X-Ray Spectrometry

PRACTICAL SPECTROSCOPY

A SERIES

Edited by Edward G. Brame, Jr.
Elastomer Chemicals Department
Experimental Station
E. I. du Pont de Nemours and Co., Inc.
Wilmington, Delaware

Volume 1: Infrared and Raman Spectroscopy (in three parts)
edited by Edward G. Brame, Jr. and Jeanette G. Grasselli

Volume 2: X-Ray Spectrometry
edited by H. K. Herglotz and L. S. Birks

ADDITIONAL VOLUMES IN PREPARATION

X-Ray Spectrometry

EDITORS

H. K. HERGLOTZ
Engineering Department
E. I. du Pont de Nemours and Company
Wilmington, Delaware

L. S. BIRKS
X-Ray Optics Branch
U. S. Naval Research Laboratory
Washington, D. C.

MARCEL DEKKER, INC.   New York and Basel

Library of Congress Cataloging in Publication Data

Main entry under title:
X-Ray spectrometry

  (Practical spectroscopy ; v. 2)
  Includes indexes.
  1. X-ray spectroscopy. I. Herglotz, Heribert.
II. Birks, L. S. III. Series. (DNLM:
1. Radiation, Ionizing. 2. Spectrum analysis.
QC482.S6 X6)
QC482.S6X18   537.5'352   77-18277
ISBN 0-8247-6625-3

Copyright © 1978 by Marcel Dekker, Inc.   All rights reserved

Neither this book nor any part may be reproduced or transmitted in any form or by any means, electronic or mechanical, including photocopying, microfilming, and recording, or by any information storage and retrieval system, without permission in writing from the publisher.

Marcel Dekker, Inc.
270 Madison Avenue, New York, New York 10016

Current printing (last digit):
10 9 8 7 6 5 4 3 2 1

PRINTED IN THE UNITED STATES OF AMERICA

CONTRIBUTORS

L. S. Birks, X-Ray Optics Branch, U. S. Naval Research Laboratory, Washington, D. C.

Paden F. Dismore, Jackson Laboratory, Organic Chemicals Department, E. I. du Pont de Nemours and Company, Wilmington, Delaware

Brent P. Fabbi, Branch of Analytical Laboratories, U. S. Geological Survey, Reston, Virginia

John V. Gilfrich, X-Ray Optics Branch, U. S. Naval Research Laboratory, Washington, D. C.

R. W. Gould, Department of Materials Science and Engineering, University of Florida, Gainesville, Florida

F. S. Goulding, Department of Instruments Techniques, Lawrence Berkeley Laboratory, University of California, Berkeley, California

Victor F. Hanson, Analytical Laboratory, The Henry Francis du Pont Winterthur Museum, Winterthur, Delaware

Kurt F. J. Heinrich, Analytical Chemistry Division, National Bureau of Standards, Washington, D. C.

H. K. Herglotz, Engineering Department, E. I. du Pont de Nemours and Company, Wilmington, Delaware

Joseph M. Jaklevic, Department of Instruments Techniques, Lawrence Berkeley Laboratory, University of California, Berkeley, California

Ron Jenkins, Engineering Department, Philips Electronic Instruments, Inc., Mahwah, New Jersey

Armin P. Langheinrich, Metal Mining Division-Research Center, Kennecott Copper Corporation, Salt Lake City, Utah

John Lucas-Tooth, Telsec Instruments, Ltd., Littlemore, Oxford, Great Britain

W. M. Tuddenham, Metal Mining Division-Research Center, Kennecott Copper Corporation, Salt Lake City, Utah

Richard L. Walter, Department of Physics, Duke University, Durham, North Carolina

R. D. Willis, Department of Physics, Duke University, Durham, North Carolina

Harvey Yakowitz, Institute for Materials Research, National Bureau of Standards, Washington, D. C.

Paul D. Zemany, General Electric Company, Knolls Atomic Power Laboratory, Schenectady, New York

CONTENTS

Contributors		iii
Preface		vii
X-Ray-Related Nobel Prizes		xi

METHODS AND INSTRUMENTS

1. Wavelength Dispersion
   L. S. Birks — 1

2. Energy Dispersion
   Joseph M. Jaklevic and F. S. Goulding — 17

3. Data Interpretation
   L. S. Birks — 59

4. Precision and Accuracy
   Paul D. Zemany — 69

5. Electron Excitation
   John Lucas-Tooth — 111

6. Proton and Alpha Excitation
   Richard L. Walter and R. D. Willis — 123

7. Electron Probe Microanalyzers
   Kurt F. J. Heinrich and Harvey Yakowitz — 163

8. Bonding and Electron Spectroscopy for Chemical Analysis
   H. K. Herglotz — 205

9. Selection and Safe Operation of X-Ray Instruments
   Paden F. Dismore — 225

APPLICATIONS

10. The General Service Laboratory
    Ron Jenkins — 241

11. Metals and Alloys
    R. W. Gould — 277

12.	Geology *Brent P. Fabbi*	297
13.	Mining and Ore Processing *Armin P. Langheinrich and W. M. Tuddenham*	355
14.	Microanalysis and Trace Analysis *John V. Gilfrich*	393
15.	Museum Objects *Victor F. Hanson*	413
16.	X-Ray Astronomy and Other Exotic Applications *H. K. Herglotz*	483

Author Index   497

Subject Index   515

PREFACE

The impact of Prof. Wilhelm Conrad Röntgen's discovery of a "novel type of radiation" in 1895 continues to exceed everyone's expectations. The medical-diagnostic application was immediately apparent to Röntgen and easily grasped by the audience when he reported his discovery. It has never ceased to benefit the general public. Other applications followed quickly: medical therapy; nondestructive testing of materials; diffraction and structure analysis; and spectroscopy with its consequences, atomic structure and elemental analysis. X Rays have been scratched from the list of interesting topics for scientists of all branches more than once, but they have bounced back just as often. This persistent vitality is not accidental. The events in atoms that generate X rays, and interactions of electromagnetic radiation of this energy range with matter, are so fundamental and informative that they never lose their significance. Ten Nobel prizes, listed in the table at the end of this preface, bear witness to this extraordinary distinction. G. von Hevesy is not included in this list of laureates, although his name is well known as the founder of an early center of practical X-ray spectroscopy.* He received the Nobel prize in 1956 for the invention of radioisotope tracer methods, not for the related and remarkable contributions of his school to X-ray spectroscopy.

This book describes only a modest band of the useful gamut of X-ray methods, namely, spectrometry, the analytical branch of X-ray spectroscopy. But even this relatively small field still exhibits the same fecundity and resilience. The simplicity and orderliness of the X-ray spectra (see Birks' Chapter 1) aroused enthusiasm in

---

*G. von Hevesy, J. Böhm, and A. Faessler, Zeitschrift für Physik 63, 74 (1930).

a few devoted spectroscopists, such as von Hevesy in the 1920s and 1930s, but widespread use did not follow. The adoption of external fluorescent excitation (outside the X-ray tube) and sensitive electronic detectors brought a renaissance of analytical X-ray spectroscopy in the 1950s and 1960s with general acceptance in the laboratory. Production and quality control remained essentially aloof. Solid-state detectors and nondispersive methods (Jaklevic, Chapter 2) conquered these strongholds of resistance. Hybridization with electron microscopy techniques has given us imaging elemental analysis (Heinrich and Yakowitz, Chapter 7). All these benefits will be apparent in those chapters that describe practical applications. Usefulness is not limited to the classical fields of chemical analysis, since plasma physics, astronomy, and nuclear structure also find in X-ray spectrometry a useful tool (Herglotz, Chapter 16).

Another aspect with historical undertones is radiation safety. Guido Holzknecht in 1905 became the world's first holder of a chair devoted completely to (medical) application of Röntgen's "novel rays." Holzknecht, the pioneer, was not aware that the therapeutic effect of X rays, if not controlled, can become deadly. Working without protection for himself, he contracted cancer and died early. A monument near the University Clinics in Vienna commemorates his sacrifice. Anyone who contracts cancer from accidental overexposure to X rays today deserves no monument. Effects of ionizing radiation on living organisms, particularly on the human body, are well explored. As a result, simple and effective safeguards are now available. Dismore's Chapter 9 deals in detail with the subject. Selection of a du Pont scientist to write this chapter is not accidental: the safety tradition of this company, initiated by its founder, Eleuthère Irénée du Pont nearly 200 years ago, has resulted in an outstanding record for safety in industry.

Authors and editors of this book were selected with the objective of making it easy for anyone to work in the field of X-ray spectrometry. We feel that the authors of the "Methods and Instruments" part, i.e., Chapters 1 through 9, have provided a good introduction to the physical concepts that is correct, concise, and sufficient to understand the rest of the book.

"Applications," the second part, is devoted to learning by participation. A set of authors generously share their experience, often the result of a lifelong effort, with their colleagues. The selection of topics for this part was, admittedly, arbitrary, but the general guidelines were

Significance of the application
Availability of an eminent, cooperative author
Instructive and universal nature of the applied methods
Reader "appeal"

What is meant by this last selection rule? The stories of the Liberty Bell and of the St. Christopher statue in Hanson's Chapter 15 would make interesting reading even if they were not examples of a new application of the nondestructive nature of X-ray spectrometry where the speed of the energy-dispersive method makes possible systematic acquisition of large sets of data.

It was our policy to avoid changes in the authors' manuscripts and thereby preserve their style, way of reasoning, and modus operandi. After all, what good is a multiple-author book if it does not represent this multitude of successful work styles? Even a bit of controversy, if not confusing and polemic, can be very stimulating.

We hope this book will interest another type of reader besides the newcomer to the field, namely, the established X-ray spectroscopist who wants to broaden his experience, to know the viewpoint of his colleagues, stay up to date with current methods, and have a recent literature guide.

The affiliation of authors is as follows: industry, 7; academia, 5; government, 5; private institutions, 1. The distribution assures a broad, multisided standpoint of the book.

So much about the chapter authors. The selection of L. S. Birks as editor needs no justification in view of the many valuable contributions by him and the X-ray spectroscopy branch that he heads at the Naval Research Laboratory. My own justification is perhaps derived from my "scientific grandfather," G. von Hevesy, and from his disciple, J. Böhm (see above), my thesis advisor and institute director in Prague. I am grateful to him for an early introduction into X-ray spectrometry, to which I maintained continuous interest, but not exclusive devotion.

Last, but not least, we, the editors, thank all contributors for their splendid cooperation. Particular thanks go to those who submitted manuscripts early. We exerted our best efforts to prevent obsolescence of their chapters before the procrastinators produced theirs. Our apologies go to the latter category for our repeated reminders of increasing intensity.

H. K. Herglotz

X-RAY-RELATED NOBEL PRIZES

Wilhelm Conrad Röntgen, 1901, for the discovery of X rays

Max von Laue, 1914, for the discovery of X-ray diffraction

W. H. and W. L. Bragg, 1915, for crystal structure analysis

Charles Barkla, 1917, for the discovery of characteristic X rays

Manne Siegbahn, 1924, for outstanding contribution to X-ray spectroscopy

Arthur Compton, 1927, for discovery of the Compton effect

Petrus Debye, 1936, for contributions to knowledge of molecular structure

Hermann Muller, 1946, for discovery of genetic effects of X rays

Max Perutz and John Kendrew, 1962, for structure analysis of globular proteins

Dorothy Crowfoot Hodgkin, 1964, for X-ray analysis of large and complex molecular structures

METHODS AND INSTRUMENTS

Chapter 1

WAVELENGTH DISPERSION

L. S. Birks

X-Ray Optics Branch
U. S. Naval Research Laboratory
Washington, D. C.

1.1	Background	1
1.2	Instrumentation	4
	1.2.1 X-Ray Tubes	4
	1.2.2 Crystals for Spectrometers	6
	1.2.3 Spectrometer Geometry	11
	1.2.4 Detectors	12
1.3	Data Interpretation	15
References		15

1.1 BACKGROUND

As it was pointed out by Moseley in 1914 [1], the characteristic wavelength for each element in the periodic table is related simply to atomic number by

$$\lambda \propto \frac{1}{Z^2} \qquad (1.1)$$

where the proportionality constant depends on the series, K, L, M, etc., and on the particular line Kα, Kβ, Lα, Lγ, etc., in that series. For instance, the Kα line of each element is given by the formula

$$\lambda_{K\alpha} \approx \frac{1300}{Z^2} \tag{1.2}$$

Figure 1.1 is a reproduction of Moseley's original plot of λ versus Z from his paper in 1914 [1]. The values for the wavelength of elements 72 and 75, as interpolated from the plot, were used as proof of the discovery of Hf and Re in the 1920s [2,3].

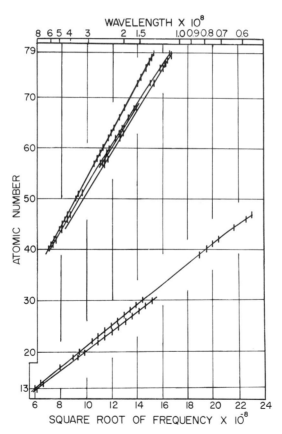

FIG. 1.1 A page from Moseley's 1914 paper showing the relationship between X-ray wavelengths and atomic number.

In wavelength-dispersion X-ray spectroscopy, we use a crystal spectrometer to separate the characteristic lines of the different elements spatially, much as a diffraction grating separates the lines in the visible spectrum. For X rays, the Bragg law tells us that the angle θ for diffraction of wavelength λ is given by

$$n\lambda = 2d \sin \theta \qquad (1.3)$$

where n is the order of diffraction (n = 1,2,3,...) and d is the spacing between planes of atoms in the crystal. Thus a particular crystal will diffract wavelength λ at several angles θ_1, θ_2, θ_3, ...; the corollary to this occurrence is that each angle θ_i corresponds to first-order diffraction of wavelength λ, second-order diffraction of $\lambda/2$, third-order diffraction of $\lambda/3$, etc. We show in Sec. 1.2.3 how the higher orders are eliminated by the detector and electronics.

The simplest geometry for wavelength dispersion is shown schematically in Fig. 1.2. Primary X rays from the X-ray tube excite the characteristic lines from the elements in the specimen by

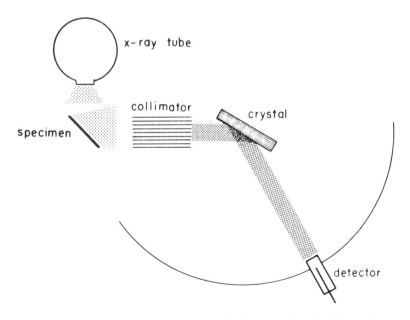

FIG. 1.2 Crystal spectrometer used for wavelength dispersion.

fluorescence. The characteristic lines emerge from the specimen in all directions, but they are limited to a nearly parallel beam by the collimator. Each characteristic wavelength is diffracted at the crystal setting corresponding to Bragg's law for that wavelength. As the crystal is rotated through a range of θ, the detector is rotated twice as fast so that it is at angle 2θ, and therefore in position to intercept the diffracted line.

As can be seen in Fig. 1.1, it seems likely that the β line of one element will overlap the α line of a neighboring element, and also that L or M lines of high atomic-number elements will overlap the K lines of lower atomic-number elements. In fact the lines do not overlap exactly and the resolution of a crystal spectrometer is generally good enough to separate such lines and allow positive identification of each line from each element.

1.2 INSTRUMENTATION

1.2.1 X-Ray Tubes

The X-ray tubes used in wavelength-dispersion systems are usually sealed, high-vacuum tubes. Figure 1.3 shows the tube and power supply schematically, and considerably simplified. The *cathode* is generally a hot filament and the *anode* is a metal such as Cr, Cu, Mo, or W. Electrons emitted by the hot filament are accelerated by a potential of up to 50 or 75 kV and strike the anode. Most of the energy is dissipated as heat; less than one part in a thousand goes into the production of X rays. Modern tubes operate at a power of

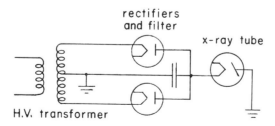

FIG. 1.3 Schematic of X-ray tube and high-voltage supply.

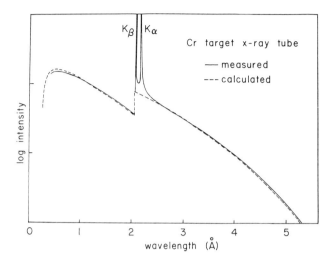

FIG. 1.4 Spectral distribution of radiation from Cr-target X-ray tube.

2-3 kW which means 50-75 mA at 50 kV (some special tubes with rotating anodes operate at higher power but are not commonplace). The target is water-cooled to prevent melting. Tubes emit about 10^{14} X-ray photons/sec/steradian. The X rays emitted by the tube consist of a continuum (bremsstrahlung) as well as the characteristic wavelengths of the target element (see Fig. 1.4).

When the primary radiation of the X-ray tube strikes the specimen, only those photons (continuum or characteristic lines) with wavelengths shorter than the characteristic absorption-edge wavelength of a given element can ionize that element. (The absorption-edge wavelengths correspond to the ionization potential of the particular element and shell, e.g., one K shell, three L shells, five M shells. The reader is referred to Compton and Allison [4] for details.) Any absorption-edge wavelength for the K shell of an element is shorter than all the absorption-edge wavelengths for the L shells, so that a different portion of the primary spectrum will excite the K lines in the sample than will excite the L lines. It should be noted that X-ray photons cannot excite bremsstrahlung; therefore only characteristic lines are excited in the sample (the background observed in X-ray fluorescence analysis is due to scattered primary radiation). For the case of a copper specimen excited by a W-target X-ray tube, Fig. 1.5 illustrates how the primary spectrum varies with tube voltage and how the intensity of Cu characteristic radiation excited by such a primary spectrum varies correspondingly. If V is the X-ray tube voltage and V_K, V_L, and V_M are the ionization potentials of the K, L, and M shells, respectively, a simple rule of thumb states that the intensity of K fluorescent radiation increases approximately as $(V - V_K)^2$ for $V \leq 3 V_K$, and approximately as $(V - V_K)$ for $V > 3 V_K$. A tube potential of 50 kV will efficiently excite characteristic K radiation for elements up to Sn and characteristic L radiation for all higher atomic numbers.

One more consideration about X-ray tubes should be mentioned, namely the choice of target material. The bremsstrahlung portion of the primary spectrum is more intense the higher the atomic number of the target. Therefore a W tube is more efficient than a Cr tube for general use. However, the characteristic lines from a Cr-target tube are so strong and so efficient in exciting elements below titanium in the periodic table, that a Cr tube is generally preferred for aluminum alloys, chlorine compounds, and the like. Special targets such as Mo or Ag have specific uses because Mo lines are especially efficient in exciting the bromine K-series lines or the lead L-series lines in a sample, and silver L lines are especially efficient in exciting sulfur.

1.2.2 Crystals for Spectrometers

The heart of the crystal spectrometer is the crystal itself. Each crystalline material has four important characteristics which must be considered in determining its usefulness as a spectrometer crystal.

1.2.2.1 Diffraction Efficiency

Some crystals will diffract a greater fraction of incident radiation than will other crystals. The term commonly used for diffraction efficiency is *integral reflection coefficient* R, and it varies

WAVELENGTH DISPERSION

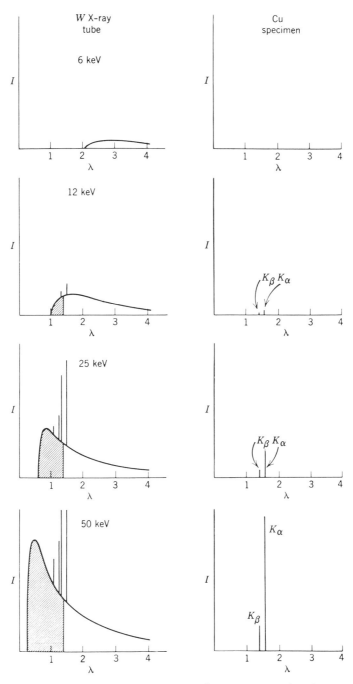

FIG. 1.5 Output of a W-target X-ray tube at selected voltages and the corresponding fluorescent intensity excited in a Cu specimen.

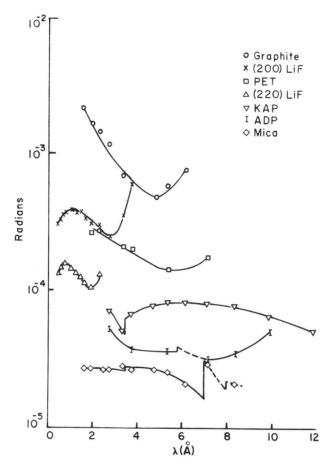

FIG. 1.6 Integral reflection coefficient R measured for several crystals.

with wavelength as well as with crystal composition and structure. Figure 1.6 shows measured R values for several commonly used crystals [5]. Nowadays we have formulas which can be used to predict the diffraction efficiency of crystals [6,7], but slight changes in crystal perfection change the value of R correspondingly, thus requiring experimental measurement for absolute calibration. Note that R is expressed in radians for historical reasons because angle was expressed in radians; it could just as easily have been expressed in seconds of arc and would have led to more tractable units.

1.2.2.2 Resolution

The resolution of a crystal depends on the perfection of the crystal lattice over its entire diffraction surface. Local distortions in the lattice cause diffraction to occur over a range of θ angle and result in a finite line width W for each X-ray line (see Fig. 1.7). Depending on the crystal spacing d and the diffraction angle θ, the value of W can be expressed in terms of Δλ, the resolution in angstroms, or as ΔE the resolution in energy (usually as electron volts). Resolution for crystal spectrometers (including the contribution due to collimator divergence discussed in Sec. 1.2.3) commonly ranges from a few eV at wavelengths greater than 2 Å to perhaps 100 eV at a wavelength of 1 Å. It is this kind of number which should be compared to 150-200 eV resolution for energy-dispersion systems discussed in Chap. 2.

1.2.2.3 Crystal Spacing (d)

As can be observed by referring to Bragg's law, the spacing d limits the maximum wavelength which can be diffracted to λ_{max} = 2d because sin θ cannot exceed unity. Likewise, the value of d controls the

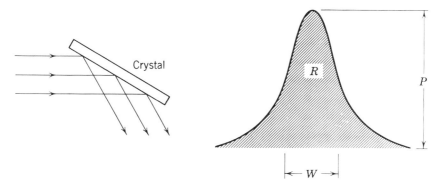

FIG. 1.7 Crystal rocking curve for parallel monochromatic radiation.

angular dispersion, $d\theta/d\lambda = 1/(2d \cos \theta) = \tan \theta/\lambda$. Dispersion is therefore better for crystals with small spacings.

1.2.2.4 Physical Size

To be useful in a spectrometer it must be possible to find in nature or to grow in the laboratory nearly perfect crystals at least 2 × 5 cm in area.

The combination of the four requirements on crystal characteristics limits our choice to a relatively small number of materials. Table 1.1 lists a few of the common spectrometer crystals along with their characteristics.

TABLE 1.1

Some Useful Analyzer Crystals

Crystal	Planes	2nd Spacing (Å)
LiF	420	1.79
LiF	220	2.84
LiF	200	4.02
Quartz	10$\bar{1}$1	6.70
Graphite	002	6.70
Quartz	10$\bar{1}$0	8.50
PET[a]	002	8.76
(ADP)[b]	110	10.64
Mica	002	19.8
KAP[c]	001	26.6
LSD[d]	layers	100.2

[a] Pentaerythritol.
[b] Ammonium dihydrogen phosphate.
[c] Potassium acid phthalate.
[d] Lead stearate decanoate.

1.2.3 Spectrometer Geometry

Two types of geometry are commonly used for X-ray fluorescence analysis, namely flat crystal and curved crystal. Flat crystal geometry was already shown in Fig. 1.2 and requires no further explanation except to say that the angular divergence allowed by the collimator, B_c, is generally larger than W, the rocking-curve breadth of the crystal (Fig. 1.7). The total angular divergence allowed by the crystal and collimator combination is B, where $B^2 = B_c^2 + W^2$. It is this B value which limits the overall spectrometer resolution to the values given in the previous section.

Curved-crystal geometry is illustrated for cylindrically curved crystals in Fig. 1.8(a) and (b) and for logarithmically curved crystals in Fig. 1.8(c). We will consider the cylindrical curvature

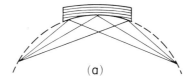

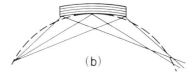

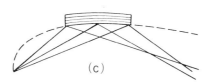

FIG. 1.8 Curved-crystal geometry: (a) Johansson focusing optics, (b) Johann optics, (c) logarithmically curved crystal.

first. If the source slit is on a circle of radius r (the Rowland circle), polychromatic radiation will diverge to the crystal, but only one wavelength will be diffracted at a given angle provided that the crystal planes are curved to a radius 2r and then the surface ground to a radius r [Fig. 1.8(a)]. In the equatorial plane, the diffracted radiation of wavelength λ will converge to the image slit as shown, and be measured by the detector. This is called *Johansson optics* [8] and can be achieved for those crystals which are not permanently damaged by abrading the surface. For those crystals which are permanently damaged by abrading, we are limited to curved but not ground crystals, *Johann optics* [9] [Fig. 1.8(b)]. For Johann geometry, the diffraction conditions are not completely satisfied for the portion of the crystal away from the center, and a broadening of the diffracted line by 0.1 to 0.2 degrees may result; this is not serious for general applications.

Figure 1.8(c) shows the logarithmically curved crystal first proposed by de Broglie [10] but not put to use until the advent of the multiple spectrometers for simultaneous measurement of 10 or 20 elements. A ray from the source makes a constant angle to every position along the spiral; this allows us to satisfy Bragg's law exactly for curved but not ground crystals. It should be noted, however, that the diffracted radiation does not converge to a line image, but this means only that the detection slit must be broader than would be necessary for Johansson optics. It should also be noted that a different spiral is needed for each θ, that is, a different crystal is needed for each wavelength. Hence the use of logarithmic crystals is limited to multiple spectrometer equipment, where several crystals are arranged around the sample.

1.2.4 Detectors

Two basic types of detectors are used in wavelength-dispersion instruments, namely gas-proportional detectors and scintillation detectors. Figure 1.9 shows them schematically.

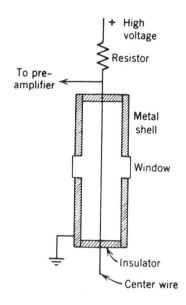

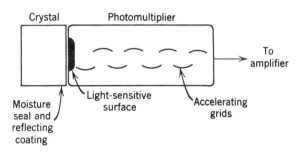

FIG. 1.9 Schematic of gas-proportional and scintillation detectors.

In the gas-proportional detector, an X-ray photon entering through the window may be absorbed by a gas atom, in which case one or more energetic photoelectrons will be generated. These photoelectrons lose energy by ionizing other gas atoms; the total number of ion pairs (atoms ionized) depends on the original X-ray photon energy:

$$\text{Ion pairs} = \frac{\text{photon energy}}{2 \times \text{first ionization potential}}$$

The electron from each ion pair is accelerated toward the (+)-charged wire and ionizes a number of other gas atoms along the way (the number of electrons in each avalanche depends on the voltage on the wire and the wire diameter since these two factors determine the field strength). These avalanches all strike the wire within 0.1 μsec after the X-ray photon is absorbed and generate a pulse which is read out on the detector electronic circuits. The important point to be understood is that different X-ray photon energies (different wavelengths) will generate pulses of different amplitude. Unfortunately, the starting number of ion pairs and the number of electrons per avalanche show a statistical spread about the average value; this results in a pulse-amplitude distribution rather than a single pulse amplitude for each characteristic X-ray line. Fortunately, when used with a crystal spectrometer the gas-proportional detector easily distinguishes between the first-order diffraction of wavelength λ and the second- or third-order diffraction of $\lambda/2$ and $\lambda/3$, and also scattered bremsstrahlung of longer or shorter wavelength than the desired signal. *This is all that is required of a detector in a crystal spectrometer.* As an aside, it should be mentioned that gas-proportional detectors have often been used without a crystal spectrometer, i.e., in the energy-dispersion mode (see Chap. 2), but energy resolution is far inferior to the solid state detector.

For $\lambda > 2$ Å, it is advantageous to use a thin window on the gas-proportional detector in order to reduce X-ray absorption by the window material. The detector gas tends to diffuse out through such a thin window and air tends to diffuse in. Therefore it is necessary to replenish the gas constantly through an inlet and outlet pipe; hence the term *flow-proportional detector*.

The scintillation detector of Fig. 1.9(b) operates on a somewhat different principle than the gas-proportional detector. When an X-ray photon is absorbed in the scintillator (usually NaI activated with Tl), the photoelectrons generate a number of visible-light photons (scintillations); the number of visible photons is proportional to the X-ray photon energy. These visible-light photons activate the photomultiplier and result in a pulse to the detector electronics. For the scintillation detector, the pulse-amplitude distribution is relatively three times broader than that from a gas-proportional detector but still more than adequate to discriminate against higher order diffraction.

Detector electronics will not be discussed in detail here because they are covered in Chap. 2. Suffice it to say that gating circuits are used to pass the pulse-amplitude distribution corresponding to first-order diffraction lines but to reject the pulse-amplitude distributions of higher order diffraction lines. Data readout is in the form of pulses per second, i.e., counts per second. Strip chart recorders plot the counting rate vs. θ to give the wavelength and intensity of the elements in the sample.

1.3 DATA INTERPRETATION

Qualitative identification of elements is achieved by the simple strip-chart recording mentioned at the end of Sec. 1.2.4, and intense lines correspond to major components. But generally the goal is quantitative analysis and this requires either calibration standards or mathematical formulations to relate intensity to composition. It does not matter whether the intensity is obtained directly with a crystal spectrometer or by unfolding overlapping energy spectra (see Chap. 2), the same calibration or correction techniques apply. They are discussed fully in Chap. 3.

REFERENCES

1. H. G. J. Moseley, *Phil. Mag.*, *27*, 703 (1914).
2. D. Coster and G. von Hevesy, *Nature*, *111*, 79, 182 (1923).
3. W. Noddack, I. Tacke, and O. Berg, *Naturwissen*, *13*, 567 (1925).
4. A. H. Compton and S. K. Allison, *X-Rays in Theory and Experiment*, D. van Nostrand Co., New York, 1963.
5. J. V. Gilfrich, D. B. Brown, and P. G. Burkhalter, *Appl. Spect.* (in press) July, 1976.

6. D. B. Brown and M. Fatemi, *J. Appl. Phys.*, *45*, 1544 (1974).
7. D. B. Brown, M. Fatemi, and L. S. Birks, *J. Appl. Phys.*, *45*, 1555 (1974).
8. T. Johansson, *Naturwissen*, *20*, 758 (1932).
9. H. H. Johann, *Z. Physik*, *69*, 185 (1931).
10. M. de Broglie and F. A. Lindemann, *Compt. Rend.*, *158*, 944 (1914).

2.1 INTRODUCTION

In energy-dispersive X-ray fluorescence spectrometry identification of fluorescence radiation is performed using detectors which directly measure the X-ray energy. Such detectors normally operate by measuring the consequence of interactions of the incident X rays in the detector medium. In the simplest case, an electron is ejected from an atom in the detector by photoelectric absorption of the incident X ray. The final result is then produced by the slowing down of this primary electron resulting in showers of electron-ion pairs in the case of a proportional counter, optical excitations in the case of a scintillation detector, or showers of electron-hole pairs in a semiconductor detector. The output from the detector is the integral of such effects which for our present purposes we can regard as proportional to the energy of the original incident X ray. This method is in contrast to wavelength dispersion in which the Bragg reflecting properties of a crystal are used to disperse X rays at different reflection angles according to their wavelength.

Although energy-dispersive detectors generally exhibit poorer energy resolution than wavelength-dispersive methods, they are capable of simultaneous detection of a wide range of X-ray energies and have inherently higher detection efficiencies. In order to operate effectively as a detector for X-ray fluorescence applications the energy resolution must be adequate to resolve the characteristic X rays of interest from interfering lines and background. The recent development of semiconductor detectors with their excellent energy resolution has made energy-dispersive X-ray fluorescence analysis a powerful analytical tool whose applications are rapidly expanding. Spectrographic applications of the other types of energy-dispersive detectors are now limited to special cases where certain features of semiconductor detectors are not acceptable, such as in space, where cryogenic cooling is difficult.

Most semiconductor detectors used for X-ray analysis consist of a single crystal of either silicon or germanium, typically 1 cm or less in diameter and 0.3-1 cm thick. Free charge produced by

Chapter 2

ENERGY DISPERSION*

Joseph M. Jaklevic and F. S. Goulding

Department of Instruments Techniques
Lawrence Berkeley Laboratory
University of California
Berkeley, California

2.1 Introduction 18
2.2 Basic Methods 23
2.3 Instrumental Considerations 36
    2.3.1 Detectors 36
    2.3.2 Signal-Processing Electronics 42
2.4 Calibration Methods 46
2.5 Spectral Analysis 52
2.6 Applications 55
2.7 Conclusion 56
References 56

---

*This work was performed under auspices of the United States Energy Research and Development Administration.

the interaction of each incident X ray in the detector is collected and accurately measured. To obtain the energy resolution required for X-ray spectroscopy, it is necessary to operate the device at a low temperature and a liquid nitrogen cryostat is normally used to maintain the detector at 77 K (-196°C).

Figure 2.1 illustrates energy resolution capabilities as a function of X-ray energy over the range of interest in fluorescence spectroscopy. Figure 2.1(a) shows the variation $\Delta E/E$ with energy for typical semiconductor detectors and proportional counters. Scintillation detectors would exhibit still poorer energy resolution. Also shown is the performance of a typical wavelength-dispersive crystal spectrometer, illustrating the improved energy resolution achieved by this method in the lower energy range. Figure 2.1(b) shows ΔE (FWHM)* for a typical state-of-the-art lithium-drifted silicon-detector spectrometer operating at 77 K. The energy resolution at low energies [120 eV in Fig. 2.1(b)] is determined by electronic noise. Small lithium-drifted or high-purity germanium detectors exhibit comparable resolution.

The dominant importance of good energy resolution in analytical X-ray spectroscopy is illustrated in Fig. 2.2a which shows the energies of the characteristic K X rays of the low-Z elements and the L X rays of the heavy elements. Figure 2.2b shows the separations of the $K\alpha$ lines of adjacent elements as Z changes, and also the separation of the $K\alpha$ line of an element from the $K\beta$ line of the next lower Z element. When those results are compared with the energy-resolution curve of Fig. 2.2a, it becomes obvious that separation of the $K\alpha$ lines of adjacent elements is impossible for the very light elements (Z < 6); even more important, interference between the $K\beta$ line of an element and the $K\alpha$ line of the next higher Z element is serious for much higher values of Z. Fortunately, computer analysis of the whole spectrum, taking into account the known X-ray line structures of the various elements, can accommodate a reasonable

---

*FWHM: full width at half maximum; corresponds to B value for crystal spectrometer in Sec. 1.2.3.

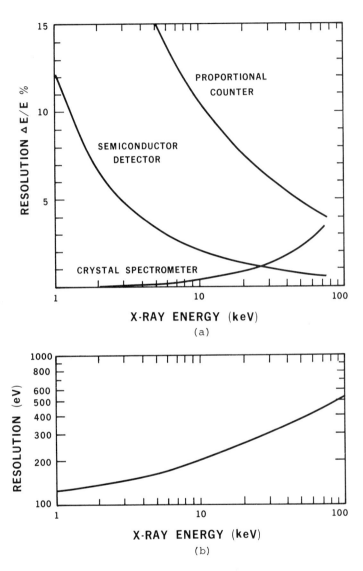

FIG. 2.1 Energy resolution capabilities of X-ray detectors: (a) comparison of typical semiconductor detector with a proportional counter and a Bragg crystal spectrometer; (b) resolution expressed as the full width at half maximum (FWHM) for a typical semiconductor system as a function of energy.

ENERGY DISPERSION

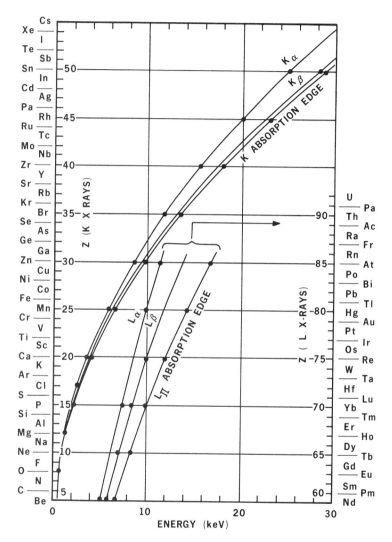

FIG. 2.2a The energies of the principal X-ray absorption edges and emission lines as a function of atomic number.

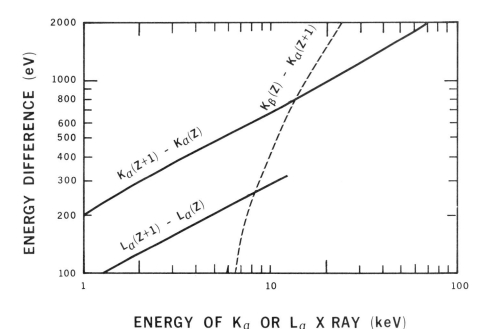

FIG. 2.2b Energy differences between adjacent X-ray lines of the elements. These curves should be compared with the resolution curve in Fig. 2.1(b).

level of such interference. These techniques are also necessary to overcome interferences between the L and M lines of heavy elements and the K lines of light elements. These problems are reduced by using detector systems with improved energy resolution.

Since the excellent energy resolution of wavelength-dispersive system avoids much of this interference, it would seem that there would be little use for energy-dispersive systems. However, the need for precise collimation in wavelength-dispersive systems results in a greatly reduced efficiency for X-ray detection with the result that the analytical sensitivity for the detection of elements is often comparable for the two methods [1]. The simultaneous multielement capability of energy-dispersive systems therefore gives them an important role in X-ray analysis.

2.2 BASIC METHODS

The basic elements of an energy-dispersive X-ray fluorescence spectrometer are illustrated in Fig. 2.3. Excitation can be any form of radiation capable of producing vacancies in the inner shells of the atoms of interest in the sample. The vacancies are then filled by transitions of electrons from higher energy atomic orbitals resulting in the emission of characteristic X rays. Measurement of the energies and intensities of these X rays is the basis of X-ray fluorescence spectroscopy. A competing process whereby vacancies fill involves the Auger effect in which deexcitation proceeds by emission of an outer-shell electron instead of an X-ray photon. This process becomes more likely than X-ray emission for the light elements where the fraction of vacancies producing X-ray emission (defined as the fluorescence yield ω) may be less than 10% [2].

The detector shown in Fig. 2.3 is normally collimated to accept X rays which originate at the sample and have sufficient energy to penetrate the thin entrance window to the detector. The observed energy spectrum contains not only the characteristic X rays of interest but also those from other elements in the sample. An additional background caused by other types of interactions, such as

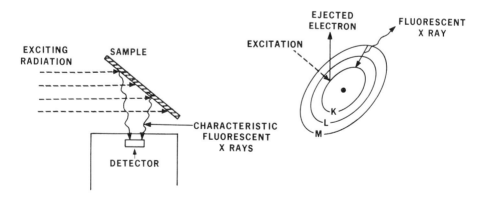

FIG. 2.3 Schematic of the X-ray fluorescence process.

scattering of the excitation radiation from the sample, can be observed. The magnitude and spectral shape of the background depends on the type and energy of the excitation and on the composition of the sample.

Various types of exciting radiation may be employed; the choice depends both on fundamental physical processes, which determine the ultimate performance potential in a given application, and on very practical factors such as cost and availability. Performance factors which must be considered include background, which establishes the limit of detectability and sensitivity, the accuracy of the quantitative information obtained, and the range of elements over which optimum sensitivity is achieved. The available options for excitation are electrons, positive ions, and photons (i.e., X rays or γ rays). The relative cross sections for inner-shell vacancy production for the three methods are similar enough that the choice is usually based on other factors.

The use of electrons for excitation has the advantage that they can be conveniently generated in large quantities and that their energy and intensity can easily be regulated. They have the disadvantages that the sample must be enclosed in the same vacuum as the source and that the ratio of fluorescent signal to background is relatively poor due to the large amount of continuum bremsstrahlung produced in the sample. This radiation is caused by the decelerations which electrons undergo as they interact with the positively charged nuclei in the sample, producing a continuous X-ray spectrum with a maximum energy equal to that of the incident electrons. The portion of this radiation seen by the detector sets a fluorescence analysis detection limit considerably higher than when using other excitation methods. However, the ability to focus electrons into precisely controlled beams of submicron size has led to extensive applications of energy-dispersive detectors in electron microprobe analysis and in scanning electron microscopes. These applications are discussed in Chap. 7. Other uses of electron-excited X-ray fluorescence are described in Chap. 5.

Since positive ions are much more massive than electrons, they produce less bremsstrahlung, and background in the fluorescence spectrum is a much smaller problem. Although not as convenient as electrons to generate, proton and α-particle beams are available at nuclear research accelerators with the result that much recent work has been done in this area. As with electrons, samples must be contained in a vacuum shared with the particle accelerator. The basic limit to the detectability of this method is the continuum bremsstrahlung produced by energetic secondary electrons resulting from collisions of the incident ions with electrons in the sample. The resulting radiation background has an endpoint energy which is directly related to the energy of the original ion beam. Consequently optimum sensitivies for most analytical applications are achieved with beam energies below 16 MeV for α particles and 3 MeV for protons corresponding to a bremsstrahlung background endpoint energy of 6-8 keV. Excellent detection limits have been achieved under these conditions particularly in situations where the thickness and size of the sample is optimum for this type of analysis [3]. A more complete discussion of the capabilities of excitation by charged particles heavier than electrons is presented in Chap. 6.

Photon-excited X-ray fluorescence analysis will be discussed in more detail in the present chapter. This approach has been extensively used and appears to be the most useful for general analytical applications, partly based on excellent performance, but also because it is very convenient to use compared with other excitation methods.

In passing through matter, photons do not undergo continuous energy-loss processes, as do charged particles. The absorption obeys an exponential attenuation law:

$$I = I_0 \, e^{-(\mu/\rho)\rho x} \qquad (2.1)$$

where I_0 is the initial intensity of the beam in photons/sec, ρ is the density of the material in gm/cm^3, I is the intensity transmitted through a thickness $x(cm)$, and μ/ρ is the mass attenuation coefficient expressed in units of cm^2/gm. The mass attenuation coefficient µ/ρ can be broken down into components associated with the various absorption processes. In considering fluorescence, the portion associated with photoelectric absorption is the important part as far as excitation of characteristic X rays is concerned. The photoelectric absorption cross section is zero for incident photon energies less than the binding energy of the particular atomic shell of a given element and jumps to a maximum immediately above that energy. As the energy is increased still further, the cross section decreases as approximately $E^{-3.5}$. This is an important factor in the choice of excitation energy since the sensitivity for detection is greatest for those elements whose absorption edges are just below the incident photon energy. Figure 2.4 is a plot of the relative photoelectric cross section per individual atom as a function of energy for several elements [4].

Photons also interact by scattering from the sample, so that, in the general case, µ is the sum of the scattering and absorption components. Furthermore, since the typical sample of interest in fluorescence spectroscopy contains atoms of many elements, the quantity µ/ρ must be replaced by a summation over all the elements and their relative concentrations:

$$\frac{\mu\rho}{\rho} \to \Sigma_i \frac{\mu_i}{\rho} \rho_i \qquad (2.2)$$

where the subscript i refers to element i in the sample.

The relative magnitude of the various unwanted scattering and absorption processes as compared with the desired vacancy production by photoelectric absorption determines the sensitivity of photon-excited fluorescence analysis. If the sample contains mostly elements of high atomic number, photoelectric absorption is dominant and the concentrations are determined by the relative magnitudes of the fluorescence peaks in the spectrum. On the other hand, if the elements of interest are contained in a low-Z matrix, the background produced by scattering processes affects the limit of detection.

The characteristics of photon-induced fluorescence and detection limits are best discussed in terms of a hypothetical

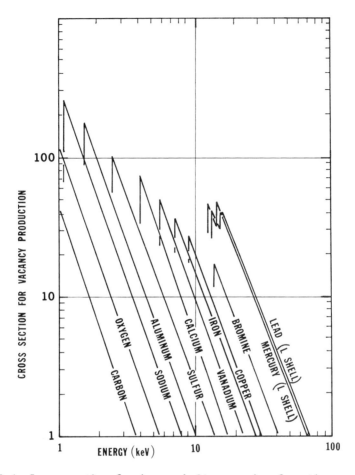

FIG. 2.4 Cross section for inner-shell vacancies formation as a function of exciting photon energy.

fluorescence spectrum. To simplify the problem, excitation will be considered to be in the form of monoenergetic photons and the sample will be assumed to be sufficiently thin that the probability of multiple interactions is small. Extensions to broad energy-range excitation and to thicker samples involves obvious integrations over the incident spectral distribution and over the sample thickness. The three modes of interaction of importance for photons in the energy range of interest are *Rayleigh scattering*, *Compton scattering*, and *photoelectric absorption*. In Rayleigh scattering an incident photon interacts with a tightly bound atomic electron and thereby is elastically scattered with no loss of energy. In Compton scattering the interaction takes place with a loosely bound electron, resulting in an energy loss in the scattered photon governed by the well-known Compton equation of atomic physics. Figure 2.5 shows the resulting energy loss as a function of scattering angle and incident photon energy. The photons scattered by either of these processes can reach the detector and contribute unwanted counts in the spectrum.

Figure 2.6 shows the idealized spectrum produced by a semiconductor detector system in a fluorescence measurement using

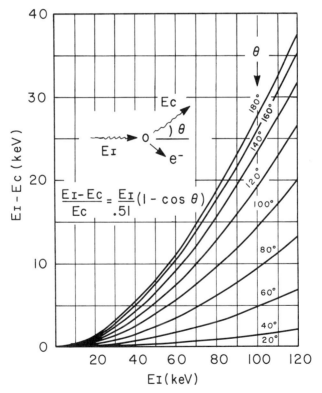

FIG. 2.5 Energy and angle dependence of the Compton effect (inelastic scattering).

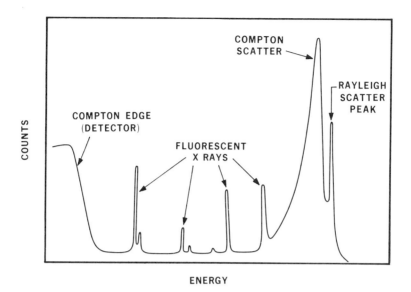

FIG. 2.6 Idealized X-ray fluorescence spectrum resulting from excitation of trace elements in a low atomic-number matrix using monoenergetic photons.

monoenergetic photon excitation. The plot represents the number of events analyzed by the detector as a function of the size of the pulse from the detector. The highest energy peak in the spectrum occurs at the energy of the incident excitation and is due to X rays that Rayleigh-scatter from the sample into the detector. The broad peak at slightly lower energy is due to the inelastic Compton scattering. The width of this peak reflects the spread of angles over which the photons are scattered and the corresponding energy losses according to the Compton relationship. The low-energy continuous distribution is due to Compton scattering in the detector whereby high-energy X rays reaching it are scattered out, leaving only that fraction of their energy which had been imparted to electrons in the detector. Fluorescence signals are represented by the small peaks occurring at energies well below the high-energy scatter peaks. The magnitude of each peak is dependent upon the concentration of a particular element in the sample and on the relative cross section for the K or L shell ionization of the element as shown in Fig. 2.4. Background beneath the peaks is due to a number of complex interactions by which the high-energy scattered X rays can produce a continuum of pulse amplitudes in the detector. These can be the result of poor charge collection in the detector or they can arise from the bremsstrahlung produced in the sample by slowing down of secondary electrons. Both these effects are serious in trace element applications. For the purpose of the present discussion, it is sufficient to observe that the detection limit is set by this background which, in turn, is proportional to the total scattered intensity.

The idealized spectrum of Fig. 2.6 represents the case of trace elements in a low-atomic-number matrix such as hydrocarbons. In this simplified example, most of the atoms in the sample are of low-Z elements that contribute most of the scattered X-ray intensity. If the average atomic number of the matrix is increased, the relative magnitude of Rayleigh vs. Compton scattering changes since the approximate Z^2 dependence of elastic scattering predominates over the $\propto Z$ dependence of the Compton effect. As the average atomic number is increased photoelectric absorption and the resulting fluorescence of the atoms in the matrix become important; analysis of the spectrum is then complicated by interelement absorption and enhancement effects. Since the heavier elements then absorb a significant fraction of the beam photoelectrically, the fluorescence spectrum consists predominantly of the X rays emitted by atoms of these elements. If the energies of these intense fluorescent X rays are above the absorption edge of other constituents in the sample, they can be reabsorbed and contribute to an enhancement of the X-ray yield of the lower Z elements. Both the absorption and enhancement effects contribute to a nonlinear relationship between the concentration and the measured X-ray intensity. These problems are well known in X ray spectroscopy of specimens such as alloys and are discussed in Chap. 3.

Practical X-ray sources often do not provide perfectly monochromatic excitation. A close approximation can be achieved by using characteristic X rays generated directly by a X-ray tube or by a radioactive source, or indirectly by a secondary fluorescence system in which the X rays from the primary source fluoresce a secondary pure-element target. The interpretation of the spectrum is then as previously discussed. In some cases it proves more convenient to use a continuous bremsstrahlung spectrum for excitation. The spectrum in the detector is then a superposition of spectra corresponding to each energy in the continuum (i.e., it is an integral over the continuous excitation distribution function). Since the scattered radiation extends over the whole spectrum, the background below the characteristic X rays is higher than with monochromatic excitation. However, detectability may be less important in some cases than the fact that the wide-band excitation gives a more uniform sensitivity over a wider range of elements.

The methods of photon excitation illustrated in Fig. 2.7 include radioisotope and X-ray tube sources used in direct or secondary fluorescence configurations. Direct excitation can employ either a broad-band or monochromatic radiation. The principal difference between radioisotope sources and X-ray tubes is the large difference in available intensity. The radiation hazard associated with the use, handling, and storage of a very active source imposes a practical upper limit of 100 mCi (1 Ci = 3.7×10^{10} disintegrations/sec), whereas an X-ray tube, which need only be turned on when required, can easily be equivalent to 10-1000 Ci. To achieve reasonable detectability using radioactive sources the total sample mass should be large and long analysis times must be tolerated. Monochromatic radioisotope sources are generally used although continuous sources are also available. Radioisotope sources are used mainly because of their low cost, small size, and simplicity, and can be used successfully where the analysis of large numbers of samples is not required. Apart from the lack of adequate intensities, another disadvantage arises from the relatively small number of available isotopes. Table 2.1 is a list of radioisotope sources commonly used for X-ray fluorescence; most are commercially available

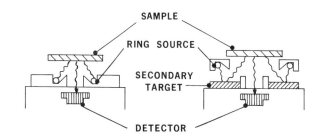

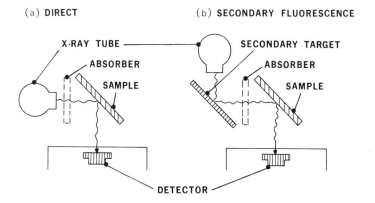

FIG. 2.7 Schematics of typical radioisotope and X-ray tube fluorescence geometries in both direct and secondary fluorescence configurations.

in packages convenient for either direct or secondary fluorescence excitation. A more complete description of possible radioisotope options is given in Ref. 5.

X-Ray tubes provide controllable photon energy and intensity but are more costly than radioisotopes and require a high-voltage power supply and associated controller. Therefore they are used in large-scale analytical programs where the speed of analysis more

TABLE 2.1

Radioisotopes for X-Ray Fluorescence

Nuclide	Half-life	Emission energies
^{55}Fe	2.7 years	5.9 keV
^{109}Cd	453 days	22.1 keV, 87.7 keV
^{125}I	60 days	27 keV
^{241}Am	458 years	12-17 keV, 60 keV
^{57}Co	270 days	6.4 keV, 122 keV, 144 keV
^{238}Pu	86.4 years	12-17 keV

than compensates for the initial cost. The direct output of a typical X-ray tube consists of the electron-excited characteristic X rays of the anode material together with a continuous bremstrahlung spectrum. The output can be filtered with an appropriate absorber, normally the same material as the anode, to provide a source predominately composed of the characteristic X rays of the anode material. A better method of creating monochromatic radiation employs the secondary target system shown in Fig. 2.7. The spectral output in this case has much less continuum radiation but at a considerable sacrifice in intensity compared with the case of direct excitation.

The use of monochromatic X-ray excitation results in a very high sensitivity for detection of elements with absorption edges slightly lower in energy than the excitation, and a decreasing sensitivity for lower Z elements. For adequate coverage of a wide range of elements, the sample must be exposed sequentially to a number of excitation energies. This is in contrast to the case of continuous excitation where the detection sensitivity is poorer but a uniform coverage of a wide range of elements is achieved in one exposure (see also Chap. 15).

One set of excitation energies suitable for analysis of a wide range of elements corresponds to the characteristic X rays of elements Cu, Mo, and Tb at 10, 20, and 44 keV, respectively. Figure 2.8 is a plot of the calculated relative fluorescence excitation

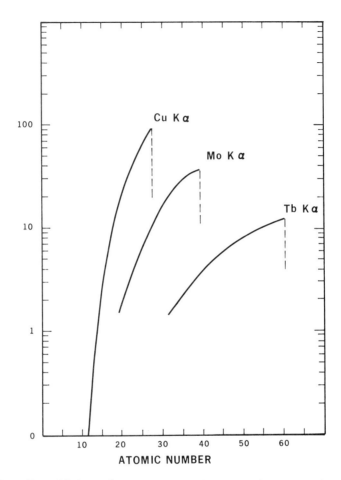

FIG. 2.8 Probability of X-ray production as a function of energy of the characteristic X rays for three different exciting radiations.

probabilities as a function of Z for this set. The rapidly decreasing values for low-Z elements result from the decrease in fluorescence yield (i.e., a larger probability of vacancy filling being accompanied by Auger-electron emission). This choice of elements for excitation is not unique and optimum detection of specific elements may require intermediate excitation energies. A practical restriction is that the secondary target should be able to be produced in a form appropriate for the geometry employed and in a chemical state not easily damaged by the X-ray radiation flux. The choice of Cu, Mo, and Tb is dictated to a large extent by their availability as stable metallic foils.

R. Giauque, et al. have reported on the single-element detectable limits obtained in a series of different types of samples using Mo K X ray for excitation [6]. Using a criterion for minimum detectable defined as a signal 3σ above background, and a counting time of 1000 sec, measured limits were below 1 ppm for elements with $24 < Z < 37$ and also for Hg and Pb contained in a biological sample (i.e., a hydrocarbon matrix). A detectable limit of 1 ppm was observed for elements in the Cd region using terbium K X-ray excitation. Russ has reported detection limits in the 1-10 ppm range in aqueous solutions over the same region using continuous bremsstrahlung and comparable counting times [7]. Similar detectability can be achieved using radioisotope excitation but counting times of the order of 10-100 times longer must be based.

Commercial X-ray fluorescence equipment manufacturers produce systems based on the various options already discussed. Since each approach has particular merits for a given class of applications, no simple conclusion regarding the superiority of any approach can be given (see Chap. 9).

A discussion of X-ray excitation sources would not be complete without mention of some other aspects of the generation of X rays which may prove useful for certain applications. Recent papers have discussed the benefits of polarized X-ray beams and means of generating them [8-10]. The advantage of polarized excitation is that no scattering is observed parallel to the direction of polarization of the incident radiation, resulting in a drastically reduced background. However, current techniques for generating polarized X-ray beams involve multiple scattering with a prohibitive reduction in the usable intensity. Alternative polarized X-ray sources, such as the synchrotron radiation produced by very high-energy electron storage rings offer high intensities but are not available for general use.

2.3 INSTRUMENTAL CONSIDERATIONS

2.3.1 Detectors

As noted earlier, the principal advantage of semiconductor detectors over others is their improved energy resolution. Secondary factors which influence their application to a given problem include efficiency, background characteristics, and counting-rate capabilities. In the following paragraphs the general feature of semiconductor-detector spectrometers will be discussed.

Figure 2.9 is a cutaway view of a typical lithium-drifted silicon X-ray detector. The device is known as a p-i-n structure referring to the p-type contact on the entry side, the intrinsic active volume, and the lithium-diffused n contact. The active volume consists of a region in which the donor lithium has been drifted under the influence of an electric field to precisely compensate p-type impurities in the original silicon. When reverse bias is applied

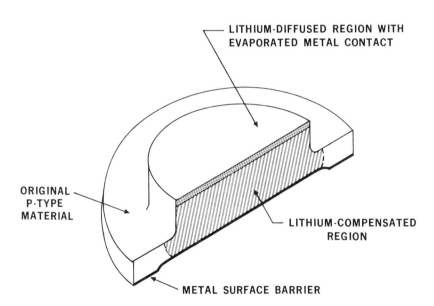

FIG. 2.9 Cross section of a typical lithium-drifted silicon-semiconductor detector.

to the device the drifted region acts as an insulator with an electric field throughout its volume. At liquid nitrogen temperature, it exhibits very low leakage current, so signals produced by radiation-induced ionization are easily detected. Germanium detectors may also be used in some X-ray spectrometers. The lithium-drifting process has been employed in the past in germanium as well as silicon, but high-purity germanium detectors are now available. Since no lithium compensation is present these detectors can, when absolutely necessary, be temperature-cycled to room temperature--unlike lithium-drifted germanium detectors. Since the mobility of lithium in silicon is much lower than in germanium, lithium-drifted silicon detectors can also be temperature-cycled.

The p-type surface barrier contact (a metal evaporation) is made very thin; entrance windows below 0.2 μm are typical. The intrinsic region is the sensitive volume of the detector where photoelectrons are produced. These lose energy by producing ionization in the form of free holes and electrons. The free charges are swept away by the applied bias and collected. Since a well-defined average energy is required to produce an electron-hole pair in a particular semiconductor material, the total number of charges produced is directly proportional to the energy of the absorbed X ray (assuming total X-ray absorption). In an ideal detector the total charge is completely collected for each detector event and the response of the system to monochromatic photons is a single peak with zero counts elsewhere in the spectrum. However, observed spectra obtained with real semiconductor detectors exhibit a continuous background extending from zero energy to that of the principal peak. Investigations of the nature of this background have established that its origin is the distortion of internal electric fields caused by the detector surfaces present in most geometries [11,12]. Figure 2.10(a) shows a cross section of a typical detector. The shaded portions indicate areas in which the internal field lines do not terminate at the electrical contact formed by the metal surface barrier but, instead, terminate on the ill-defined surfaces. Events occurring within this

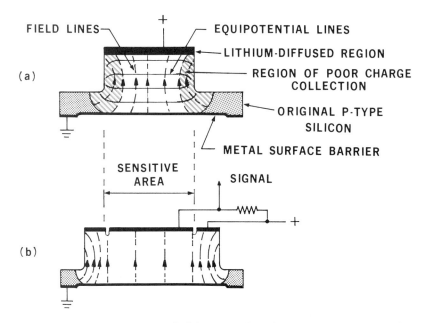

FIG. 2.10 Cross sections of (a) a top-hat detector geometry and (b) a guard-ring detector showing the internal field distributions.

region experience incomplete collection of the charge signal resulting in a smaller signal than should be produced. To the extent that such events are continuously distributed up to the amplitude corresponding to the full-energy peak, they account for much of the observed continuum background.

The magnitude of the continuum relative to the peak can be reduced by using a larger detector and collimating the incoming radiation to avoid the shaded regions. Another approach is to use the guard-ring detector illustrated in Fig. 2.10(b). Here, an additional annular contact or guard ring is added on the lithium-diffused side and maintained at the same dc potential as the central region. The signal is taken from the central area while the outer ring serves to maintain uniform internal field lines in the detector. Such guard-ring detectors show greatly reduced background. Further reduction can be achieved by extracting a signal from the guard ring and, if an event occurs sufficiently close to the edge of the sensitive region that a portion of the charge is collected in the guard ring, the coincident signal in the active region is rejected. This is termed the *guard-ring reject* method.

Assuming that the appropriate measures are taken to reduce such detector artifacts as these, then the fundamental limit in detector background below the principal peak is established by the escape of secondary particles or photons from the detector surfaces. Photoelectric absorption of an X ray in the detector ejects an electron from one of the shells of Si or Ge atoms creating a vacancy which subsequently fills with accompanying emission of Si or Ge characteristic X rays or an Auger electron. To produce a "full-energy" signal, the whole energy of the initial photoelectron and that of the secondary radiation produced when the Si or Ge atom deexcites must be absorbed in the sensitive region of the detector. Any loss of energy will cause a signal smaller than desired resulting in detector background. This type of detector background is fundamental and depends only on the characteristics of the detector material and on its geometry. It should be noted that even in the ideal case where continuous background due to detector effects could be reduced to zero there would still be significant background contributions due to bremsstrahlung produced in the sample by the electrons ejected from atoms in the sample.

The escape of a characteristic X ray of the detector material subtracts a definite energy from an event and such events produce discrete "escape" peaks in the background distribution at energies slightly below the main peaks. In the case of Si, the K escape causes a peak at an energy 1.74 keV less than the full-energy peak. Due to the large penetration depth of X rays in Si and the small fluorescence yield of Si vacancies, these peaks are important for photon energies below 6 keV and cannot achieve a size of more than 2% of the full-energy peak. In Ge, K-escape peaks occur at an energy of 9.87 or 11.0 keV below the main peak depending on whether the Kα or Kβ X rays escape. Due to the large fluorescence yield in Ge and the penetration of these high-energy X rays, these peaks can be as large as 20% of the full-energy peak intensity. Since the energy of these peaks occurs in a very awkward range from the point of view of fluorescence analysis, Ge detectors find only limited applications here.

The efficiency of Ge detectors is much higher than that for Si detectors at high energies due to the Z^3 dependence of photoelectric cross sections. Figure 2.11 shows the efficiency variation of various detectors with energy. Note that the Ge detector efficiency changes rapidly in the 10-keV region near the K-shell absorption edges. In practical systems the fall off in efficiency at low energies is set by the thickness of Be window on the cryostat-vacuum enclosure. The curves of Fig. 2.11 are for a 25-μm window but Be windows as thin as 7 μm can be fabricated where good low-energy efficiency is essential. Air absorption is also important in such

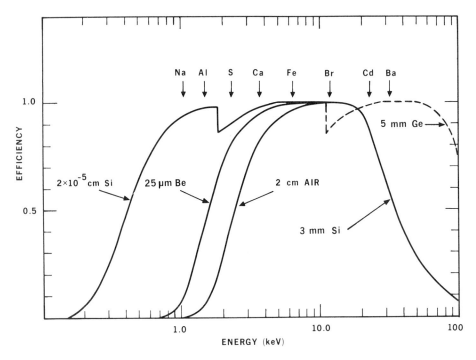

FIG. 2.11 Calculated efficiencies of Ge and Si detectors as a function of incident X-ray energy.

systems, so helium or vacuum paths are provided between the sample and the detector system. In special applications windowless systems are sometimes used; the detection of boron K X rays at 185 eV has been reported [13] in a system of this type. Problems associated with operating windowless systems make their use difficult for general purposes. Although the intrinsic efficiency of semiconductor detectors makes them sensitive to characteristic (either K or L) X rays of all elements from boron to uranium in the period table, the characteristics of the excitation and the energy resolution normally restrict the use of a spectrometer to a much smaller range of elements.

The energy resolution of a semiconductor detector spectrometer is determined partly by electronic noise but also by statistical fluctuation in the number of electron-hole pairs generated by monochromatic radiation. The root-mean-square (RMS) deviation of a monoenergetic X ray line caused by these two sources of fluctuation can be expressed as the quadratic sum of two components:

$$\sigma_{RMS} = \sqrt{\sigma_e^2 + \sigma_s^2} \qquad (2.3)$$

where σ_e is the electronic noise contribution and σ_s represents the statistical spread in the number of free charges produced in the detector. The electronic noise contribution is determined by the input amplifier stage and by detector leakage current. Operating the detector at low temperatures is essential to reduce the latter component. The statistical spread σ_s can be represented by

$$\sigma_s = \sqrt{F\ E\ \varepsilon} \qquad (2.4)$$

where E is the X-ray energy, ε is the average energy required to produce an electron-hole pair, and F is the Fano factor which corrects for the departure of the energy-loss process from a Poisson distribution. Typical values are ε = 3.7 eV and F = 0.12 for Si, and ε = 3.0 and F = 0.10 for Ge. Comparison of these with the value of ε = 30 eV for gas detectors illustrates the inherent resolution advantage of semiconductor detectors. The lower value of both F and ε for Ge relative to Si accounts for slightly better resolution at high energies for Ge relative to Si. Figure 2.1(b) is a plot of Eq. (2.2) with appropriate values for ε and F for Si, allowing also for the contribution of electronic noise.

2.3.2 Signal-Processing Electronics

In order to achieve the minimum possible resolution contribution due to electric noise sources it is necessary to filter the output signals to enhance the signal relative to noise. Figure 2.12 is a schematic of a typical signal channel employed with a semiconductor detector. The input preamplifier stage is mounted with the detector and maintained at low temperature in the cryostat. It integrates each detector charge signal to produce a voltage step proportional to the charge. This pulse is then amplified and shaped in a series of integrating and differentiating stages to achieve the optimum shape to maximize the signal-to-noise ratio. The resulting shaped pulse is then processed and encoded in a multichannel pulse-height analyzer which accumulates the X-ray spectrum.

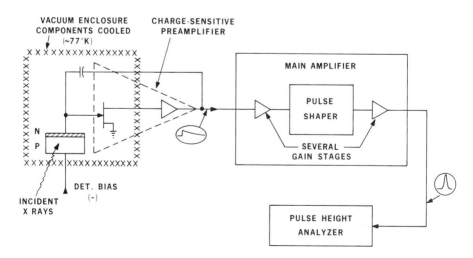

FIG. 2.12 Schematic of complete X-ray spectrometer system.

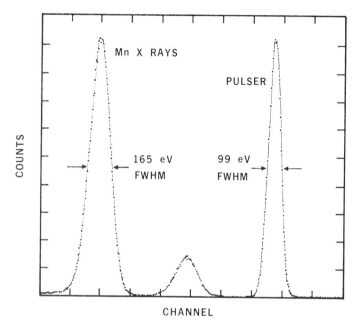

FIG. 2.13 Spectrum of Mn K X rays together with a pulser peak.

Figure 2.13 is an example of a spectrum obtained from a typical Si detector spectrometer showing the K X-rays of Mn together with a peak generated by a pulser. The resolution of the pulser peak is a measure of the electronic noise σ_e. The difference between the 99-eV pulser and the 165-eV Mn Kα widths is due to the statistics of charge collection as expressed in Eqs. (2.3) and (2.4). Electronic resolutions as low as 60 eV have been achieved in special systems.

As compared with γ-ray spectrometers, X-ray spectrometers are characterized by the large values of signal-shaping times used to optimize energy resolution. Since these long pulse-shaping times place a severe limit on the maximum allowable counting rate, a shorter shaping time than the optimum (from the point of view of resolution) is used. Figure 2.14 is a plot of the counting-rate characteristics of a typical semiconductor system operated at various pulse-shaping times. The characteristic shaping time referred to in Fig. 2.14 is the time required for a gaussian-shaped pulse to

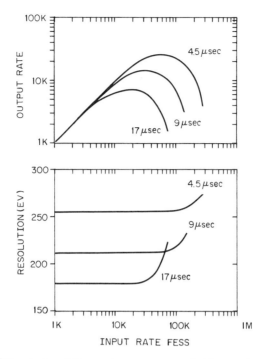

FIG. 2.14 Output-counting and energy-resolution rate as a function of input counting rate for a pulsed optical-electronic processor employing pile-up rejection.

reach its maximum value and must be distinguished from the characteristic shaping time constant which can be considerably shorter. For the cases shown in Fig. 2.14, the best resolution is achieved at 17 μsec peaking time. However, the output rate at high input counting rates is higher for the lower values of peaking times. Consequently a compromise value is often employed depending on the application.

The shape of the input/output data curves of Fig. 2.14 illustrates the effects of loss of pulses due to pulse pile-up. Since the signals arrive at random times, the time interval between pulses obeys a Poisson distribution. Following one event, the probability that a second event will occur within a time t is given by

$$P(t) = 1 - e^{\lambda t} \tag{2.5}$$

where λ is the average rate in count/sec. At counting rates of 10^4 count/sec there is a 40% probability of two events occurring within 50 μsec of each other. Such pulses overlap and produce erroneous pulse amplitudes. To eliminate such ambiguous events, most spectrometer systems employ circuitry which detects such pulse pile-up and eliminates any signals contaminated in this way. The drop in output rate for high input rates as shown in Fig. 2.14 reflects the effect of pulse pile-up rejection. If lower counting rates are normally encountered, such as may be the case in radioisotope excitation, pile-up rejection may not be necessary.

Since pile-up rejection eliminates some of the valid events, an absolute measurement of the intensity of X-ray lines requires a method of correcting for these losses in the final data. Fortunately, the loss fraction is the same for all amplitudes in the spectrum so compensation can be affected by extending the total counting interval by the appropriate factor. The normal method is to use a live time clock system in which the counting time is determined by counting a predetermined number of periodic clock pulses which pass through the same rejection gate as the signals. If this system is properly designed the clock pulses undergo the same probability of rejection as the X-ray events.

Another method of overcoming pulse pile-up and dead time losses is possible if a pulsed X-ray tube is available. By shutting off the X-ray tube excitation immediately an event is detected the probability of additional events occurring during the pulse-processing time is reduced essentially to zero. At the end of the pulse-processing time the tube is then turned on again and the next event is awaited. In such a system the output rate equals the input rate, no pile-up rejection is required, and higher counting rates can be achieved.

Following the analog signal processing just discussed, signals pass into an analog-to-digital converter and thence are stored in a multichannel analyzer or small computer. The result is a spectrum of the number of detected events as a function of X-ray energy.

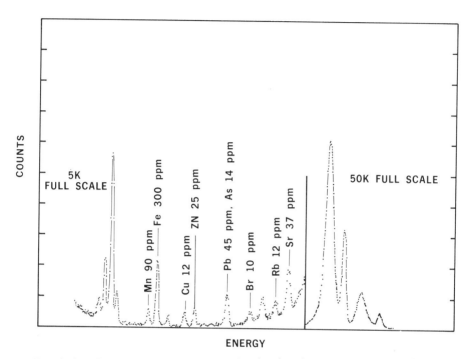

FIG. 2.15 Fluorescence spectrum obtained using Mo K X-ray excitation on a National Bureau of Standards Orchard leaf specimen (30 mg/cm^2).

This is the information which must be used to derive elemental concentrations.

Figure 2.15 is an example of a pulse amplitude spectrum obtained using Mo K X-ray excitation of an National Bureau of Standards orchard leaf sample. Elemental concentrations are indicated on the plot. The peak-to-background ratio for the fluorescent peaks is representative of current techniques.

2.4 CALIBRATION METHODS

In this section we will analyze factors which determine the relationship between the concentrations of elements in the sample and the counting rate for the characteristic X rays in the detector.

Only the simplest case will be analyzed but the generalizations required to extend the analysis to other classes of samples will be indicated.

For the purpose of illustration we chose a sample in the form of a homogeneous slab of thickness d (cm) which is irradiated by photons of energy E_0 and intensity I_0 as illustrated in Fig. 2.16(a),

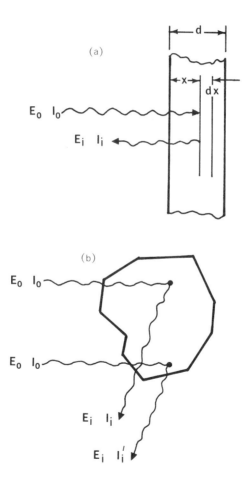

FIG. 2.16 Diagrams showing the mechanism X-ray absorption corrections for (a) a uniform layer and (b) an irregular particle. In the latter case, I_i and I_i' are due to the difference in absorption paths.

To avoid cumbersome trigonometric functions, we will assume that the incident and fluorescent X rays have directions normal to the surface. To extend the analysis to other angles the appropriate distances can be divided by the cosines of the angle of incidence.

From the absorption expression of Eq. (2.1) we can write the intensity of fluorescence K or L radiation detected by the detector from the atoms of element i contained in a layer of thickness dx located at distance x below the surface of the sample as follows:

$$dI_i = [I_0 e^{-(\mu/\rho)_0 \rho x}]_1 [\tau_i \rho_i \omega_i \varepsilon_i dx]_2 [e^{-(\mu/\rho)_i \rho x}]_3 \quad (2.6)$$

The brackets logically separate the three components as follows:

1. The intensity of the incident flux transmitted by a thickness x, density ρ, and average total mass attenuation $(\mu/\rho)_0$ for the incident radiation.

2. The fraction of the X rays reaching the element of thickness dx which interact in the element to produce vacancies in the K or L shells of element i followed by X-ray emission. The number of vacancies is $\tau_i \rho_i dx$. The fluorescence yield is ω_i. For convenience ε_i, the detector efficiency which is a function of the energy E_i, is also included in this term.

3. The probability that the fluorescence X rays of energy E_i will escape back to the surface of the sample from a depth x in a medium with average mass absorption coefficient $(\mu/\rho)_i$.

Integration of Eq. (2.6) over the sample thickness d gives the following expression for the total intensity of fluorescence radiation observed by the detector:

$$I_i = I_0 \, G \, \tau_i \rho_i \omega_i \varepsilon_i \, \frac{1 - e^{-[(\mu/\rho)_i + (\mu/\rho)_0]}}{[(\mu/\rho)_i + (\mu/\rho)_0]} \quad (2.7)$$

where the factor G is an efficiency term determined by the effective solid angle from the sample to the detector.

Two limiting cases of practical importance can be calculated.

For very thin samples in which absorption and interelement effects are negligible $[(\mu/\rho)_i + (\mu/\rho)_0]\rho d \ll 1$, and the expression reduces to

$$I_i = (I_0 \, G \, \tau_i \omega_i \varepsilon_i) \rho_i d \tag{2.8}$$

where the quantity $\rho_i d$ is the concentration of the element of interest in gm/cm^2.

The quantity in parentheses relates the concentration of the sample to the observed counting rate in the simplest of cases. Calculation of this "relative X-ray excitation probability" can be performed using published values for τ_i and ω_i [14-16]. The results of these calculations were shown in Fig. 2.8. The fact that these curves are smoothly varying functions of atomic number has important consequences for quantitative multielement analysis. If the system has been carefully calibrated for a given element, then an approximate result for a neighboring element can be easily inferred from a crude plot of Eq. (2.8). R. Giauque et al. [22] have refined the technique using a semiempirical approach to the evaluation of this function and have obtained accurate calibrations extending over a wide range of atomic numbers. Table 2.2 is a comparison of measured and calculated results for thin-film standards. These authors have

TABLE 2.2

Relative Excitation Efficiencies

Peak	Calculated intensity	Standard value
Cr Kα	0.381	0.370 ± 0.011
Mn Kα	0.450	0.435 ± 0.003
Fe Kα	0.587	0.599 ± 0.003
Ni Kα	0.884	0.882 ± 0.007
Cu Kα	1.000[a]	1.000 ± 0.015[a]
As Kα	1.653	1.660 ± 0.083
Se Kα	1.776	1.753 ± 0.057
Pb Lα	0.804	0.774 ± 0.019

[a]Values are normalized for Cu Kα.

also extended the method to include an attenuation correction of Eq. (2.6) for moderately thin homogeneous samples in which the exciting radiation is not severely attenuated. This calibration technique will permit analysis of a wide class of samples such as air pollution filters and many biological specimens.

The second special case of Eq. (2.7) involves very thick samples in which $\rho d \to \infty$. In this case the expression reduces to

$$I_i = \frac{I_0 \, G \tau_i \omega_i \varepsilon_i}{(\mu/\rho)_i + (\mu/\rho)_0} \left(\frac{\rho_i}{\rho}\right) \tag{2.9}$$

This equation can be applied to samples in which the absorption of the matrix is known beforehand (i.e., $(\mu/\rho)_i$ and $(\mu/\rho)_0$) and the concentration ρ_i is not large enough to significantly alter the matrix absorption. This is normally so in trace element analysis (see Chap. 14).

The extension of these ideas from monoenergetic excitation to broad-energy excitation involves a second sum or integration of Eq. (2.7) over the incident energy distribution. This can be done using parametric expressions for μ/ρ and τ together with measured X-ray distributions. Alternatively a totally empirical approach can be used. The qualitative features of a calibration curve generated experimentally are similar to those discussed for monoenergetic excitation (see Chap. 15).

In order to use these semiempirical formulas for accurate analytical purposes, it is necessary to prepare samples in the right form. Ideal samples are homogeneous and must be dilute to avoid interelement X-ray enhancement effects. Techniques for preparing such samples are discussed elsewhere in this book. In situations where these requirements are not met and where interelement and absorption effects may be important, a more elaborate calibration procedure must be undertaken. In its most general formulation, the intensity of X rays produced by element i depends on the concentration of all other elements in the sample. This can be expressed as follows:

ENERGY DISPERSION

$$\rho_i = \frac{I_i}{\sum_j a_{ij} I_j} \tag{2.10}$$

The quantitative a_{ij} represent a relative measure of the absorption of enhancement effect of element j on the observed intensity I_i. The problem of calibration then involves the determination of the coefficients a_{ij}. A number of methods of doing this have been developed [17,18]. This problem is also present in the case of wavelength-dispersion analysis, but the spectrum obtained from an energy-dispersive spectrometer can give a simultaneous measurement of I_i for all elements of interest. These and related topics are discussed in the chapter on data interpretation (Chap. 3). These complications are avoided by preparing standards which closely approximate the form of the unknown sample; this method becomes even more desirable as the sample form departs far from the simple case discussed earlier. Where interelement and matrix effects are important, and the number of constituent elements is not large, the solution of Eq. (2.7) can be effectively accomplished by using calibration curves. As the sample departs far from uniformity and homogeneity, it becomes essential that a standard sample closely replicating the main constituents of the unknown be prepared.

The question of sample homogeneity closely relates to the particle size problem often encountered in X-ray fluorescence analysis. With the increasing importance of energy-dispersive analysis applied to air particulate analysis an understanding of this effect is desirable. Figure 2.16(b) illustrates the problem of measuring the X-ray intensities from an irregularly shaped particle. Defining a characteristic absorption length $[(\mu/\rho)\rho]^{-1}$, if the average diameter of the particle is comparable to or greater than this length, there will be a large difference between the output intensities I_i and I_i' arising from different parts of a particle. In this case the measured intensity may be a function of the distribution of particle sizes in powder specimens, or of surface irregularities in solid samples. Rhodes has treated the question of a generalized particle size distribution and derives models for various distributions [19].

T. Dzubay has derived correction factors to be used for air particulate analysis assuming that the particles are spherical [20]. Criss has calculated the particle size correction for a variety of shapes and compositions [21]. In any of these approaches, assumptions must be made regarding the particle composition and shape before a reasonable correction can be applied. To overcome this, R. Giauque et al. have used a variable excitation energy to measure the fluorescence output as a function of the penetration depth of incident X-ray beam [22]. Since these corrections are only approximate, the particle-size effects always limit accuracy, particularly when analyzing for low-Z elements.

2.5 SPECTRAL ANALYSIS

The earlier discussion of calibration techniques assumes that an accurate measure of X-ray line intensities could be derived from the multichannel fluorescence spectrum. This is clearly easy for high concentration elements where a simple summation over the channels in the peak is adequate. However, in the general case, the size of the peak of interest must be derived despite the presence of continuous background and interference from lines due to other elements in the sample. The accurate subtraction of continuous background is particularly important in trace analysis applications where the peaks are small. In large-scale analysis programs, this must be done accurately and rapidly.

Figure 2.17 helps illustrate the complexity of the problem of spectral analysis. This example is from a typical air pollution sample consisting of the collected particulates on a 5-mg/cm^2 cellulose backing. The sample was irradiated with Mo K X rays. The logarithmic plot emphasizes the number of details which must be included in a successful analysis procedure. For example the continuous background beneath the peaks in the region between Ca and Br is very difficult to define since the density of peaks obscures any reference level which might exist. The number of elements present requires careful consideration of the possible overlap of lines from adjacent

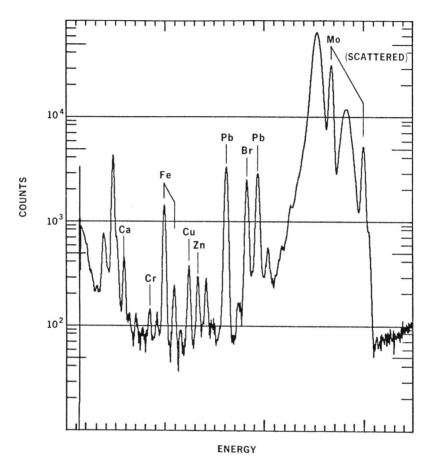

FIG. 2.17 A typical fluorescence spectrum obtained with an air pollution sample. Note the density of peaks and the subsequent lack of regions from which a smooth background might be estimated.

elements and of their K X rays with the complex L X-ray spectrum of Pb. Often, the spectra to be analyzed are much simpler than this but a general purpose computer program should be capable of handling all types of problems.

The simplest method for extracting peak areas is to sum the data over a selected channel interval, typically the Kα or Lα X ray, after subtraction of a continuous background. Interferences between elements can be handled by a sequential subtraction of the overlapping lines using the known ratios between Kα and Kβ to determine the amount of overlap. In cases where the number of interelement interferences are few and the background can be easily determined, this simple method can be used effectively.

The next level of sophistication involves the use of fitting routines in which the background and peaks are compared to reference standards on a point-for-point basis. The comparison can be performed either by a simple sum over the peaks of interest or by least squares techniques.

The background is usually estimated either by interpolation over a limited region or by relating the background at a given point to the intensity of scattered radiation at the high-energy region of the spectrum. The standard peak shapes can be generated mathematically using gaussian peak shapes of appropriate width and locations and intensities specified by previously known data. Difficulties with the mathematical method of peak generation arise where the characteristic X-ray lines become complex, such as the L X rays of heavy elements or the partially resolved Kα1-Kα2 doublet for $Z > 40$.

An alternative method for generating peak shapes is to experimentally measure the spectrometer response to X rays of a given element and store that spectrum in the computer [23]. This has the advantage that multiple structure and nongaussian peak shapes are automatically determined. The cost of the added storage capacity required is offset by the simplifications involved in programming. The advantages of using standard shapes and more complicated fitting routines are a slight increase in the statistical accuracy of the measurement and a more flexible approach to the problem of interelement interferences. However, the program can be extremely complex particularly if a large number of elements are to be simultaneously analyzed and do not fit easily into a small computer.

The next level of mathematical complexity involves the use of transformation techniques to enhance the peak-to-background ratio prior to spectrum unfolding. Three examples of interest are Fourier analysis [24], digital filtering [25], and correlation techniques

[26]. Although each is slightly different they are all basically methods whereby the mathematical transformation filters out the smoothly varying background. These algorithms can be incorporated into small computer systems. On the other hand, the problems of multiple lines and interelement interference must be handled separately.

As yet no completely general method of X-ray spectral analysis has been developed. Various combinations of the above techniques are used by a number of laboratories and commercial manufacturers with varying success depending on the type of application. It is probable that all the available programs achieve the same level of accuracy for the more straightforward problems and fail equally in the more complicated situations. The reasons for this lie not only in the complexity of multiple element spectra but also in instrumental effects which are difficult to compensate mathematically. These effects include small gain shifts in the spectrometer as a function of time, changes in peak shape and location as the count rate varies, and variations in the relative intensities of the characteristic X ray for a particular element due to absorption effects. Since these problems are not easily compensated for in typical programs, the subsequent spectral analysis is subject to large errors particularly in the case of the less intense peaks. The net effect is to limit the dynamic range over which elemental concentration can be determined.

2.6 APPLICATIONS

Areas in which energy dispersive analysis has been successfully employed are numerous and have been reported extensively in the literature. In most respects the applications duplicate those of conventional X-ray fluorescence. Applications range from on-line process control, where only one or two elements must be continuously maintained, to environmental analysis applications, where a complete list of elements and their absolute concentrations must be determined. Some specific areas of interest are discussed in Chaps. 10 to 16.

2.7 CONCLUSION

The technique of energy-dispersive photon-excited X-ray fluorescence analysis has been developed to the point where it constitutes a potent analytical tool for accurate and sensitive nondestructive multiple-element analysis. A number of instruments based on this method are now commercially available. As they become more widely used the full range of applications of the technique will no doubt be expanded.

REFERENCES

1. J. V. Gilfrich, P. G. Burkhalter, and L. S. Birks, Anal. Chem., 45, 2002 (1973).
2. E. H. S. Burhop and W. N. Asaad, Advan. At. Mol. Phys., 8, 163 (1972).
3. J. W. Cooper, Nucl. Instrum. Methods, 106, 525 (1973).
4. W. H. McMaster, N. Kerr Del Grande, J. H. Mallet, and J. H. Hubbell, Lawrence Livermore Laboratory Report UCRL-50174 (1970).
5. J. R. Rhodes, Design and Application of X-Ray Emission Analyzers Using Radioisotope X-Ray or Gamma Ray Sources, in Energy Dispersion X-Ray Analysis: X-Ray and Electron Probe Analysis, ASTM Special Technical Publication 485, 243 (1970).
6. R. D. Giauque, F. S. Goulding, J. M. Jaklevic, and R. H. Pehl, Anal. Chem., 45, 671 (1973).
7. J. C. Russ, A. O. Sandborg, M. W. Barnhart, C. E. Soderquist, R. W. Lichtinger, and C. J. Walsh, Advan. X-Ray Anal., 16, 284 (1973).
8. T. G. Dzubay, B. V. Jarrett, and J. M. Jaklevic, Nucl. Instrum. Methods, 115, 297 (1974).
9. L. Kaufman and D. C. Camp, Advan. X-Ray Anal., 18, 2 (1975).
10. R. H. Howell, W. L. Pickles, and J. L. Cate, Jr., Advan. X-Ray Anal., 18, 265 (1975).
11. F. S. Goulding, J. M. Jaklevic, B. V. Jarrett, and D. A. Landis, Advan. X-Ray Anal., 15, 470 (1972).
12. J. M. Jaklevic and F. S. Goulding, IEEE Trans. Nucl. Sci., NS-19, No. 3, 384 (1972).
13. J. S. Hansen, J. C. McGeorge, D. Nix, W. D. Schmitt-Ott, I. Unus, and R. W. Fink, Nucl. Instrum. Methods, 106, 365 (1973).

14. W. J. Gallagber and S. J. Cipolla, *Nucl. Instrum. Methods*, *122*, 405 (1974).
15. Bambynek, *Rev. Mod. Phys.*, *44*, 716 (1972).
16. J. S. Hansen, H. V. Frevnd and R. W. Fink, *Nucl. Phys.*, *A142*, 604 (1970).
17. L. S. Birks, in *X-Ray Spectrochemical Analysis*, Vol. 11, Chemical Analysis, John Wiley, New York, 1969.
18. R. O. Muller, in *Spectrochemical Analysis by X-Ray Fluorescence*, Plenum Press, New York, 1972.
19. J. R. Rhodes and C. B. Hunter, *X-Ray Spectrometry*, *1*, 113 (1972).
20. T. G. Dzubay and R. O. Nelson, *Advan. X-Ray Anal.*, *18*, 630 (1975).
21. J. W. Criss, *Anal. Chem.*, *48*, 179 (1976).
22. R. D. Giauque, R. B. Garrett, L. Y. Goda, J. M. Jaklevic, and D. F. Malone, *Advan. X-Ray Anal.*, *19*, 305 (1976).
23. J. M. Jaklevic, B. W. Loo, and F. S. Goulding, Photon Induced X-Ray Fluorescence Analysis Using Energy Dispersive Detector and Dichotomous Sampler, (to be published) in *X-Ray Fluorescence Analysis of Environmental Samples*, 1976. Lawrence Berkeley Laboratory Report LBL-4834.
24. N. G. Volkov, *Nucl. Instrum. Methods*, *113*, 483 (1973).
25. T. Inouye, T. Harper, and N. C. Rasmussen, *Nucl. Instrum. Methods*, *67*, 125 (1969).
26. A. Robertson, W. V. Prestwich, and T. J. Kennett, *Nucl. Instrum. Methods*, *100*, 317 (1972).

Chapter 3

DATA INTERPRETATION

L. S. Birks

X-Ray Optics Branch
U. S. Naval Research Laboratory
Washington, D. C.

3.1	Introduction	59
3.2	Dead-Time Correction	60
3.3	Background	61
3.4	Line Interference	62
3.5	Converting X-Ray Intensity to Composition	64
	3.5.1 Types of Samples	64
	3.5.2 Simple Analytical Expression	65
	3.5.3 Regression Equations	66
	3.5.4 Fundamental-Parameter Equations	67
References		68

3.1 INTRODUCTION

An X-ray analyst quickly develops the ability to recognize elements and to make rough estimates of composition merely from the appearance of the energy-dispersion display or the strip-chart recording from a crystal spectrometer. To go from the easy rough estimates to quantitative analysis requires more than additional practice; it requires mathematical data treatment, perhaps as simple as preparation of a calibration curve or as complex as computer calculation of elaborate matrix effects. Data interpretation can be optimized if the analyst understands why it is necessary and how detailed it needs to be for different kinds of applications or degree of reliability. We shall consider individually the several factors which the analyst needs to account for.

3.2 DEAD-TIME CORRECTION (SEE ALSO SEC. 2.3.2)

The X-ray analyst should be aware of the two kinds of dead-time problems and what to do about them. The simplest kind of dead time is in the multichannel analyzer for energy dispersion where a few μsec are required to store a pulse in the proper energy channel. While the pulse is being stored the analyzer is electronically gated so it will not accept another pulse and the "off" time is dead time. However, one mode of reading the counting time only records while the analyzer is "on," so the dead-time correction is accounted for automatically. The second, and more important, kind of dead time t_D is in the detector itself or the amplifier circuits. This is the recovery time for the detector or amplifier to return to equilibrium after a pulse is generated. It cannot be accounted for automatically. Several years ago Beaman [1] published an elaborate treatment for dead-time correction under various detector and amplifier situations. Such elaboration is unnecessary, however, and one can use the simple expression

$$I_{true} = \frac{I_{meas}}{1 - t_D I_{meas}} \tag{3.1}$$

if the counting rate is kept low enough so that $t_D I_{meas}$ is less than 0.1 ($1 - t_D I_{meas}$ approximates the series expansion of Beaman's exponential term because the third term is negligible when the second term is less than 10%).

3.3 BACKGROUND

The signal we want to relate to composition is the characteristic line for each element. For either wavelength or energy dispersion there is always some background intensity, and this must be subtracted from the total intensity measured at each line position. Direct manual subtraction of the background under each line is probably still the most common practice although automated methods are becoming more popular. Figure 3.1 illustrates some of the comments to be made about background subtraction: it shows line A on a low constant background, line B on a sloping background, and line C on the side of a strong adjacent peak. The signal may be specified as the intensity at the peak or the integrated intensity under the full peak or the integrated intensity within the full width at half maximum (called B in Sec. 1.2.3) as illustrated for line A in Fig. 3.1. Whatever wavelength or energy interval is integrated for the peak, the same interval must be integrated for the background.

The analyst can estimate the background by fairing in a curve under the peaks as shown in the figure or, in the case of microsamples,

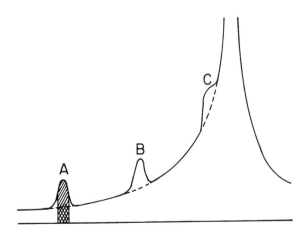

FIG. 3.1 Three types of background situation: A, a low, nearly constant background; B, a sloping background due to a nearby peak; C, a severe background slope on the side of a strong peak.

he can run a blank which does not contain the elements of interest. Or he can express the background intensity as a parametric or analytic function of energy or wavelength and let a computer make the background subtraction automatically. (The reader should turn to more detailed literature [2-4] for specific instruction on background curve fitting.)

One thing that should be recognized by the analyst is that for trace elements the lines are hardly distinguishable from the background. It then becomes necessary to know the background quite accurately because its statistical uncertainty limits the precision and detectability for trace constituents.

3.4 LINE INTERFERENCE

Whenever instrumental resolution is not adequate to separate neighboring X-ray characteristic lines the analyst must correct for the interference. The simplest situation is shown in Fig. 3.2 where the β line from element B interferes with the α line from element A, but there is no interference with the Bα line. The analyst simply measures the fractional contribution of Bβ at the Aα position using a sample of element B only. In the figure the fraction is 10% of Bβ or what is more important, 2% of Bα. Then in any sample containing both A and B, the Bα is measured and 2% of its intensity is subtracted from the measured intensity at position Aα. (Note, in all of these evaluations the dead-time and background corrections are made before operating on the line intensities.)

The next more complex situation occurs when Bα as well as Bβ is interfered with by some other element line, say Cγ (not shown in Fig. 3.2). Then it may become easier to determine Aα and Bβ by an iterative solution of the equations

$$I_{T\alpha} = I_{M\alpha} - K_{\alpha\beta} K_{T\beta} \tag{3.2a}$$

$$I_{T\beta} = I_{M\beta} - K_{\beta\alpha} I_{T\alpha} \tag{3.2b}$$

where $I_{T\alpha}$ and $I_{T\beta}$ are the true intensities at the Aα and Bβ positions,

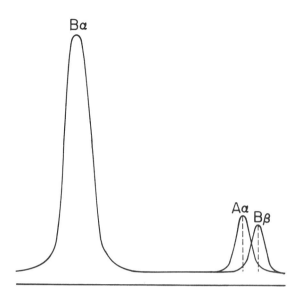

FIG. 3.2 Simple line interference. The Bβ line interferes with Aα, but there is no interference with Bα.

$I_{M\alpha}$ and $I_{M\beta}$ are the measured intensities at the Aα and Bβ positions, and $K_{\alpha\beta}$ and $K_{\beta\alpha}$ are the fractional interferences at the Aα line by Bβ and vice versa (for symmetric line shapes, such as gaussian, $K_{\alpha\beta} = K_{\beta\alpha}$). For the first iteration the measured value $I_{M\beta}$ is substituted for $I_{T\beta}$ in Eq. (3.2a) and likewise $I_{M\alpha}$ is substituted for $I_{T\beta}$ in Eq. (3.2b). The values for $I_{T\alpha}$ and $I_{T\beta}$ calculated in the first iteration are employed in the next iteration, etc. Usually three or four iterations are adequate to give unchanging values of $I_{T\alpha}$ and $I_{T\beta}$.

Somewhat more elaborate treatment of line interference is desirable in energy dispersion, where interference is more common and the interference coefficients much larger. The test method is unfolding the overlapping lines with a computer program which either assumes the lines are gaussian or which stores measured line shapes. Because the position of each line is known the unfolding is rapid but, as always, subject to large uncertainty when a weak line must be unfolded from a strong line close by.

3.5 CONVERTING X-RAY INTENSITY TO COMPOSITION

All of the factors discussed up to this point are merely to eliminate interferences so that the true intensity of a given characteristic X-ray line can be stated unambiguously and quantitatively. We now come to the important matrix corrections which must be made so that X-ray intensity can be converted to chemical composition.

3.5.1 Types of Samples

Three types of samples must be considered: thin layers, intermediate-thickness layers, and thick (bulk) samples.

3.5.1.1 Thin Samples

For very thin layers such as evaporated films or pollution samples, the intensity from element I is linear with mass of element i, and concentration C_i is simply the measured intensity I_i divided by the measured sensitivity S_i in count/seg/g.

$$C_i = \frac{I_i}{S_i} \tag{3.3}$$

Note that particle size corrections are necessary for some thin samples such as particulate pollution, which is discussed in Chaps. 2 and 14.

3.5.1.2 Intermediate Thickness

Strangely enough samples of intermediate thickness are more difficult to treat than either thin samples or bulk samples. Some attempts have been made to write expressions for intermediate thickness [5,6], but no general formulations can be presented at this time.

DATA INTERPRETATION

3.5.1.3 Bulk Samples

For bulk samples when there is negligible secondary fluorescence the expression for concentration of element i in a multicomponent sample looks deceptively simple and nearly the same as Eq. (3.3) for thin samples:

$$C_i = \frac{I'_i \mu'_i}{S'_i} \quad (3.4)$$

where μ'_i represents absorption in the composite sample.

$$\mu'_i = \mu_P \csc \phi_1 + \mu_F \csc \phi_2 \quad (3.5)$$

where μ_P is the linear absorption coefficient for primary radiation in the composite sample, and μ_F is the linear absorption coefficient for the emitted characteristic radiation from element i by the composite sample, ϕ_1 and ϕ_2 are the incident and take-off angles. S' in Eq. (3.4) corresponds to sensitivity in Eq. (3.3), but is the intensity from a bulk sample of pure element i times a μ' term for the pure element i?

Unfortunately, Eq. (3.5) cannot be evaluated directly because μ_P and μ_F depend on total sample composition (which we are trying to determine) and on the spectral distribution of the primary incident radiation. In addition, if secondary fluorescence of one element by other elements within the sample occurs, the additional contribution I_i depends on the concentrations of the secondary fluorescer elements and their fluorescent yields, and the expression becomes much more complex.

The remainder of this chapter describes the various approaches used for analysis of bulk samples.

3.5.2 Simple Analytical Expression

In X-ray analysis, dependence on intensity of all elements present precludes preparation of a simple analytical curve except for binary samples or when the variation of all components is within a very limited range. A good example of the latter is in the steel industry where for each alloy type (say 316 stainless steel), an expression of the algebraic form y = ax + b can be used,

$$C_i = aI_i + b \tag{3.6}$$

where a and b are constants determined explicitly for only the 316 alloy. For each other nominal composition different values of a and b must be determined.

Equation (3.6) is widely used in large-scale industrial analysis *because it works* in spite of its formal shortcomings.

3.5.3 Regression Equations

The next most complicated expression, and the one most commonly used in general X-ray analysis, is a set of simultaneous linear regression equations which employ empirically determined interelement coefficients α_{ij} and include the concentration of each element explicitly. Each equation in the set is of the form

$$\frac{C_i}{I_{r,i}} = \alpha_{i0} + \alpha_{ii}C_i + \sum_{j \neq i} \alpha_{ij}C_j \tag{3.7}$$

where α_{i0} corresponds to an intercept, $\alpha_{ii} \equiv 1$, and α_{ij} is the effect (either + or -) of element j on the intensity of element i; $I_{r,i}$ is the relative X-ray intensity of element i in the sample, i.e., the measured intensity divided by the intensity from pure element i. The best way to obtain the numerical values for the α_{ij} values is to measure as many known standards as there are elements of interest and substitute the measured $I_{r,i}$ and known C_i values into the set of simultaneous equations, Eq. (3.7), so they can be solved for each α_{ij} value.

Regression equations of the form of Eq. (3.7) have advantages over other regression equations in that (1) they are nonlinear in $I_{r,i}$, which allows a better approximation to a real analytical curve; (2) the curved line which results for a local range of C_i

DATA INTERPRETATION   67

is not constrained to pass through the origin because of the α_{10} term; (3) it is easy to solve for the α_{ij} values because for standards where $I_{r,i}$ and C_i are known, the simultaneous equations are linear in the α's and hence easy to solve, and (4) the equations are linear in evaluating an unknown because the $I_{r,i}$ values are measured and hence constant for each equation.

The α coefficients are always determined for a particular type of sample and are valid for only that type of sample (i.e., iron-base alloys, copper ores, etc.). Nevertheless the regression equations allow much more latitude in composition than the simple analytical expression described in the preceding section.

3.5.4 Fundamental-Parameter Equations

It has always been possible to write exact analytical expressions for the intensity-concentration relationship [7], but until the advent of large computers they could not be solved in practice. The formulations use the spectral distribution of primary radiation [8] and tabulated values for mass absorption coefficients [9], fluorescent yield [10], and jump factors [11].

The most practical form of a computer program for analytical use of the fundamental-parameter equations is available to analysts from COSMIC [12]; the documentation is available separately from the listing. Use of the fundamental-parameter approach allows complete freedom in type of sample with only pure elements as standards (or one known compound for each element). Recent, soon-to-be-published, tests of the fundamental-parameter program on data from Rasberry and Heinrich [13,14] and NRL indicate accuracy to within 2% relative for up to nine elements in iron-base and nickel-base NBS certified samples. The cost of analyzing data by the fundamental-parameter program is less than ten cents per element and competitive with other X-ray methods when time sharing to a large computer is available.

REFERENCES

1. D. R. Beaman, J. A. Isari, H. K. Birnbaum, and R. Lewis, *J. Phys.*, *E*, *5*, 767 (1972).
2. R. L. Walter, R. D. Willis, W. F. Gutknecht, and J. M. Joyce, *Anal. Chem.*, *45*, 843 (1974).
3. J. C. Russ, *EXAM Methods*, Edax International Inc., Prairie View, Illinois.
4. C. E. Fiori, R. L. Myklebust, and K. F. J. Heinrich, *Anal. Chem.*, *48*, 172 (1976).
5. J. W. Criss, unpublished.
6. A. W. Witmer and M. L. Verheijke, *Proc. 18th Colloq. Spect. Int.*, *3*, 658 (1975).
7. J. Sherman, ASTM Spec. Tech. Pub. #157, June, 1953.
8. D. B. Brown, J. V. Gilfrich, and M. C. Peckerar, *J. Appl. Phys.*, *46*, 4537 (1975).
9. W. H. McMaster, Lawrence Radiation Laboratory Report UCRL 50174 (1969).
10. W. Bambynek, *Rev. Mod. Phys.*, *44*, 716 (1972).
11. *Handbook of Spectroscopy* (included X-ray tables), CRC Press, Cleveland, 1974.
12. *COSMIC* (Computer Software Management and Information Center), University of Georgia, Athens, Georgia 30601.
13. S. D. Rasberry and K. F. J. Heinrich, *Anal. Chem.*, *46*, 81 (1974).
14. K. F. J. Heinrich and S. D. Rasberry, *Advan. X-Ray Anal.*, *17*, 309 (1974).

Chapter 4

PRECISION AND ACCURACY

Dr. Paul D. Zemany

General Electric Company
Knolls Atomic Power Laboratory
Schenectady, New York

4.1	Introduction	70
4.2	Types of Errors	71
4.3	Enumeration: Coincidence Loss	74
4.4	Random Sample: Binomial Distribution	77
4.5	Poisson Distribution	79
4.6	Standard Deviation	83
4.7	Gaussian Distribution	84
4.8	Counting Compared to Other Measurements	86
4.9	Estimate of Variance and Standard Deviation	93
4.10	Net Line Intensity: Comparison of Intensities	93
4.11	Minimum Detectable Amount	96
4.12	Measurement of Peak Intensity: Goniometer Setting Error	99
4.13	Background Measurements	100
4.14	Measurement Strategy	102
4.15	Estimating Composition: Calibrations	103
4.16	Sample Configuration	107
4.17	Light Elements	109
References		110

4.1 INTRODUCTION

All measurements of continuous parameters such as mass, duration, or length are made with reference to an arbitrary standard or unit. These measurements are not exact and are only estimates of the true value. Only the standard kilogram, second, and meter defined by the National Bureau of Standards are exact. Carbon 12 is the reference isotope for atomic weights and is exact by definition; all other atomic weights are measured values and subject to refinement. The assigned values of standard reference materials are slightly in error despite the fact that the greatest effort possible is used to assure results of the highest quality. The amount by which a measured value differs from the true value is the error of the result. However, the magnitude of the error is not known exactly--otherwise the exact, true value could be deduced. The best that can be done is to state the *probability* that the error does not exceed certain limits.

X-Ray spectroscopy and other methods that use counting techniques are unique in that it is fairly easy to state the minimum error, or the best result that can be expected in the measurement.

The main functions of X-ray spectroscopy are to identify the various chemical elements present in a sample of material (qualitative analysis) and estimate the amount of each (quantitative analysis). This chapter will discuss the quality or reliability of the results of these analyses. The discussion will consider errors of measurement in general, counting statistics, the application of statistical techniques to measurement data, and the errors made in using these data to derive concentration or amount.

This discussion will be concerned mainly with single channel X-ray spectrographs that use crystals for dispersing the spectrum and scintillation or proportional detectors with scalers for measuring intensity. The principles and equations presented can usually be applied to other types of equipment and data collection methods.

The concentration of an element in an unknown is calculated by comparing the intensity of a characteristic line of that element with the intensity of the same line in a standard or set of standards. The precision with which the intensities can be measured is ultimately limited by counting statistics. The first step in minimizing errors is to assure that deficiencies in the equipment or in manipulations on it do not cause this limit to be exceeded significantly. The efficiency of the determination can be optimized by choosing the best method for sample preparation and measurement.

The accuracy of the analytical result depends on how well all factors that could bias the determination are minimized or accounted for, and is ultimately limited by the accuracy with which the standards are characterized. The comparison will not be valid unless standard and unknown microscopic and macroscopic geometries are similar; the backgrounds are properly treated and the total composition of standard and sample are considered. Interelement effects complicate the calculation of composition from intensity data. The use of an incorrect form of equation, inaccurate coefficients, or neglecting the effects of an element in either specimen or standard will result in an incorrect interpretation of the data.

4.2 TYPES OF ERRORS

Errors can be classified as the random errors of measurement, biases or systematic errors, and blunders or mistakes. Biases in a comparative measurement can be eliminated, at least in principle, thus the random error of measurement ultimately limits the quality of the final result. Ideally, an X-ray spectrographic method should be developed to the point that the magnitude of the error of the final result is no greater than that contributed by the random error of measurement. This ideal is too seldom attained and too often assumed, in practice.

Because of the random error of measurement, no matter how carefully the same quantity is measured, different values for the result will be obtained each time the measurement is repeated. When a group of such results are treated by the methods of statistics the *precision* of the measurement method can be determined and the probability of deviations of various magnitudes from the *mean* can be estimated. When the precision of an analytical method is high the result is very sharply defined. The results on replicates are nearly alike, with minor variations in the final digits.

A second type of error is a bias, or a consistent or *systematic* error which is added to the random error. A bias in a measurement may be inferred only if there is already some estimate of the precision and if the difference between the expected or true result and the measured result is greater than the limits of error set by the precision.

Accuracy is concordance of the results of measurements with the known or true value. An accurate result has high precision and is free of bias. A measurement or analysis may be highly precise but not necessarily accurate unless systematic errors have been corrected.

A measurement or analysis may be subject to several different biases. For example, a marksman usually fires a few rounds at a target to determine the biases of the sights for the prevailing conditions. If the sighting-in shots are close together (high precision) but high and to the right of the target center (true value), the biases of elevation and windage are compensated by adjustment of the sights. At longer range the pattern of a group spreads out--the random error is greater--and readjustment of the sights is required since the biases change with range. If the sighting-in shots scatter too much the sight adjustment cannot be made. At a range of a foot or so precision is excellent, there is no need for a wind correction, but since the rifle bore and sights are not collinear, a large elevation correction is required to correct for a third bias.

The procedure just described is an example of a *calibration*. Calibrations or standardizations are needed for practically every chemical analysis. One purpose of calibrations is to estimate biases so that there will be guidance in correcting for or eliminating them. The other purpose is to establish the relation between the measured results and concentration, i.e., the scale factor. Calibrations are done by measuring standards, or known targets which are accepted as valid. Standards should be available for the entire measurement range. The biases important in trace determinations may be different than those at higher concentrations.

Inasmuch as X-ray spectrographic methods are calibrated by comparisons to standards the accuracy of the method is of no concern except that it is limited by that of the standards. The precision is limited by the random error of measurement. Precision is best treated by the methods of statistics.

Since accuracy is limited by the accuracy with which the composition of reference materials are known, it is essential that they be carefully prepared or obtained from reliable sources. The best standard for a determination is one that matches the sample in all pertinent details, e.g., composition, surface finish. Such standards are seldom available and the sample composition is computed by interpolation between standards or by more involved procedures as in the fundamental-parameters method [1], where the "standards" are the fundamental parameters.

A third type of error that occurs too often, which is frequently large and frequently neglected or dismissed in a discussion of errors in measurement, is a mistake or blunder. Mistakes (as distinct from random errors or biases) can be attributed to equipment malfunction or misadjustments, errors in sampling or identification, operator error in performing the measurement, arithmetic mistakes, faulty standards, erroneous assumptions, neglect of a bias, and a great many other factors. When discovered the result is discarded, the error corrected, and the incident forgotten. If a result is very far from that expected it is likely to be recognized as a mistake, but if there is no prior knowledge or expectation, it may be undetected. A record of the frequency and cause of those detected would be useful in estimating how often they can be expected in analyzing unknown materials or in giving guidance to prevent their occurrence.

A faulty result due to a blunder is not considered in discussions of the accuracy or precision of a method. The most frequently recommended method for discovering the occurrence of a mistake where it is not likely to be recognized is to repeat the analysis; by an entirely different method if possible. This is also the procedure used to detect bias in the absence of reliable reference materials.

It is probably true that the most frequent cause of large discrepancies between the reported and true values are due to blunders rather than lack of precision in the measurement.

4.3 ENUMERATION: COINCIDENCE LOSS

Counting the number of objects or events is an entirely different operation than a measurement of a continuous property of a material. Biases and random error are not involved, in fact have no meaning in this context. The Bureau of Standards does not provide a "standard dozen." Yet mistakes and blunders do occur. Legend has it that the Bakers' Dozen (13 units) was a response to medieval consumer protection legislation which stipulated that the punishment for a short count was amputation of a hand. Evidently bakers were not much more skilled than their customers in the art of counting! The legislation did not protect the customers from a second kind of mistake--a mistake in identification. The baker could slip in a few stale rolls when the customer expected them to be fresh.

X-Ray emission spectroscopy uses counting for the estimation of intensity. Intensities are expressed as discrete values, the number of events counted in a given time interval. The usual unit of intensity I is counts per second. With modern equipment the error in timing is so small that it can be neglected; it is not likely to exceed a few parts per million provided the proper operation of the timing device is verified.

Mistakes in counting can and do occur in the measurement of intensity. The two kinds of mistakes mentioned above have their counterparts in counting quanta. The events counted can be incorrectly identified since the detector does not distinguish events

PRECISION AND ACCURACY

due to the analytical line from background, interfering lines, or other causes. At low count rates the total number of events counted is exactly correct, but at higher count rates a pair of nearly simultaneous events may be counted as one. This type of mistake is a *coincidence error or loss*, and becomes apparent at count rates over 10,000 count/sec with proportional, scintillation, or solid state detectors and at much lower rates for geiger counters.

An approximate correction can be made to compensate for coincidence loss. If two quanta cause ionization in the detector in the interval before charge collection the ionization will be the sum of the ionization caused by each. (See also Sec. 2.3.2.) When using pulse-height selection neither will register as a count, without pulse-height selection a single pulse is registered. With solid state detectors using multichannel analyzers, a single pulse is recorded at an energy equal to the sum of the energies of the two quanta. Immediately after registering a pulse the detector system gain is greatly reduced and a closely following ionizing event will not be registered. The recovery of the ability to detect a second count, or of the gain, takes place exponentially with an average half-time in the order of a few microseconds. Quanta absorbed during the end of the recovery period yield a pulse reduced in size which results in an amplitude shift, that is lower average pulse size, or it may be missed because it is not great enough to pass the lower level discriminator. The inactive period is lost from the total counting time, and occurs after each pulse. Thus, the total effective counting time T_e is

$$T_e = T_T - NT_d \tag{4.1}$$

where T_T is elapsed time, N the number of pulses recorded, and T_d the dead or recovery time of the detector. For example, if $T_T = 100$ sec, $N = 10^6$, $T_d = 2 \times 10^{-6}$ sec, then $T_e = 100 - 10^6 \times 2 \times 10^{-6} = 98$ sec, and the intensity is $10^6/98 = 10204.08$ count/sec.

The detector system average dead time for a particular energy can be estimated experimentally by setting the spectrometer to count a high intensity line ($I > 10^5$ count/sec) and adding thin absorbing foils, one at a time, to the beam. The count obtained initially and for each additional foil is measured. A semilog plot of intensity vs. number of foils will be a straight line with many foils in place, but the extension of the line to the unfiltered beam will extrapolate to a higher count than that recorded. The number of lost counts can be estimated and the detector system dead time calculated. For example [2, p. 188], if the extrapolated rate I_{true} is 1.5×10^5 count/sec and the measured rate I_{obs} is 2.4×10^4 with all foils removed, then

Coincidence loss correction:

$$I_{true} = \frac{I_{obs}}{1 - I_{obs}T_d} \qquad (4.2)$$

or $I_{true} = 2.4 \times 10^4 / (1 - 2.4 \times 10^4 \, T_d)$ and $T_d = 3.5 \times 10^{-5}$ sec. Equation (4.2) is the one usually used to calculate the coincidence loss correction.

Another method to determine T_d is to plot a characteristic line intensity as a function of tube current. They are proportional. At high tube currents and count rates the linear relationship will not seem to hold since the observed intensity is subject to coincidence loss. This method can only be used if the current can be measured precisely and if the wave form or "ripple" of the constant-voltage X-ray tube power supply does not change with tube loading.

A third method [3, p. 105] of estimating dead time is to measure the intensities of a pair of lines as excitation is increased. Suitable line pairs are the Kα and Kβ lines of an element or the same line in the first and second order, or the same line with or without a primary filter. The true ratio of intensities R can be measured at low excitation where coincidence losses are negligible by counting for a long time. Then, since

$$\frac{I_a}{1 - I_a T_d} = \frac{RI_b}{1 - I_b T_d} \qquad (4.3)$$

at all intensities, T_d can be calculated from measurements at high intensities:

$$T_d = \frac{I_a - RI_b}{(1-R)I_a I_b} \tag{4.4}$$

Another method is to calculate both R and T_d by substituting I_a and I_b measured at different intensities in Eq. (4.3). T_d is sometimes a function of energy and of the setting of the electronics. At very high count rates other perturbations occur so it is poor practice to use the correction if the coincidence loss exceeds about 10-15%. If a direct comparison is being made, that is the standard and sample are virtually identical, there is no need for a coincidence correction (or any other correction). The coincidence loss correction is made only when the actual count rate is required for the calculation.

Some detector electronics, particularly those used with the solid state detectors, minimize the coincidence loss errors. The counting time may be extended by the amount of the dead time and the circuit can be designed so that recovery time is very short. Another complication that is corrected to some extent by more sophisticated electronic circuits is that caused by the fact that pulses occurring before complete recovery of the amplifier gain are, without the correction, smaller than normal and would be registered as pulses of lower energy in a multichannel analyzer.

4.4 RANDOM SAMPLE: BINOMIAL DISTRIBUTION

Except for coincidence loss the number of counts recorded is exactly the number of events affecting the detector and passed to the scaler. The reason different totals are recorded on each repetition is that each counting interval is for a different *random sample*. ("Sample" is used here in its mathematical sense, not as "sample" of material to be analyzed.) The count is correct, exactly correct for each repetition. The variation is in the individual random samples. In contrast, if some continuous parameter is measured repeatedly, for example the length of an object, the length remains exactly the same for each repetition, the variation in results is due to random variations in the measurement process.

To illustrate the counting of random samples suppose it is required to determine the relative amounts of elements in a piece of a binary alloy that contains 2×10^y atoms of element A and 8×10^y atoms of element B. If it were possible to count the individual atoms of each kind the result would be a "perfect" analysis and the fraction of each kind could be stated exactly. An approximation of the relative amounts could be obtained from the number of each kind when a "sample" consisting of a few atoms, for example n = 10, were counted. If the entire piece was used up by removing 10 at a time, the sum of the results would be the same as that obtained when the whole sample was analyzed by counting. Individual estimates f_A of the fraction $F_A = (2 \times 10^Y)/(2 \times 10^Y + 8 \times 10^Y) = 0.2$ could have any of the values 0, 0.1, 0.2, ..., 1.0 and $\Sigma f_A/N = F_A = 0.2$, where N is the total number of samples, (10^Y). The average of all results gives the correct value, the individual random samples only approximate it.

How are the individual results distributed? How many of the random samples of 10 atoms had 0, 1, 2, ..., 10 A atoms? The answer to this question is given by one of the statistical distribution functions. If the total number of atoms is limited, the removal of 10 atoms might change the fraction of A atoms in the remainder. For this case, the hypergeometric distribution applies. If the samples are returned after counting or the total number of atoms is large, the *binomial distribution* applies. The binomial distribution is the expansion of $[F + (1 - F)]^N$. The general form for the individual terms is

$$f_X = \frac{N!}{X!(N-X)!} F^X (1-F)^{N-X} \qquad (4.5)$$

when N is the number of units sampled (10 in the present case), X is any of the integers 0, 1, 2, ..., N, and f_X is the fraction of the total number that occurs for any particular value of X. F is the average value, 0.2 in this case. Figure 4.1a shows the distribution when F = 0.2 and N = 10. Tables [4, p. 183] of values of f_X for various N and X eliminate the need for calculation.

4.5 POISSON DISTRIBUTION

While the binomial distribution applies to samples of a given size (N fixed), the results of counting quanta, where N can have any positive integral value, are described by the *Poisson distribution*. The value of individual terms is given by

$$f_X = \frac{F_A^X e^{-F_A}}{X!} \tag{4.6}$$

which can be derived as a limiting case of the binomial distribution with undefined sample size. Figure 4.1b shows the Poisson frequency distribution for each integer when the average count is two units. It expresses the probability of obtaining any given result for the enumeration of events in a given time interval when the events occur randomly in time. It has only one parameter F_A, the average count. In contrast, the binomial has two parameters, the number of units in the sample and the mean value.

To demonstrate that the Poisson distribution is applicable to the results of counting X-ray quanta, an X-ray spectrometer was set up to take a large number of counts of a low intensity. The average count in the first example, shown in Table 4.1, was 0.525 for 1,049 repetitions. Run 2 had an average of 3.71 for the 544 counting periods, and 830 repetitions averaged 27.97 in run 3. These last results are also displayed in Figure 4.1c. The good agreement between the observed and expected distributions is a verification of the hypothesis that counting results obtained by measuring the intensity of an X-ray line have a Poisson distribution provided that the equipment is working properly and that the intensity is *constant* during the period. If the agreement is poor, it is evidence that some disturbing influence is present.

This is a most important conclusion. Demonstrating that counting results do have a Poisson distribution is one of the most important uses of statistics in assessing the quality of X-ray spectrochemical results. Note, however, that the 830 measurements used to draw the histogram in Fig. 4.1c are scarcely enough to demonstrate

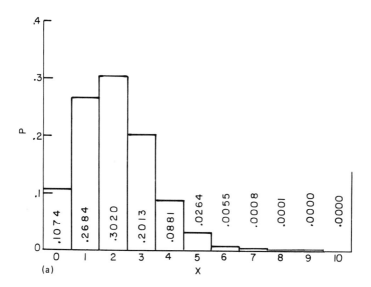
(a)

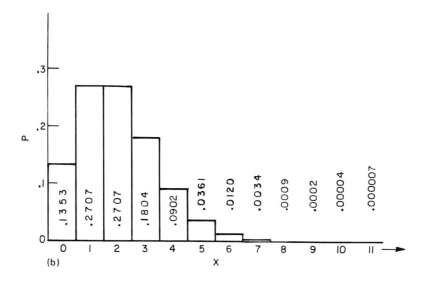
(b)

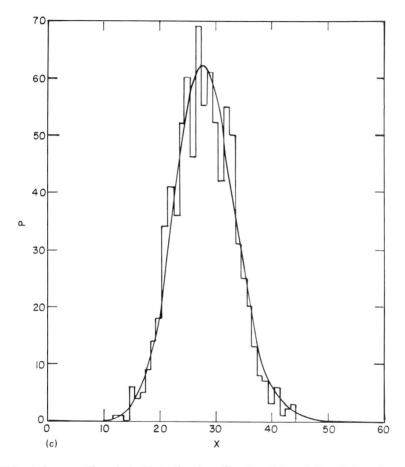

FIG. 4.1 a. Binomial distribution for $N = 10$ and $F = 0.2$. (See Sec. 4.4.) b. Poisson distribution for $F_A = \bar{X} = 2$. (See Sec. 4.5.) c. Poisson distribution for "Run 3" of Table 4.1. P is the probability that a sample will contain x units. The sum of the probabilities is one in 4.1a and 4.1b, but 830 in 4.1c. The smooth curve 4.1c represents the Poisson distribution, the histogram experimental results.

TABLE 4.1

Test of Poisson Distribution for X-Ray Spectrometer Counting Results

	Run 1			Run 2	
	Mean value = 0.525			Mean value = 3.71	
	Variance = 0.513			Variance = 3.90	
	Standard deviation = 0.716			Standard deviation = 1.97	
	Number of occurrences:			Number of occurrences:	
Count	Observed	Calculated	Count	Observed	Calculated
0	621	621.1	0	15	13.3
1	320	324.9	1	51	49.3
2	93	85.4	2	95	91.6
3	15	15.1	3	110	113.3
4	0	2.0	4	100	105.1
5	0	0.3	5	71	78.0
	1049		6	52	48.3
			7	29	25.6
			8	14	11.9
			9	5	4.9
			10	1	1.9
			11	1	0.6
			12	0	0.2
				544	

	Run 3				
	Mean value = 27.97				
	Variance = 28.30				
	Standard deviation = 5.323				
	Number of occurrences:			Number of occurrences:	
Count	Observed	Calculated	Count	Observed	Calculated
11	0	0.12	29	61	60.18
12	1	0.28	30	52	56.11
13	1	0.60	31	42	50.63
14	0	1.21	32	55	44.26
15	6	2.26	33	50	37.52
16	4	3.96	34	31	30.86
17	5	6.52	35	25	24.67
18	9	10.13	36	20	19.17
19	14	14.91	37	13	14.49
20	18	20.86	38	8	10.66
21	34	27.79	39	7	7.65
22	41	35.33	40	3	5.35
23	36	50.08	41	6	3.65
24	52	56.03	42	1	2.43
25	60	60.28	43	2	1.58
26	46	62.45	44	3	1.01
27	69	62.39	45	0	0.62
28	55	60.18	46	0	0.38
				830	

the agreement. Perfect agreement would require an infinite number of measurements. Obviously it is impractical to make as many measurements as in the examples given when testing the agreement, and a more efficient statistical method is needed.

4.6 STANDARD DEVIATION

Table 4.1 lists values for the *standard deviation* S. It is the square root of the sum of the squares of the deviation of each result from the average of all results divided by one less than the number of results.

Standard deviation:

$$S = \sqrt{\frac{\Sigma(X_i - \bar{X})^2}{n-1}} = \sqrt{\frac{\Sigma X_i^2 - n(\bar{X})^2}{n-1}} = \sqrt{\frac{n\Sigma X_i^2 - n(\Sigma X_i)^2}{n(n-1)}} \qquad (4.7)$$

The last two of the three equivalent forms of Eq. (4.7) are more convenient for the calculation of S than is the first.

In any actual case, the value calculated for S by the equations above is only an estimate of the true standard deviation σ which could be calculated only from an infinite number of random samples, i.e., the entire *population*. In the same way the average $\bar{X}$ is an estimate of the true mean μ, the average of the entire population. If a second set of measurements of intensity of the same population were made, slightly different values of $\bar{X}$ and S would be obtained, since these are random samples of σ and μ. Generally, the greater the number of random samples used to calculate $\bar{X}$ and S, the more closely they agree with μ and σ. The square of the standard deviation S^2, called the *variance*, is close to the average $\bar{X}$ for the data shown in Table 4.1. This is true for random samples from a population having a Poisson distribution. Conversely, the square root of an average value, or even of a single measurement, of random samples from a Poisson distribution is an estimate of the standard deviation. The significance of these statements will be developed later.

The variance of a few measurements can be calculated and compared to the mean to test if the population sampled has a Poisson distribution. This procedure is very much easier than to gather enough data to display in tabular form, as in Table 4.1, to perform the test. The χ^2 (chi square) goodness-of-fit test is a more powerful test, but requires more calculations [5].

4.7 GAUSSIAN DISTRIBUTION

Figure 4.1b with $\bar{X} = 2$ is noticeably skewed, while Fig. 4.1c is nearly symmetrical. As $\bar{X}$ increases, the Poisson distribution becomes more symmetrical. Also, as $\bar{X}$ increases, the arithmetic of calculating the individual terms becomes more difficult.

In the limit, as $\bar{X}$ increases, the Poisson distribution $f_X = F_A^X \exp(-F_A/X!)$ can be expressed by

$$f_X \simeq \frac{1}{\sqrt{2\pi}\, F_A} \exp\left[\frac{-(X - F_A)^2}{2F_A}\right] \simeq \frac{1}{\sqrt{2\pi \bar{X}}} \exp\left[\frac{-(X - \bar{X})^2}{2\bar{X}}\right] \quad (4.8)$$

In the more general case, where the mean μ and the standard deviation σ are independent variables, the expression

$$f_X = \frac{1}{\sigma\sqrt{2\pi}} \exp\left[\frac{-(X - \mu)^2}{2\sigma^2}\right] \quad (4.9)$$

gives the density function for the *gaussian distribution*, also called the *normal distribution*.

The gaussian is a continuous function in contrast to the binomial or Poisson in which the data points are positive integers. The gaussian extends from $-\infty$ to $+\infty$. With the Poisson distribution approximated as the unique gaussian for which $\sigma^2 = \mu$, numerical analysis is facilitated. For the approximation to apply reasonably well, $\bar{X}$ should exceed about 100. Some useful relations can be derived from this gaussian, and they can be applied to the count data obtained in X-ray spectrometry. Experimental verification of the fact that X-ray counting results conform to this unique gaussian are given in Ref. 6.

PRECISION AND ACCURACY   85

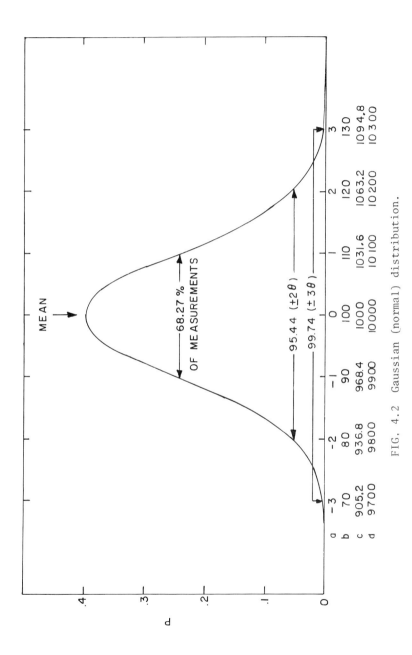

FIG. 4.2 Gaussian (normal) distribution.

The total area under the gaussian distribution curve as shown in Fig. 4.2 is one unit. It represents the entire population, or 100% of all possible results. The abscissa is first marked off in σ or S units, with the origin located at the mean value. The other three calibrations of the abscissa represent the distribution of X-ray counting results when the averages are $\overline{100}$, $\overline{1000}$, and $\overline{10,000}$. For the case where $\bar{X} \approx \mu = 10,000$, 68.27% of a long series of counts would lie between 9900 and 10100 (or ± 1S), while 95.44% (≈ 95%) would be between the 2S limits of 9800 and 10200. Five percent would exceed these limits.

A common way to express the precision of a result is to refer to the 2S limits as the *95% confidence limits*. (In strict statistical terms confidence limits apply to intervals containing parameters only [that is, μ]. The term is used here (loosely) to be a probability statement about the random variable X.) Expressed in a different way, 19 of 20 measurements would be included by the 2S limits, i.e., the chance that a particular result deviates from the mean by more than 2S units is about 5%. The *relative standard deviation* is the ratio of the standard deviation to the count obtained (or $\sigma/\bar{N}$). If $\bar{N} = 100$, the *coefficient of variation* is 10%; if $\bar{N} = 10,000$, the coefficient of variation is 1% provided the precision is limited by counting statistics (or $S = \sqrt{\bar{N}}$).

4.8 COUNTING COMPARED TO OTHER MEASUREMENTS

If a weight or a volume is measured (as is often done in analytical chemistry) on "exact" replicates, the results will vary about some mean value. Given a sufficient number to construct a histogram, it is likely that a smooth curve drawn through it would approximate a gaussian curve (Fig. 4.2). A standard deviation could be calculated which would be entirely independent of the mean value. Usually there is no single influence responsible for the variation of the results, many contributing factors can be named. When a series of counting results on replicates of a material are obtained in X-ray

PRECISION AND ACCURACY

spectrometry, it may be found that the standard deviation is not equal to the square root of the mean. On rare occasions it may be found that $S^2 < \bar{N}$. If the scaler skips the first decade giving a count of 10 for each pulse detected, then S^2 can be less than $\bar{N}$. If the counting circuit picks up a steady pulse rate, such as the power-line frequency, in addition to the pulses actually detected, then a low value of S will be obtained. In some counting systems, a low value of S is observed when coincidence losses are high. All three of these effects have been observed and all are indications of a defect in the counter electronics.

The usual concern is that $S^2 > \bar{N}$. Again, various random influences could be named responsible for the variation, such as errors in settings of the equipment, irregularities in the samples, and instability of the apparatus. When $S^2 > \bar{N}$, the precision of the measurements is not as good as is possible. Each of the variables contributing to the total S_T has a standard deviation S_i or variance S_i^2. The total variance of the measurement S_T^2 is the sum of variances of all factors involved including that arising from counting statistics S_c^2. Thus,

$$S_T^2 = S_c^2 + S_1^2 + S_2^2 + \cdots + S_N^2 \qquad (4.10)$$

and the standard deviation of the measurement is

$$S_T = \sqrt{S_c^2 + S_1^2 + S_2^2 + \cdots + S_N^2} \qquad (4.11)$$

S_c, the *standard counting error* [7, p. 331], is the square root of the count obtained in a determination of N.

The total variance S_T^2 is the only one estimated by direct computation for a series of measurements. Sometimes it is possible to estimate the variance due to a particular variable (say the error of resetting a goniometer, S_A^2) by making a series of counts when all but the one being investigated are kept constant. Then,

$$S_A^2 = S_T^2 - S_c^2 \quad \text{and} \quad S_A = \sqrt{S_T^2 - \bar{N}} \qquad (4.12)$$

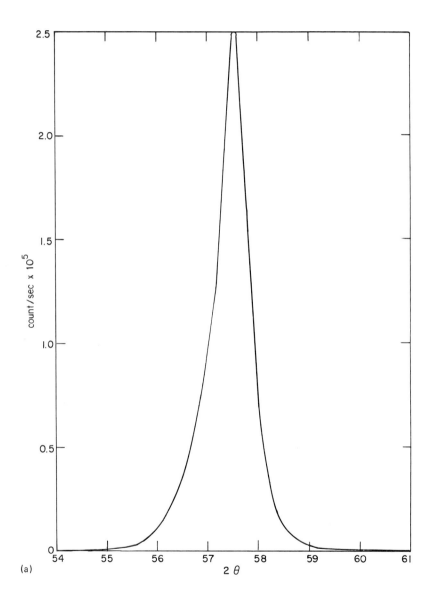

(a)

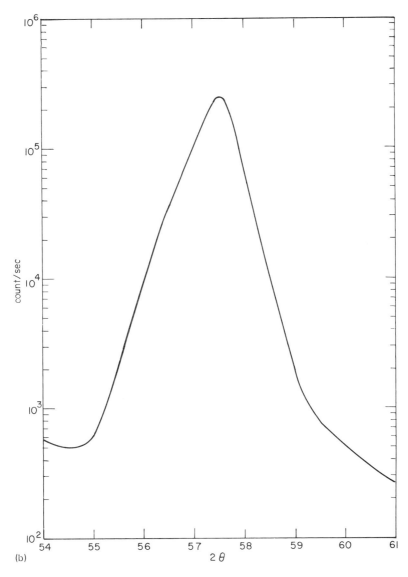

FIG. 4.3 Count rates from step scanning through an Fe Kα peak: (a) linear scale and (b) logarithmic scale.

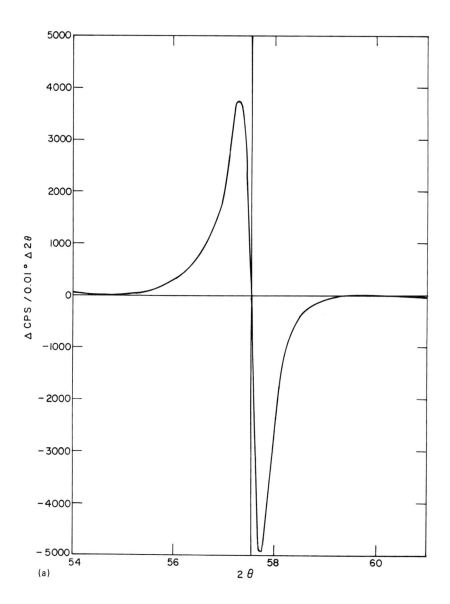

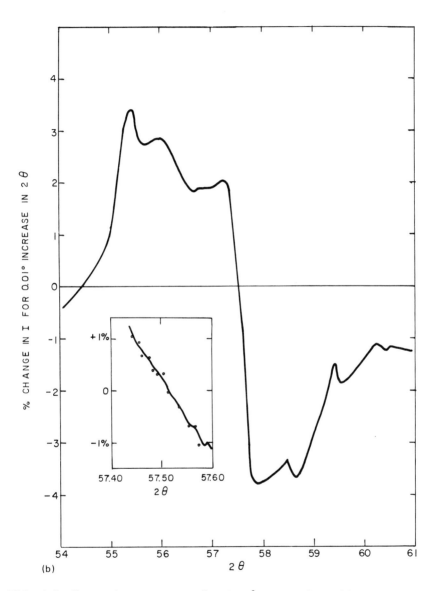

FIG. 4.4 Change in count rate for 0.01° versus 2θ: (a) in counts and (b) in percent.

If, in the preceding example, S_A accounted for the major part of the increase of S_T over that predicted from S_c, it may be possible to improve the techniques of resetting the goniometer, so that S_A becomes insignificant. Figure 4.4 shows that a change of 0.01° can cause a change of more than 3% in intensity when measured near a characteristic line peak. With high resolution, the change can be as great as 20% per 0.01°. If S_A is determined at the point of maximum percent rate of change with 2θ, S_A can be expressed in degrees 2θ. The voltage output stability and reset error of the X-ray tube power supply can be investigated by making measurements a kilovolt or two above the absorption edge of a characteristic line, for example, 52 kV for Gd K lines or 39 kV for Ba K lines, since a change of voltage of 1% changes the intensity by 10, to 20%.

Other variables suspected of contributing to loss of precision can be investigated in similar fashion. Sometimes $S_T^2 > \bar{N}$ even if nothing is changed between measurements. This may indicate instability or drift of the generator or detector, and steps can be taken to correct such a defect. X-Ray generator or detector instability can be distinguished by substituting a radioactive source, such as iron 55, for the X-ray tube and noting any change in the ratio of $S^2/\bar{N}$.

At some point, as $\bar{N}$ is increased by longer counting intervals, S_c^2 will exceed $\bar{N}$. This is the limit of the stability of the generator-detector system. With modern equipment S_c should not greatly exceed 10^3 when $\bar{N}$ is 10^6. The system short-term stability can reasonably be expected to have a coefficient of variation of 0.10%.

There are several types of system instability or variation of intensity with time. The 120-Hz variation in X-ray output is usually of no concern since it is regular, and short compared to normal counting times. Random short- and long-term changes in rate, spikes, and gradual drift are all forms of instability that cause errors in the interpretation of counting data if the change occurs during the time that related rate measurements are made.

The main uses of the statistical tests outlined are to estimate the precision of the equipment, or certain manipulations on it or on the material being examined. Identification of a source of variability is a first step in improving the analytical technique. Note that accuracy, or lack thereof, or the existence of biases is not dealt with. But accuracy cannot be achieved in a measurement until the precision is adequate.

4.9 ESTIMATE OF VARIANCE AND STANDARD DEVIATION

While the variance S^2 is an unbiased estimate of σ^2, S calculated by Eq. (4.7), is a biased estimate of σ. If S is calculated from two, three, or four values for N, it should be multiplied by the factors 1.253, 1.128, or 1.085 to get a corrected, unbiased estimate. The correction factors decrease as the number of values of N increases. An easier way to calculate S, and nearly as efficient for a few values of N, is to use the range, i.e., the difference between largest and smallest value. If the range for 2, 3, or 4 N values is multiplied by the factors 0.886, 0.591, or 0.486, the result is an unbiased estimate of S. Reference 4, p. 349, gives factors for other cases.

In addition to being biased, a standard deviation based on a few measurements of N is not likely to be a good estimate. A guide in identifying significant differences is useful. Chi square over degrees of freedom tables provide the desired guide [4, p. 295]. These tables list confidence limits for S, depending on the number of values of N used to estimate $\bar{N}$ and S, in terms of S^2/σ^2 ($\approx S^2/\bar{N}$). Usually the concern is that $S^2 > \bar{N}$, so only the upper limit is used. Table 4.2 lists a few values of the confidence limits.

4.10 NET LINE INTENSITY: COMPARISON OF INTENSITIES

A simple quantitative determination is the measurement of the thickness d of a film or a very low concentration of an element in some other material. The line intensity I_e increases linearly with thickness or concentration (up to a point). In the linear region,

TABLE 4.2

Chi Square Over Degrees of Freedom (χ^2/df)

df^a	95^b	99^b
1	3.841	6.635
2	2.996	4.605
3	2.605	3.782
5	2.214	3.017
9	1.880	2.407
15	1.666	2.039
19	1.587	1.905
49	1.354	1.529

[a] df (degrees of freedom) is one less than the number of replicates.

[b] The value of the ratio $S^2/\bar{N}$ is likely to be less than that given in the table 95 or 99% of the time if the counting results follow a Poisson for the given number of degrees of freedom. For example, if 10 replicate counts are taken (df = 9) there is one chance in 20 that $S^2/\bar{N}$ would equal or exceed 1.880 if the counting results had a Poisson distribution. A larger value for the ratio means that it is doubtful that the counts have a Poisson distribution.

$$I_e = Kd \qquad (4.13)$$

The proportionality constant K has to be determined for each case by the use of standards, and I_e for the sample determined by counting. I_e cannot be measured directly; it is the difference between the measured intensity I_t at the line position and the background intensity I_b. So

$$d = \frac{I_t - I_{b1}}{K} \qquad (4.14)$$

and

$$K = \frac{I_c - I_{b2}}{D} \qquad (4.15)$$

where I_c is the intensity for the standard of known thickness or concentration D, and I_{b1} and I_{b2} are the two measured backgrounds.

Then the concentration (in linear range) by ratio method, when compared to the standard of concentration D, fixed time T, and background correction becomes

$$d = \frac{D(N_T - N_{b1})}{N_c - N_{b2}} \quad (4.16)$$

where the N values are counts for a given time interval. Each of the intensities are subject to a random counting error.

The random error in the estimate of d depends on how the errors of the intensities interact; the greater the number of factors subject to independent random error required to calculate an analytical result, the greater the random error of the result. Equation (4.15) can be expressed in the form $KD = N_c - N_{b2}$. Then the variance of KD is $N_c + N_{b2}$, and $S_c = \sqrt{N_c + N_{b2}}$.

The *relative error*:

$$\varepsilon = \frac{\sqrt{N_c + N_{b2}}}{N_c - N_{b2}} \quad (4.17)$$

Since D is known, this is the relative error in the estimate of K. The error of $N_e = N_T - N_{b1}$ can be expressed in a similar fashion. When the ratio of net counts is used to determine d by Eq. (4.16), the relative error is the square root of the sum of the squares of the relative errors.

$$\varepsilon_T = \sqrt{\varepsilon_1^2 + \varepsilon_2^2} = \sqrt{\left(\frac{(N_T + N_{b1})^{1/2}}{N_T - N_{b1}}\right)^2 + \left(\frac{(N_c + N_{b2})^{1/2}}{N_c - N_{b2}}\right)^2} \quad (4.18)$$

Two background measurements are indicated. If the background is measured only once, the error of the result is increased due to the propagation of error [5], even if they measure the same thing.

To illustrate the application of the preceding equations and how the error grows as more measurements are combined to calculate a result, assume that the counts observed for background, standard, and sample are 2000, 10000, and 5000, respectively. The standard counting errors ($\sqrt{N}$) are 44.7, 100, and 70.7; the relative errors ($\sqrt{N}/N$) are 0.022, 0.010, and 0.014. The net counts for standard and sample after subtracting background are 8000 and 3000. The corresponding standard counting errors are 109.54 and 83.67 while the relative errors increase to 0.0136 and 0.0279. If the four counts (two backgrounds) are used to calculate concentration or thickness, the relative error of the result is 0.0310. The probability that the true value is within 6.2% of the value calculated is 95%.

In the previous discussion, it was assumed that counting statistics alone determine the error, that a proper standard was available, and that the background could be measured. These conditions are not always met.

4.11 MINIMUM DETECTABLE AMOUNT

A frequently occurring analytical problem is that of deciding if a sample of material contains a particular element. The simplest procedure for making the determination by X-ray spectrometry is to scan over a characteristic line of the element while making a strip-chart recording of rate meter output. If the element is present at a concentration equal to or greater than the minimum detectable amount, the scan will show a positive deflection at the line position. The height of the peak will be roughly proportional to the concentration. At low concentrations the peak height will approach the background rate, and the problem is that of distinguishing a true peak from the random variations or noise. It has been found [8, p. 54] that a deflection that is three standard deviations above the background average is a positive indication of the presence of the element. For example, if the background rate is 100 count/sec, the recorder time constant 1 sec, and the scan rate is 1° (2θ) per minute, a deflection of the recorder to 130 count/sec (background + 3σ) at the peak position is a positive indication of the presence of the element. A calibration with a suitable standard containing a known, greater amount of the element is required to calculate a numerical value for the minimum detectable amount or concentration. For example, if 1% of the element gives a net count rate of 3000 count/sec, then 0.01% would, in the present example, give a count rate of 30 count/sec, and 0.01% is the minimum detectable concentration. Experience has shown that samples containing the minimum detectable amount, defined in this way, will give a positive result in 90% of all trails.

If counting at the peak and background positions is done and a different set of conditions postulated, a more rigorous definition of the minimum detectable amount (MDA) can be stated. It is that amount or concentration of an element that can be detected with some specified confidence level. Thus, if the confidence level chosen is 95% and if a set of blanks and samples containing the MDA were measured, 19 of 20 of both blanks and those containing the MDA would be correctly identified. The determination is qualitative in the sense that the element is either found to be present or is not detected; it is the limiting amount for qualitative analysis.

The factors that determine the MDA are the specific count rate for the element (either counts per second per microgram for small samples or counts per second per % for bulk samples), the background rate, the time available, and the confidence level desired. The confidence levels need not be symmetrical. It is also assumed that a background rate can be determined, say by counting at an adjacent wavelength where the background does not change with wavelength. See Sec. 4.12.

The MDA_{95} (the subscript denotes the confidence level) is derived from counting statistics. It is assumed that no other significant error sources are present. Measurements of N_B and N_T are made on each sample for equal times. The net count $N_E = N_T - N_B$ is proportional to the amount present. The standard counting error for N_E is $S_c = \sqrt{N_T + N_B} \simeq \sqrt{2N_B}$. The error introduced by this approximation is small and can be neglected since the difference between N_T and N_B is small. $N_E = 0$ for blanks but individual results are distributed on the gaussian for which the standard deviation is S_c. Five percent of the individual values will exceed $+1.66 S_c$ (or $2.35 \sqrt{N_B}$). The usual $2 S_c$ limit for 95% confidence includes deviations above and below the mean, here only the 5% exceeding the mean value are significant. 1.66σ limits are used to define the upper (or lower) 5%. A sample for which $N_E < 2.35 \sqrt{N_B}$ is reported "element sought not detected," while a sample for which $N_E > 2.35 \sqrt{N_B}$ is reported "element present."

Sample containing the element at a concentration that has $N_E = 2(2.35)\sqrt{N_B} = 4.70 \sqrt{N_B}$ will have 5% of individual values of N_E less than $2.35 \sqrt{N_B}$. This concentration is the MDA, and a net count of $2.35 \sqrt{N_B}$ is the *decision line*. Samples containing the MDA and blanks can be distinguished with 95% confidence that the two possible assignments "not detected" and "present" are correct.

Samples containing less than the MDA will be reported "not detected" more frequently. For example, those whose concentration corresponds to the count for the decision line will be reported "not found" with a probability of 50%. Concentrations below the MDA are in a gray zone. The probability of a positive judgment increases with the concentration. The width of the gray zone increases with the confidence limits chosen.

The concentration or amount corresponding to the count required for the MDA is a function of background rate R_B (count/sec), counting time T, and the sensitivity S.

Minimum detectable amount (MDA):

(a) The rate meter deflection at line position equals or exceeds 3σ above the adjacent background if wavelength scan mode is used.

(b) $$MDA_{95} = 4.70 \frac{\sqrt{R_B T}}{ST} \qquad (4.19)$$

and the decision line

$$DL_{95} = 2.35 \sqrt{R_B T} \qquad (4.20)$$

when peak and background are counted for time T.

The factors 4.70 and 2.35 are adjusted for other confidence limits.

If the background rate is 200 count/sec, counting time 1000 sec for each, and the sensitivity 1000 count/sec per %, then N_B = 200,000, S_c = 632.5, and the net count for the MDA_{95} is 2102 which corresponds to 0.002102% ~ 0.002%. The decision line is at a net count of 1051.

The experimental verification reported in Ref. 9 of an equivalent definition of MDA shows that it applies to actual samples. Computer simulations using 10,000 pairs of random normal numbers to represent counting results gave 4.88% "present" if none of the element was added, and 94.82% "present" for a simulated addition of the MDA_{95}.

4.12 MEASUREMENT OF PEAK INTENSITY: GONIOMETER SETTING ERROR

The intensity observed from a sample is a function of wavelength or 2θ. Ordinarily, a characteristic line intensity is measured at the angle which gives the highest count rate; this is not essential. The angle for measurement might be that which gives only half the maximum. All that is required is that the measurements always be made at the same angle. How precisely can the peak be located, and subsequently relocated? Figure 4.3 shows count rates observed on step scanning through an iron $K\alpha$ peak on both linear and logarithmic scales, and Fig. 4.4 shows the change in count rate observed for a 0.01° displacement or error of the goniometer setting in counts per second, and as a percentage of the count. The irregularities in rate of change of count rate, most apparent in the curve showing percent change in rate, are probably due to irregularities in the collimator [7, p. 204].

An error of 0.005° in the resetting of the goniometer causes a significant error in the count rate except at the maximum. With higher resolution peak location is even more critical. Precise goniometer adjustment is also particularly important when one of a pair of unresolved lines is being measured. The contribution of the second line can vary significantly for a small error in setting.

4.13 BACKGROUND MEASUREMENTS

If only the linear plot of the Fe Kα line in Fig. 4.3 is considered, it would appear that the background count rate could be measured 3° above or below the peak. The logarithmic plot indicates that a precise estimate is not so simple. The average of the two rates (360 count/sec) would be composed of the detector noise and the scattered continuous spectrum from the sample which together make up the "true" background and the tails of the peak itself. The peak tails are exaggerated by the low resolution and collimator defects. The error in the estimate is the contribution of the peak tails, which is about 100 count/sec. Such an error is negligible in the case of a pure element, being much less than the standard counting error of the peak rate.

Another, more common, practical example of the problem of estimating background is shown by the counting results illustrated in Fig. 4.5. The Kβ line of a trace of chromium in an iron sample produced the counting result shown. The dotted line represents adjusted counting results on a pure iron sample. A simple interpolation of rates a few degrees above and below would not give a good estimate of the peak height. In cases where a small peak is on the shoulder of a large peak, or in a region of rapidly changing background, the background can be determined on a blank at the line position and a degree or two above or below. The ratio of rates can then be used as a factor to calculate the background at the line position from a background measurement on the sample at the reference angle. When a suitable blank is not availabe, graphical interpolation on a plot of count rates vs. 2θ in the region of the line will permit a good estimate of the blank.

Figure 4.5 is a typical example of a sloping background. A constant background is the exception rather than the rule.

Background measurements are not always necessary to estimate composition. If one variable element is to be determined in an otherwise constant matrix, and suitable standards provide calibration data, subtracting the background usually does nothing to

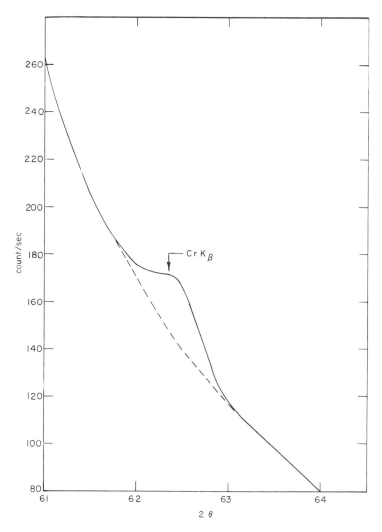

FIG. 4.5 Background estimating on Kβ line of trace amounts of chromium in iron.

improve the reliability of the determination but does increase the standard counting error. On the other hand, a measurement of the background does provide assurance that the equipment is operating properly, and that a significant, and unsuspected, matrix variation is not present. The decision as to whether the net count is necessary to calculate composition has to be decided for each particular case.

4.14 MEASUREMENT STRATEGY

Measurement strategy is concerned with choosing that procedure which will give the desired result with the greatest precision and reliability in the shortest time.

For example, if it were required to determine a few micrograms of cobalt in a liter of aqueous solution with a precision of a few percent, it would be possible to make the determination by counting Co Kα and background directly on the liquid sample. Counting would require several days. On the other hand, the cobalt could be concentrated on ion-exchange paper [10]. The separation would take less than an hour, and the counting could be accomplished in a few minutes. The choice of the best method for a particular determination is discussed in other chapters, and is probably the most important choice that has to be made. The concern here is with the optimum procedure for counting a particular sample [11, p. 93].

After the sample has been prepared for measurement by the best method, and the 2θ settings for peaks and backgrounds identified, the problem is to determine either how many counts on each are required if a fixed count mode is chosen, or how long to count each if a fixed counting time is used. The fixed time mode is usually used, where the choice can be made, despite the fact that the counting precision is determined by the number of events counted.

The relative error $\varepsilon = \sqrt{N}/N$ desired determines the net count in the simple case where a single peak is measured without background correction and also determines the time required $T = N/R$, where R is the count rate (count/sec). In the more complex case where a background subtraction is required:

Relative error (statistical):

$$\varepsilon = \frac{\sqrt{N}}{N} \quad \text{(for a single count)}$$

$$= \frac{\sqrt{N_T + N_B}}{N_T - N_B} \quad \text{(background correction)} \quad (4.21)$$

The recommended procedure [7, p. 347] for a sample about which nothing is known is to make estimates of R_T and R_B by a very brief (a few seconds) count, and calculate the counting time that will give the desired precision. Counting times usually can not be preset closer than by a factor of two. The longer interval should be used when the calculated time falls between two possible counting times, and all counts are for the same duration. This recommendation does not result in the shortest possible counting time when widely different rates are used to establish a net rate or ratio of rates. In such a case,

$$\frac{T_T}{T_B} = \sqrt{\frac{R_T}{R_B}} \qquad (4.22)$$

should be solved for the ideal ratio of times; it may be appropriate to calculate these optimum times where many similar samples are routinely measured, but if only a few samples that may have very different composition are to be determined, the time saved in counting may be lost in the calculation.

When long counting times are required, the error added by instrument instability or other causes will become significant. In those cases, it is best to count at each 2θ setting for a short time, repeating the sequence to get the desired total counting time.

4.15 ESTIMATING COMPOSITION: CALIBRATIONS

High precision in the primary data, i.e., in the net count or ratio of rates, does not assure that the composition will be determined accurately. The accuracy depends on how well the sample-standard comparison is made and on the validity of the calibration. Counting results on standards are subject to the same counting errors as are the counting results on samples. The variance of the final result is equal to the sum of the variances due to all causes. Any bias in the standards will be transferred to the analytical result. Far too often both the standard counting error on standards and the estimate of analytical error or bias of standards are neglected in estimating the reliability of an analytical result. A calibration curve based on a number of standards is not necessarily unbiased or free of random error.

The net intensity of a characteristic line of an element in a constant matrix is a linear function of the weight fraction or amount of the element over a narrow range of composition. Within the linear range the net or total intensity I, in count/sec, is given by

$$I_i = a + bx_i \tag{4.23}$$

where x_i is the weight percent or amount of the element in the ith standard. If several standards are available within the linear composition range, the coefficients a (the intercept) and b (the slope of the straight line) can be calculated and a least square fit obtained by solving the following equations.

$$\Sigma I_i = na + b\Sigma x_i \tag{4.24}$$

$$\Sigma x_i I_i = a\Sigma x_i + b\Sigma x_i^2 \tag{4.25}$$

The coefficients of least square straight line $I = a + bx$.

$$b = \frac{n\Sigma x_i I_i - (\Sigma x_i)(\Sigma I_i)}{n\Sigma x_i^2 - (\Sigma x_i)^2} \tag{4.26}$$

$$a = \bar{I}_i - b\bar{x} = \frac{\Sigma I_i}{n} - b\frac{\Sigma x_i}{n} \tag{4.27}$$

The standard error, analogous to a standard deviation, of I is given by

$$S_E = \sqrt{\frac{\Sigma[I_i - (a + bx_i)]^2}{n - 2}} \tag{4.28}$$

The equation $I = a + bx$ is analogous to the average of n measurements of intensity on a single sample. It can be used to plot a calibration line, or the equations used directly in calculating composition from counting results. Note that in these calculations it is assumed that there are no errors in the standards and that the probable counting error of each is the same.

The linear relation between intensity and composition is an

approximation that is adequate for a limited range of composition. Usually a more complex function is required for an extended range. Calibration data can be fitted to the more complex function by the method of least squares, and the best values for the coefficients in the mathematical expression relating intensity and composition can be determined. The calculation of the coefficients becomes more complex as the number increases.

If the spectrometer is operating properly, the error of I of any one of the standards should be limited by counting statistics. The more complex the standards, the less likely is this to occur in practice. If one or more of the standards deviates from the least squares line by more than the statistical limits, the assumption of a linear, or other function, is incorrect or there is a defect in the standard. Concentrations of other elements may vary resulting in unequal interelement effects that bias the count rate, the assumed composition may be in error, the standards may be inhomogeneous or may have different surface finishes or particle size.

If defects of the standards are perturbing the expected count rates and cannot be remedied or explained, then the precision of the calibration will be limited by this factor and may exceed the standard counting error. In the cases where standards' defects limit the precision rather than the counting statistics, the standard error of composition should be calculated by interchanging the dependent and independent variables I and x. There are statistical methods for treating errors of both variables simultaneously but they are complex and an unnecessary refinement where errors of one parameter are significantly greater than those of the other. The standard error of composition (of the standards) is a measure of the reliability of the calibration. If the analytical samples are similar to the standards they are also subject to the same error. In addition, there will be a counting error of the samples. The random error of the composition of the samples is the square root of the sum of the squares of these errors, all expressed in the same terms.

Absorption and enhancement effects can be considered as biases or systematic errors. Their magnitude can be estimated by two very different approaches. The fundamental-parameters method [1] calculates the intensities to be expected from a sample of a given composition or the composition from observed intensities using some parameters depending on sample geometry and three sets of basic data: the X-ray tube spectral distribution, mass absorption coefficients for each of the elements in the specimen, and fluorescent yields. These data serve as the "standards." Standards similar to the samples are not required. If all elements in the sample are considered, the primary data correct, and the equations relating intensity to composition valid, the composition can be calculated from observed intensities.

The second method is an empirical one exemplified by the multiple regression treatment of Lucas-Tooth and Pyne [12]. Factors for calculating the influence of each element on the element being determined in a complex sample can be obtained from intensities measured on at least X + 2 analyzed standards where X is the number of elements in the sample. The reliability of the analytical results in both cases is limited by the standard deviations of the intensity measurements on samples and the quality of the coefficients used in the calculations.

There are many other methods for estimating interelement effects: empirical, semiempirical, or those involving only fundamental data. The empirical methods apply only over concentration ranges covered by the standards; the fundamental parameters methods require a great many items of input data and long calculations.

Another reason that high precision in the measurement of intensities does not assure equally precise analytical results can be illustrated by the data [7, p. 389] on the weight percent Fe and relative intensity of Fe Kα in FeO (77.8%, I = 95.7) and Fe_2O_3 (70.0%, I = 95.0). The relative change of composition is 14.5 times as great as the relative change in intensity. The precision of an iron analysis would be reduced by about that same factor. The very low absorption of oxygen causes the low sensitivity or ratio of rates of change in intensity to rate of change in composition. The remedy in such a case is to dilute the highly absorbing element with an absorbing material. The loss in intensity will be more than compensated by the gain in ratio or slope of the calibration curve.

4.16 SAMPLE CONFIGURATION

The topics considered under the heading sample configuration are sample size, surface finish, heterogeneity, sample location in holder, thickness, and particle size. Variation between sample and standard of any of these may introduce a bias in the calculation of compositions.

The intensity observed from samples that fill the plane of the sample holder represents the integration of all of the area. But the sensitivity varies over the area [7, p. 308]. The intensity from a small sample will vary depending on the exact location. Displacement of a 2-mm diameter sample by a few millimeters may change the response by a factor of two. If quantitative results are required for small samples, a contour map of the response of the spectrograph should be made by recording the count rates observed when a small sample (i.e., 2-mm square foil of nickel) is placed in different regions of the sample holder, or different distances from the center if samples are rotated. To make analytical determinations on small samples, the standards and samples should have the same area and be placed in the region of maximum response. This can be facilitated by masking.

Sensitivity also depends on the elevation of the sample in the sample plane. The intensity from a bowed or dished liquid sample will differ from that obtained from a liquid sample with a flat surface.

The intensities obtained from a heterogeneous sample differ from that obtained from a homogeneous sample of the same average composition in part due to the variations in sensitivity with location in the sample holder and in part to variations in interelement effects. Spinning a heterogeneous sample (or a small sample) may reduce or average out the error due to sample location to a small extent, but will not help the error arising from interelement effect variations. These variations can be eliminated only if the heterogeneities are much smaller than the critical thickness.

Interelement effects and efficiency of excitation are functions of the angle of incidence of the exciting radiation and the take-off angle, hence a sample with a rough surface will give different count rates than a sample of the same composition that has a smooth surface. Also, loose powders with low bulk density deviate from the compacted material. The ideal sample is homogeneous with a smooth, flat surface.

The intensity I_d for a film of a single element is related to I, the intensity from an infinitely thick sample by

$$\frac{I_d}{I_\infty} = 1 - e^{-ad} \qquad (4.29)$$

where a is a function of absorption coefficients and spectrometer geometry [7, p. 291]. For thin samples, where ad is small and e^{-ad} is almost one, I_d increases linearly and $1 - e^{-ad} \approx ad$. For thick samples, $I_d \approx I$. Self-absorption and interelement effects in multi-element samples are absent in thin samples, and fixed in thick samples, but there is an intermediate region, below the (minimum) critical thickness for thick samples, and above the maximum critical thickness of thin samples where these effects are variable. In this intermediate thickness range calculation methods and experimental data have not been presented in great detail, except for the case of pure-element thickness measurements. The *critical thickness* can be defined as that thickness at which a thin sample departs perceptibly from the linear relationship, or, for thick samples, I_d is detectably less than I_∞. "Perceptible deviation" can be arbitrarily given the value of 1%; for thick samples, $I_d/I_\infty = 0.99$, while for thin samples, $ad/(1 - e^{-ad}) = 1.01$ at the critical thickness. The corresponding values of ad are 0.19934 and 4.6051. Samples for which ad is between these two values are in the intermediate region where self-absorption and interelement effects are significant but less than in thick samples. Heterogeneous sample particles are apt to be in this size range. An indication of the magnitude of self-absorption effects can be had by comparing the intensity I_d of a film at the upper critical thickness with the extrapolation of the thin film approximation; the critically thick film has an intensity that is only 22% of the extrapolation, i.e., 78% of the intensity expected is lost by self-absorption.

All of the perturbations mentioned in regard to sample configuration indicate the need for extreme care in sample and standard preparation if reliable analytical results are desired. A mismatch of sample and standard in area, location, thickness, surface finish, homogeneity, particle size, bulk density, etc., may result in erroneous conclusions regarding composition.

4.17 LIGHT ELEMENTS

It is difficult to get reliable analytical results for the light elements. The usual error sources are more prominent in measurements on the light elements. Fluorescence yields and hence count rates tend to be low. Interferences due to higher order lines and the L and M lines of heavier elements are numerous. Absorption coefficients are extremely high at the wavelengths of interest, so that surface finish and particle size effects tend to be more serious. Pulse heights are small and closer to the noise level of the detector. The high absorption coefficients result in extremely high interelement effects.

While the perturbing phenomena noted become more significant for the light elements, the light elements have several additional complications. Changes in absorption in the optical path due to a small amount of residual air in the vacuum or in the helium can effect the intensity. Crystals used for the light elements have large temperature coefficients of expansion, resulting in significant changes in 2d spacing with changes in temperature. An even more serious effect is the change in wavelength, line shape, and intensity with changes in the chemical state of the element [13].

The magnitude of the chemical effect is significant for the lighter elements, and it is necessary that the calibration standards have the element present in the same chemical state as the samples in which the element is to be determined.

REFERENCES

1. J. W. Criss and L. S. Birks, *Anal. Chem.*, *20*, 1080 (1968).
2. E. P. Bertin, *Principles and Practice of X-Ray Spectrometric Analysis*, Plenum Press, New York, 1970.
3. R. Jenkins and J. L. DeVries, *Worked Examples in X-Ray Analysis*, Springer Verlag, New York, 1970.
4. *Handbook of Probability and Statistics*, 2nd Ed. (R. C. Weast, ed.), The Chemical Rubber Co., Cleveland, Ohio, 1968.
5. J. B. Scarborough, *Numerical Mathematical Analysis*, 3rd Ed., Johns Hopkins Press, Maryland, 1955.
6. H. A. Liebhafsky, H. G. Pfeiffer, and P. D. Zemany, *Anal. Chem.*, *27*, 1257 (1955).
7. H. A. Liebhafsky, H. G. Pfeiffer, E. H. Winslow, and P. D. Zemany, *X-Rays, Electrons, and Analytical Chemistry*, Wiley Interscience, New York, 1972.
8. L. S. Birks, *X-Ray Spectrochemical Analysis*, Interscience, New York, 1959.
9. H. A. Liebhafsky, H. G. Pfeiffer, and P. D. Zemany, *X-Ray Microscopy and X-Ray Microanalysis, 2nd Int. Symp.* (A. Engstrom, V. Cosslett, and H. Pattee, eds.), Elsevier, New York, 1960, p. 321.
10. W. J. Campbell, E. F. Spano, and T. E. Green, *Anal. Chem.*, *38*, 987 (1966).
11. R. Jenkins and J. L. DeVries, *Practical X-Ray Spectrometry*, Springer Verlag, New York, 1967.
12. H. S. Lucas-Tooth and E. C. Pyne, *Advances in X-Ray Analysis*, Vol. 7, Plenum Press, New York, 1964, p. 523.
13. P. D. Zemany, *Anal. Chem.*, *32*, 595 (1960).

Chapter 5

ELECTRON EXCITATION

John Lucas-Tooth
Telsec Instruments, Ltd.
Littlemore, Oxford
Great Britain

5.1	General	112
	5.1.1 Historical Background	112
	5.1.2 Present Instrumentation	113
5.2	Specimen Requirements	114
	5.2.1 Metal Samples	114
	5.2.2 Nonmetal Samples	115
5.3	Quantitative Analysis	115
	5.3.1 Wavelength Dispersion	115
	5.3.2 Energy Dispersion	116
	5.3.3 Mathematical Formulations	117
	5.3.4 Calibration Standards	118
	5.3.5 Examples	118
References		122

5.1 GENERAL

5.1.1 Historical Background

It is interesting to realize that the very early work in using the excitation of characteristic X rays as a possible method of analysis were performed using direct electron excitation. It was noticed that the target of the X-ray tube emitted the characteristic lines of the elements of which it was composed, and it was suggested by Moseley before 1914 that this could be used as a method of chemical analysis. However, the advantages of X-ray spectroscopy as an analytical technique were only really manifest when methods of detecting X rays better than the photographic plate were available. This meant that work did not really proceed until the science of nuclear physics had produced first the geiger counter, then the scintillation and proportional counter. By this time the availability of good, sealed, medical X-ray tubes meant that a reliable source of X-ray excitation was available to experimentalists. It was perhaps because of this sequence of timing that when X-ray fluorescent spectrometers were introduced in the early 1950s, they all used sealed X-ray sources rather than direct electron excitation.

The main limitation of these early tube-excited X-ray instruments was the decreasing performance on the lower atomic number elements. In many cases analysis was just not possible for elements of lower atomic weight than calcium. Gradually, however, the air path instruments were replaced by vacuum instruments and the detector windows made thinner and thinner. Similarly, large lattice-spacing crystals became available and the range of X-ray fluorescent analysis was extended quickly down to 10 Å. It was found that the fluorescent yields for elements such as magnesium, aluminium, and silicon were much poorer than for the transition elements, and hence sensitivities and accuracies were very poor. Attempts were made at this time to improve the performance by increasing the excitation by increasing the power on the X-ray tube and also by making thinner window X-ray tubes. Special targets in the X-ray tube were also chosen which were more suited to exciting the light elements. By 1960 it became obvious that there were no more large factors to be gained by these methods, and at this time the analysis of magnesium and heavy elements was perfectly feasible although the analysis of elements of lower atomic numbers still presented very great difficulties.

In the late 1950s the advent of the microprobe analyzer (see Chap. 7) appeared to give a breakthrough for light-element excitation. For a start the dispersive system for such light X-ray spectrometers was readily available together with the electronics associated with them, and the use of the microprobe showed that the efficiency of excitation was very much higher for electron excitation than for X-ray excitation. It was also apparent about this time that a useful saving in power dissipation could be gained. It is not uncommon for the input power to a sealed X-ray tube to be in excess of 2 kW. The actual X-ray power is, of course, much lower, probably less than 1 W. As the efficiency for electron excitation is higher than the efficiency for X-ray excitation, similar count rates should be obtained for an electron beam of about 1 W. Thus, for the same X-ray performance, the input power is reduced by 1,000-fold. This could be reflected in less complexity and reduced prices.

5.1.2 Present Instrumentation

The microprobe is specifically excluded from the instrumentation discussed in this paragraph. To the writer's knowledge there are only two, commercial, X-ray fluorescent analyzers using direct electron excitation for the analysis of bulk specimens. Rank-Hilger of Margate, Kent, England, manufacture an add-on electron excitation attachment to their standard range of X-ray fluorescent analyzers (the Fluoroprint series) to enhance the light-element performance. The other instrument uses electron excitation alone and is manufactured by Telsec Instruments, Oxford, England. There are two main purposes for using electron excitation. First, the increased efficiency for the low-atomic number elements gives them some advantage over standard X-ray spectrometers for the elements fluorine, sodium, magnesium, and aluminium. For even lighter elements such as boron, carbon, nitrogen, and oxygen, they produce data which are hard to acquire in any other way. There is, however, another advantage which has been touched on previously. The fact that the power requirements are so much less than for X-ray excited spectrometers means that significant reduction in price can be achieved, and some laboratories are using electron excitation instruments for the heavier elements for work which would more normally be done by X-ray fluorescent spectroscopy in the interests of economy.

5.2 SPECIMEN REQUIREMENTS

Electron excitation is very much a surface phenomenom. This being so, it is even more important with this type of excitation to prepare a surface which is representative of the bulk of the sample. Over and above this onerous requirement, the sample must be made electrically conducting.

5.2.1 Metal Samples

These are naturally conducting and so are the easy samples to prepare. If all the metals are in solid solution, a good surface finish should not be difficult to produce and the requirements are very similar to those of normal X-ray spectroscopy. Should the metal sample be inhomogeneous, great care must be taken to ensure that a representative surface is presented to the electron beam in a precisely similar manner as is important for normal X-ray fluorescent spectroscopy. In particular, the problem of lead in copper-based alloys is well known. Smearing of the lead particles over nonlead areas can produce high spurious readings for lead, but similarly, if the lead particles are completely "pulled out" of the matrix, a low lead count rate can ensue.

5.2.2 Nonmetal Samples

Two methods have been tried in order to make the specimen conducting. As is commonplace for microprobe analysis, a thin conducting layer can be deposited on the surface of the sample. This has proved to give unsatisfactory quantitative results for electron excitation. As it is common practice to use a lower voltage beam and greater power per unit area, the absorption of the conducting film can affect the intensity of the fluorescent radiation, and if a very thin film is used, this can be physically burned off by the beam. It is also customary to be aiming at higher reproducibility than for microprobe analysis. Both these effects tend to litigate against the use of conductive coatings. The more standard method is to grind the specimen concerned with an added amount of conducting material. The most normal material to use is graphite, which tends to coat each individual grain with a conducting layer and also helps in the briquetting of the sample when pressed. In some cases (for instance when carbon itself is to be analyzed), the use of carbon can be unfortunate and other work has been carried out with both aluminium and copper powder. Since a certain amount of heat is being injected into the sample and since many of the samples to be analyzed are potentially unstable, it is worth using the lead-disk technique. This is described in various papers but essentially consists of pressing a small amount (about 200 mg) of sample into the surface of a lead disk. The softness of the lead presents an ideal binder and also a heat sink for the heat produced by the electron beam.

5.3 QUANTITATIVE ANALYSIS

5.3.1 Wavelength Dispersion

In almost every way, the spectrometers designed for electron excitation use precisely similar wavelength-dispersive systems as for standard X-ray fluorescence. It is worthwhile to state here that the background is higher with electron excitation than for standard X-ray fluorescence. This is because of the well-known bremsstrahlung

effect which is described in Chap. 2. Hence it is worth giving
some thought to increasing the resolution of the spectrometer as
much as possible. This is enhanced by the fact that the intensity
is rarely a problem, and hence it is worth improving the resolution
at the expense of the intensity. This usually means tighter collimation and the very careful choice of crystal. It is also worth
stressing that special care needs to be taken for the detection of
long-wavelength X-rays. Again there is no difference from microprobe analysis, but the use of relatively low-gain proportional
counters with charge-sensitive preamplifiers and special counter
gasses at low pressures enhances the natural performance of the
spectrometer very considerably. There is nothing exceptional in
this, merely the intelligent use of the principles of wavelength
dispersion outlined elsewhere in this book.

5.3.2 Energy Dispersion

The use of energy dispersion does not seem to have been recommended
by any of the workers in this field. It is probable that the higher
background of bremsstrahlung coupled with the inherent problems of
solid state detectors, means that so many factors are going against
obtaining good data, that it is not worth the extra convenience of
using nondispersive detectors. This is, of course, in marked contrast to the microprobe user, and it is worth digressing for a few
moments on this apparent paradox. The microprobe user tends to be
less oriented towards quantitative analysis; his problem is to identify very wide ranges of elements, any of which may be present in
considerable quantities. The typical analyst using a standard X-ray
fluorescent spectrometer is normally completely aware of what his
specimen contains. He is not usually doing semiquantitative work
on a very large number of specimens, and hence, if the specimen is
unknown, he can afford the time of an angular scan using a crystal
spectrometer and goniometer. This leads to easy identification of
the peaks, although a much longer time of analysis. The usual mode

of analysis, however, is for him to be looking at a few fixed
elements which are always present in a particular series of samples.
Many exceptions to this spring to mind, such as pollution studies
and work connected with forensic and archaeological studies, but
here the use is with a standard fluorescent spectrometer and not
electron excitation. Furthermore the characteristic wavelength of
low atomic-number elements are not ideal for nondispersive solid
state detectors.

5.3.3 Mathematical Formulations

A great deal of effort has gone into the mathematical formulation
of relating X-ray output to chemical concentration using the electron microprobe. It might be thought that the equivalent problem
of bulk samples would be well documented and well known. This is
not the case. A few moments thought will show why this should be
so. In microprobe analysis it is very rare to have a large number
of chemically analyzed standards. The microprobe is almost the
only method of analyzing submicrogram quantities of material and
the use of the microprobe is primarily aimed at nonhomogeneous specimens. Thus, one is endeavoring to analyze the exact chemical content of a few square microns of sample rather than an average bulk
analysis. These few square microns tend to have a discrete chemical formulation, and hence the use of mathematical correction procedures have tended to be related to the very wide range of concentrations which occur for many different elements. Therefore one
tends to use as standards known chemical compounds which cover the
range (albeit very widely) and work out elaborate calculations for
extrapolation. In the case of bulk analysis using electron excitation, the problem is much more similar to standard X-ray fluorescent analysis. A much narrower range of chemical concentrations
are encountered, and it is normal to use the standard techniques
rather than those in microprobe analysis.

5.3.4 Calibration Standards

With reference to a previous paragraph, it can be seen that the use of calibration standards is similar to those for standard X-ray fluorescence. One point of interest, however, is that the interelement effects are usually very much smaller. An example of this is given in the references which shows that the average interelement effect on chromium analysis in high-chromium steels is probably less than one-third that for standard X-ray fluorescence. This comes about because of two factors: The absorption curve for electrons entering the sample has a nondiscontinuous response against wavelength and hence does not produce the sort of absorption anomalies of X rays. Secondly, the emerging X-ray beam, owing to the reduced depth of penetration, is less absorbed on its emergence from the sample than in the equivalent X-ray excitation situation. However, this same effect can lead to larger discrepancies when it comes to inhomogeneity and particle-size effects. Many workers do not distinguish sufficiently between these two effects and blame (unfairly) their poor results on interelement effects rather than on the limited bulk of sample that they are analyzing. As has been said, great care has to be taken over the proper preparation of standards using electron excitation but having done this, the interelement effects are reduced considerably.

5.3.5 Examples

A few representative examples of applications of the *direct electron-excitation* technique are given for illustrative purposes.

1. Analysis of copper in brasses: Figure 5.1 shows the calibration curve for copper in brass from 50% to 90%. The conditions of analysis are as given beside the figure. The graph is self-explanatory and indicates both the reproducibility of the method and the linearity of response.

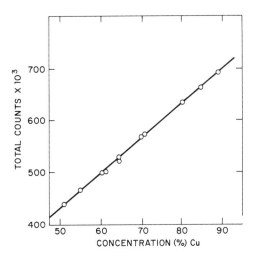

FIG. 5.1 Analysis of copper in brasses. Conditions: electron-beam voltage = 15 kV; electron beam = 0.5 mA; crystal = LiF 110; time of analysis = 20 sec.

2. The next example shows the analysis of lead in copper based alloys (Fig. 5.2). This example is given because the analysis was performed on the lead Mα line. The graph is included to show how, with appropriate sample preparation techniques, satisfactory analysis can be performed on special elements, and it also shows that the higher background of electron excitation is not necessarily a drawback.

3. The two calibration curves in Fig. 5.3(a) and (b) show the analysis of carbon in low-alloy steels and cast irons.

4. Figure 5.4 shows the analysis of oxygen in fused synthetic cryolite. One of the interesting points about this work is that no conducting agent was added to the sample. It should be noted that these were synthetic samples and that natural samples have considerably more interelement effect.

5. Figure 5.5 shows the analysis of boron in cobalt and nickel alloys. The results are self-explanatory.

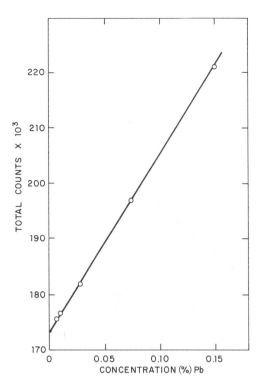

FIG. 5.2 Analysis of lead in copper-based alloys. Conditions: same as in Fig. 5.1, except for NaCl crystal.

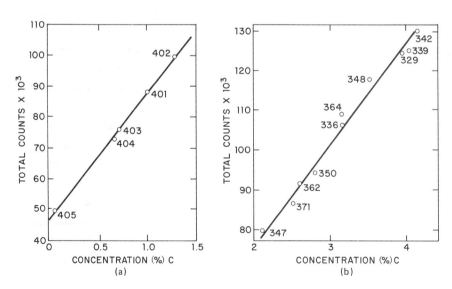

FIG. 5.3 Analysis of carbon (a) in low-alloy steels and (b) in cast iron. Conditions in both: electron-beam voltage = 15 kV; electron beam = 0.5 mA; crystal = octadecyl hydrogen maleate (O.H.M); time of analysis = 10 sec.

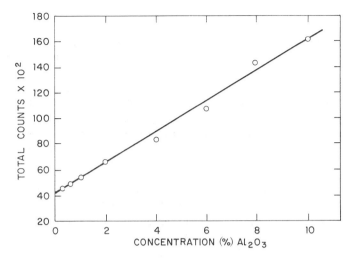

FIG. 5.4 Analysis of oxygen in fused synthetic cryolite. Conditions: electron-beam voltage = 12 kV; electron beam = 0.5 mA; crystal = rubidium acid phthalate (RbAP); time of analysis = 20 sec.

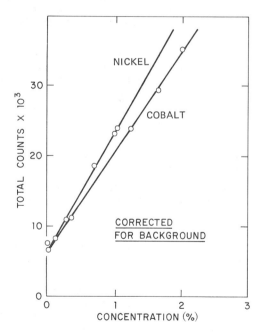

FIG. 5.5 Analysis of boron in cobalt and nickel alloys. Conditions: electron-beam voltage = 10 kV; electron beam = 0.5 mA; crystal = lead stearate; time of analysis = 50 sec.

REFERENCES

1. Banks, M. S., A Practical Direct Electron Excitation Spectrometer, *Proc. 5th Conf. on X-Ray anal. Methods*, Swansea Symposium, 1966.

2. Lucas-Tooth, H. J., and M. S. Banks, X-Ray Fluorescent Analysis Using an Electron Excitation Method, *Acta Imeko* (1967).

3. Lucas-Tooth, H. J., and M. S. Banks, The Use of an Electron Excitation X-Ray Spectrometer for the Quantiative Analysis of Elements with Atomic Number 6-11, *XIV Colloquium Spectroscopicum Internationale*, Debrecen, 1967.

4. Watling, J., H. J. Lucas-Tooth, and M. S. Banks, The Analysis of Elements of Atomic Number 5-9 by the Betaprobe, *XV Colloquium Spectroscopicum Internationale*, Madrid (1969).

5. Haylett, D. W., The Use of the Betaprobe for X-Ray Spectrometric Analysis by Direct Electron Excitation, *Metallurgia LXXX*, No. 478, August, 1969.

6. Clarke, W. E., The Application of the Telsec Betaprobe Electron Excitation X-Ray Spectrometer to the Analysis of Cast Iron, B. C. I. R. A. Report No. CQ17, August, 1969.

7. White, G, P. S. Bramhall, P. Rudd, and W. Dolbey, The Application of Direct Electron Excitation to the Analysis of Slags and Fluorspar, B. I. S. R. A. Report Ref. MG/D/470/70, March, 1970.

8. Haylett, D. W., The Application of the Betaprobe to the Analysis of Cryolite from Aluminium Extraction Baths: Progress Report, B. N. F. M. R. A., Report A1752, February, 1970.

9. Poole, A. B., E. D. Pickard, and D. Lawrence, The Determination of Carbon and Fluorine in Rocks and Minerals by X-Ray Emission Spectrometry, *Spectrochimica Acta*, *26B*, 145-150 (1971).

10. Price, B. J., Modern X-Ray Analysis by Electron Excitation, *Metals and Materials* (1973).

11. Luschow, H. M. and K. Schaefer, Aufbau und Anwendung einer Elektronenstrahlmakrosonde, *Mikrochimica Acta*, Suppl. 5, 161-180 (1974).

12. Johansen, O. O., Operation of the B200 Betaprobe in the Aluminium Industry with Special Reference to Long Wavelengths, Alcan Spectroscopy Conference, Arvida, Canada, 1975 (available as Telsec Application Report 76021).

13. Price, B. J., H. J. Lucas-Tooth, J. V. Yelland, and K. M. Field, Progress in Analysis Equipment for X-Ray Fluorescence and Direct Electron Excitation X-Ray Spectrometry, *XVIII Colloquium Spectroscopicum Internationale*, Grenoble, 1975.

Chapter 6

PROTON AND ALPHA EXCITATION

Richard L. Walter and R. D. Willis

Department of Physics
Duke University
Durham, North Carolina

6.1	Historical Background	124
6.2	Present Instrumentation	125
6.3	Sample Characteristics	131
	6.3.1 Sample Requirements	131
	6.3.2 Sample Types	131
6.4	Qualitative and Quantitative Analysis	136
	6.4.1 Calibration Methods	136
	6.4.2 Beam-Heating Effects and Volatilization	142
6.5	Data Analysis	144
	6.5.1 Typical Spectra	144
	6.5.2 Computer Fitting Codes	149
6.6	Applications of Particle-Induced X-Ray Emission Analysis (PIXEA)	152
References		161

6.1 HISTORICAL BACKGROUND

X-Ray emission analysis clearly has been developing in the last decade into a reliable method for multielemental analysis for a wide variety of sample types. Until a few years ago scientists concentrated heavily on developing X-ray fluorescence (i.e., *X-ray-induced* X-ray emission) systems for this purpose because such equipment was relatively economical, reliable, and convenient to use. In addition to X rays, charged particles can also be employed for specific purposes as a means of inducing atoms of a sample to emit characteristic radiation. Although the apparatus required to produce beams of ions, e.g., $^1H^+$ or $^4He^+$, is usually expensive and complex, there are a number of laboratories that are presently pursuing X-ray emission analyses with such beams, some on a regular basis. There are several advantages of using ion beams instead of X-ray beams to induce X-ray emission for elemental analyses, but a large number of disadvantages also exist.

In 1964, Birks [1] reported on the relative merits of electron-, proton-, and X-ray-induced emission for elemental analysis, and the authors chose X rays as being the most practical at that time. However, three important occurrences have been responsible for renewed activity in developing systems which employ proton or α beams for *Particle-Induced X-Ray Emission Analysis*, now commonly called PIXEA (and pronounced like "pixie"). These events were (1) the development of high resolution Si(Li) detectors for energy-dispersive X-ray analysis, (2) the decrease in federal support for basic nuclear studies for low-energy accelerator laboratories in the United States, and (3) a pair of publications by T. B. Johansson et al. [2] which suggested phenomenal detectability limits for 3-MeV proton-induced emission. Since 1972, nuclear scientists at more than 24 accelerator laboratories around the world have pursued some degree of multi-elemental analyses using Si(Li) detectors and either proton-, α-, or heavier particle beams.

The intent of the present paper is to review the gains made with PIXEA during the last few years and to try to put this analytical technique into a proper perspective. The latter goal is particularly difficult to achieve as many of the breakthroughs of the last two years have not been documented in accessible reports.

Lastly, it is important that the PIXEA method not be confused with proton-activation analysis. The latter technique involves nuclear excitation (similar to neutron activation) with high-energy proton beams and observation of the γ rays emitted (sometimes after a few days) from the decaying radioactive nuclides. The PIXEA method involves atomic excitation and detection of X rays that are emitted instantaneously, i.e., during the radiation interval.

6.2 PRESENT INSTRUMENTATION

Because reports of analyses using the PIXEA method are spread over a wide range of scientific and environmental literature, it is impossible to be certain that a review of this topic is complete. Many of the pertinent articles of the past few years are referenced at the end of this chapter. Most of the PIXEA work to date was conducted with protons using single-ended Van de Graaff accelerators. At several labs, tandem Van de Graaff accelerators have been employed although the larger types of tandems cost more to purchase and are more expensive to operate. A major effort has revolved around a moderately new, variable-energy cyclotron at the University of California at Davis, where a group under the direction of T. A. Cahill [3] have underway one of the most active programs of aerosol analyses. At this facility, 18-MeV alpha-particle beams are employed. Some work using ^{16}O beams have been reported [4], but these measurements were more of an exploratory nature.

The consensus of current PIXEA analysts seems to be that for most purposes of trace-element investigations with particles, proton beams with energies near 3 MeV are about the best choice and that α-particle beams about four times as energetic give nearly as good sensitivity. This conclusion is reinforced by the recent reports of Folkmann et al. [5]. These authors have made calculations of the X-ray background due to the bremsstrahlung produced in an organic matrix by the incident beam itself and due to the electrons that have been knocked out by the beam. The bremsstrahlung from the electrons, that interacts inside the matrix, is the dominant source of background X rays below about 12 keV for the 3-MeV proton case. The calculations of Folkmann et al. are represented by the solid curves in Fig. 6.1. Folkmann et al. also measured the background levels for proton, ^{4}He, and ^{16}O beams, and these results are shown by the broken curves in Fig. 6.1. Recalling that the bremsstrahlung production is proportional to the inverse square of the

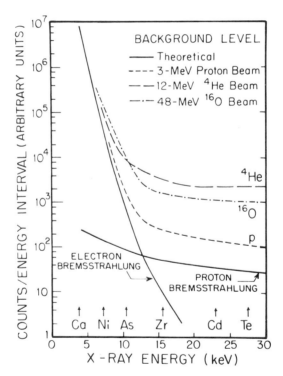

FIG. 6.1 Solid curve: Calculated detectability limit for various elements in an organic matrix for 3-MeV protons. Dashed curves: Observed detectability limits for protons, α-, and ^{16}O-particle beams. The rise at low energies is caused by bremsstrahlung from knocked-out electrons.

mass of the projectile, electron beams of the same energy will produce nearly 10^6 times more intense background radiation than proton beams. For this reason, the discussion on PIXEA using beams in the MeV range concern only high-energy proton, α, and ^{16}O beams.

A schematic view of the PIXEA system of Walter et al. [6] is shown in Fig. 6.2. Here a beam of protons of energy E_0 emerges from a particle accelerator and is directed onto a *thin*-membrane sample. An individual proton undergoes collisions with atoms of the sample, losing a small amount of energy in each collision, and eventually exits from the specimen, slightly diminished in energy from E_0. The beam then impinges upon a graphite "beam stop" which can be far removed from the specimen chamber.

Two other important aspects are also exhibited in Fig. 6.2. First, the beam of particles will excite any medium (such as air) through which it passes and thereby loses energy in the process. Therefore, the path of the beam inside the accelerator and along the beam path prior to reaching the sample is normally evacuated. This requires that the samples must also be placed inside an evacuated chamber for irradiation purposes unless one resorts to special

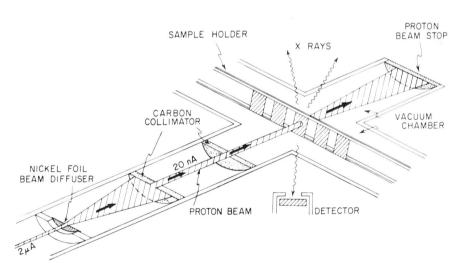

FIG. 6.2 Schematic diagram of proton beam, samples, and vacuum chamber.

tricks like bringing the beam through a thin window into a short helium-filled chamber as discussed below.

The second important aspect in Fig. 6.2 refers to the passage of the 2-µA beam through a Ni diffuser foil whose presence spreads out the narrow beam through multiple Rutherford scatterings from Ni nuclei. Only a small percentage (~1-2%) of the original beam impinges on the target after passing through a series of collimators of which only two of four are shown. For most purposes, the 20-nA shown gives a sufficiently high flux of X rays to make X-ray analysis feasible. The purpose of the foil diffuser is to provide a beam of uniform intensity and of known area. The diffuser eliminates "hot spots" due to inhomogeneities in the original beam or due to sharp focusing by the magnetic lenses of the accelerator beam-handling equipment. A profile of the diffused beam can be obtained by measuring the X-ray yield when a small point target deposited on a thin-film substrate is moved transversely across the beam axis. Figure 6.3 illustrates the beam uniformity determined in such a manner. A second method of diffusing the beam on the samples has been adopted at several laboratories (for example, see Ref. 2). This involves sweeping a narrow beam over the irradiation area with a pair of electromagnetic deflectors in a manner similar to that in which

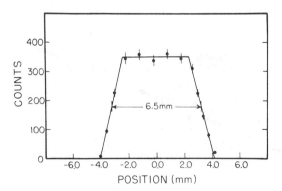

FIG. 6.3 Profile of proton-beam intensity measured along a diameter. The slope at the edges is due in part to the finite size of the point target.

the electron beam is swept across the screen of a TV cathode-ray tube.

The amount of charge that is incident on the target during a specific exposure is measured using commercially available beam-current integration circuits. Such modules are reproducible in measuring currents down to the 10^{-10} A range. They also can operate in preset modes which permit turning on and off the analog to digital converter (ADC) associated with the energy-dispersive X-ray detection after a preselected amount of charge has been accumulated. To avoid improper integration due to proton scattering, the samples, sample holder, and chamber can be connected electrically to the beam stop.

In the system shown in Fig. 6.2, the X rays that are emitted from the sample pass from the vacuum chamber, through a 25-μm Mylar film, and then through 1 cm of air before reaching the Si(Li) detector. Keeping the detector in air is a mere convenience; some PIXEA systems that were designed to measure X rays emitted from light elements operate with a Si(Li) detector which protrudes into the evacuated region. Although crystal spectrometers have been used in conjunction with X-ray production or cross-sectional measurements for particle beams, nobody has reported on the successful use of wavelength-dispersion methods coupled to a PIXEA system for applied or environmental multielemental analyses. The main reason for this is associated with the low efficiency of wavelength-dispersion systems relative to a Si(Li) detector. Because of this, the remainder of the chapter will be devoted to PIXEA measurements made using Si(Li) detectors and to problems related to this method.

The sample holder shown in Fig. 6.2 is a simple ladder design which can be either manually positioned or controlled with a computer. Automatic systems designed for analysis of aerosol specimens deposited on thin membranes have been installed inside vacuum chambers at several facilities. At least three such systems have been economically constructed by modifying commercially available 2 × 2 in. slide-changer mechanisms which are suited for up to 100 samples.

Clearly the size of the irradiation area can be conveniently adjusted by changing the beam-restricting collimators. Typical beam diameters at the sample are 6 mm. More will be said about decreasing the beam diameter in one of the last sections of this chapter where miniscans will be considered briefly.

One of the limiting factors concerning minimum detectability limits (mdl) is counting statistics. (See Chap. 4.) In a fixed time interval, the mdl can be lowered by increasing the counting rate. Two problems arise however with increasing the particle beam on the sample. For some samples, the X-ray flux reaching the detector is already large with just 20 nA if the Si(Li) detector is placed at the usual 7-10 cm from the sample and if no filter is placed over the detector (see Sec. 6.5.1). The second, and more serious, concern is dissipation of the power deposited in the sample by the beam during penetration. This effect can be quite appreciable when the specimen depth approaches a "thick sample." For a "thin sample," for the example of a beam of 3-MeV protons which penetrates 1 mg/cm^2 of organic material, the beam loses 130 keV of energy due to ionization and excitation. For a 100-nA beam this corresponds to 0.013 W of power that must be dissipated through either conduction, instantaneous X-ray or light emission, or thermal radiation. Calculations of the beam heating, or in other words, the temperature rise as a function of the time the beam is on target, are complex and to our knowledge have not been carried out accurately. Probably, heat transfer due to conduction is relatively small for organic samples or thin organic substrates and therefore one can roughly estimate an upper limit of the temperature as a function of time by assuming no heat loss except through thermal radiation to the chamber walls. For the case where a 3-MeV proton beam of 100 nA stops completely in a "thick target," i.e., where 0.3 W are deposited in the sample, the equilibrium temperature that will be reached is about 300°C if the only heat loss was that due to thermal radiation. Of course, recall that this is an upper limit which is based on the assumption of negligible X-ray radiation emission and negligible heat conduction.

6.3 SAMPLE CHARACTERISTICS

6.3.1 Sample Requirements

For *most operating systems* the samples must conform to the following requirements:

1. The element forms to be measured must be able to withstand a hard vacuum, i.e., about 5×10^{-5} torr.
2. The region to be analyzed cannot be more than 100 μm below the exposed surface.
3. Samples 25 μm or thicker should be homogeneous for quantitative analyses.
4. The cross-sectional area to be investigated (in one irradiation period) should be less than approximately 1 cm^2.
5. The sample must be able to withstand some heating, the amount ultimately being inversely related to the detectability limit being pursued in unit time.

Attempts to overcome some of these requirements have been reported but they have not been fully explored yet. For instance, Seaman and Shane [7] describe an arrangement in which the proton beam exits from the vacuum system through a thin foil into a helium atmosphere in which the samples are located. Hence, this group eliminated requirement 1 and reduced, through the conductive cooling ability of He, requirement 5. Seaman and Shane also described a thin-window cell intended for studying liquid samples. Their cell can be located inside the He atmosphere. The He method and the cells introduce new types of X-ray backgrounds and different analysis problems that are beyond the range of those which will be described in this report.

6.3.2 Sample Types

During the past few years, a wide range of samples have been subjected to PIXEA studies and considerable data on relative abundances of elements can be found in various PIXEA articles and reports.

However, very few quantitative results can be found, except possibly for air filter analyses for which specialized PIXEA systems have been constructed. In the immediate future it should be possible to give reliable, quantitative numbers for a broad range of specimens because many of the serious problems are gradually being recognized and resolved.

The categories of samples which have been analyzed are dried (liquid) deposits, filter media, thin self-supporting sections, thick pieces of materials, powdered samples that have been packaged between thin films, and pelletized substances. Specifically, the liquid deposit samples are water, wine, blood, serum, urine, fuel oil, soil extracts, and products of ashing processes. Filter media implies materials like Millipore, Nuclepore, or fiber glass which would have been exposed to a flow of air or other fluids. The thin self-supporting samples are carbon foils, microtomed tissue sections, ion-exchange membranes, and aerosol-sprayed thin substrates. Individual leaves and single strands of hair are self-supporting samples that fall into a category between thin and thick depending on the proton energy and the species from which they are collected. Thick samples are defined categorically to be those whose thickness is greater than the proton beam can penetrate. These include tissue from marine life, humans, and other animals, layered hair, seeds, some leaves, grass, pine needles, cloth, finger nails, and thick layers of residues from some liquid specimens. Packages of powdered samples have contained fly ash, coal dust, orchard leaf, chemicals, lake sediment, lyophilized tissue, and dried blood. Pelletized substances include all of the powdered materials above plus clay, humus, and porous tissue.

The manner of preparing samples in each of these categories for irradiation will be discussed next. Deposits are usually prepared by laying droplets ($\sim$20 µl) in a well controlled region of a thin substrate and then by drying them slowly in a clean atmosphere. Common substrates are either porous like Nuclepore or Millipore, or have good holding properties like (possibly) carbon foils.

The advantage of the porous membranes is that the dried residue is apt to attach itself more rigidly than it would to a substance like Mylar or Formvar. If the deposits are not entirely encompassed by the proton-beam perimeter, or if there is danger of the deposit flaking off, the best quantitative results would be achieved by proper doping of the original solution with a known amount of a standard, noninterfering element. This is described at some length in Walter et al. [6], Campbell et al. [8], and Johansson et al. [9].

For best results on air filter analyses, the thin filters of Nuclepore have proven to be more durable than Millipore which suffers either from beam heating or from radiation damage to the long-chain molecules. Recently, thin filters that have a mesh backing for strength have been used in high flow-rate filtering devices. This backing is detrimental to PIXEA analysis because the ion beam does not penetrate it and the accumulated charge causes sparking if it is not conducted away before a large voltage is built up. (Two methods for reducing charge buildup will be discussed later.) Fiber-glass filters have been standard membranes in Environmental Protection Agency (EPA) air-particulate measuring devices for many years. Such filters are not thin to 3-MeV protons and are not suited for quantitative analyses. The PIXEA emission spectrum from a typical 24-hr fiber-glass filter exhibits only a few elements which are not part of the original fiber glass or are not lost in the background. These elements are Pb, Br, and Fe.

The thin self-supporting samples can either be glued onto the target ladder directly, onto small graphite rings, or onto the conveniently available 5-cm × 5-cm slide frames. Long strands of hair can be attached lengthwise on a sample holder like the one shown in Fig. 6.2 for a scan of the elemental abundances as a function of position or, in other words, age. For one project to be discussed later, up to 20 leaves from one plant and as many as 60 strands of grass were mounted on a single frame. Thin, sheet-type ion-exchange membranes can be attached to the frame as shown schematically by the filters in Fig. 6.2. Mylar or paper tapes obtained in air filter devices that expose a new section of filter medium either continuously or stepwise in a fixed time interval can be mounted linearly on the frames, so that a longitudinal scan will give the details of the particulate levels in the atmosphere for every 2-hr interval during a period from 2 to 7 days. Such air filter mechanisms have been developed at the nuclear physics laboratory at Florida State University [9] for EPA and at the University of California at Davis [3] for the California Air Resources Board. Both devices were developed around the potential for small areal scans of PIXEA. A similar device with a mechanized feed has been proposed at our lab for monitoring streams or other solutions incorporating long strips of ion-exchange membranes.

Thick sections or thick pieces of specimens have been mounted by either attaching directly to the long aluminum frames or else by gluing onto a substrate like Mylar which had previously been attached to one of the rings or to the aluminum frame. In the latter case, unless a clean adhesive is used, one will contaminate the specimen so that only the front surface can be irradiated. In many of our studies, the glue has been sufficiently clean that the glued side is also irradiated (merely by rotating the target ladder through 180°) to obtain information about the homogeneity of the sample. In this case where the reverse side is studied, any contaminants in the thin substrate clearly must be measured and subtracted out. Vacuum-dried tissue is conveniently handled in either of these ways of mounting. In order to help guarantee that the samples do not become detached from the substrate, frequently a thin layer of plastic wrap is also attached over the front of the sample holder. (As the sample chamber must have a hard vacuum in it before the valve to the accelerator vacuum system is opened, it is frequently expedient to dry the samples in another evacuated chamber prior to insertion into the irradiation chamber.)

Earlier we mentioned that one should eliminate the buildup of charge associated with the proton-beam stopping in thick targets. If this is not done a high voltage can be developed, resulting in

a discharge which emits X rays. These X rays appear as a large
pulse in the detector and thus in the spectrum. Shabason and Cohen
[10] avoid this buildup by neutralizing the charge with electrons
emitted from a heated filament located inside the vacuum chamber
near the irradiation area. The filament acts as a cathode and the
sample as the anode of a simple diode. The other method is to coat
the front face of the sample with a pure aluminum layer or to mount
the targets on a conducting substrate like Mylar which has a thin
aluminium layer evaporated onto it. There may still be sparking in
the latter case, but at a reduced voltage.

Powdered substances can be contained between layers of conventional plastic-wrap materials, but Mylar has not proven to work successfully. Vacuum packaging between two layers of mildly heated polypropylene has proven successful to keep the material localized in the central region of the frame, so that the specimens can be manipulated without spreading. The problem with using powdered samples in PIXEA is that it is extremely difficult to account for the nonuniformity of the layer in making quantitative assignments unless the powder was uniformly doped or unless the layer is extremely thin, i.e., less than 1 mg/cm^2 at its greatest thickness. The best results have been achieved when the powder layer is thick relative to the range of the particle beam and spread over an area larger than the beam diameter or else when the powder is compressed to form a solid pellet as described in the following paragraph.

Thick samples of some substances have been made by compressing the substances into 1.5- or 3-cm diameter pellets about 1 mm thick in a 20-ton press. Soil, coal, and plant material that have been finely ground press well. Lyophilized tissue and blood, and NBS orchard leaf also form good pellets. A binder of purified paraffin of about 2% by weight is necessary to bond fly ash or starch into durable pellets. The pellets are mounted for irradiation either by using a small amount of a clean adhesive to attach them to a Mylar substrate or else they are sandwiched between layers of plastic-wrap and Mylar. In both cases the pellets may be irradiated from either face in order to obtain a more representative measure of the abundances. As is the case for other thick targets, corrections for X-ray attenuation and the decreasing X-ray production as the particles slow down in the material appear manageable.

6.4 QUALITATIVE AND QUANTITATIVE ANALYSIS

6.4.1 Calibration Methods

As is the case for XRF units, the overall sensitivity of a PIXEA system can be obtained by using uniform thin films of the elements of interest, when the beam profile is known, or by using a very small amount of an element deposited essentially in a point on a thin backing. In the first case, the areal density of the standard must be known, whereas in the latter case, only the total amount needs to be known. To date all of the difficulties in producing standards have not been sufficiently eliminated to permit PIXEA calibration curves of better than ±5%. The problems with the thin films are knowing how much material is present (the main difficulty with vacuum-evaporated films) or else manufacturing a very uniform layer (the main difficulty with depositing calibrated liquid standards). The problem with making a "point" deposit from a calibrated solution is assuring that there is good adhesion and no spreading. If one understands all of the absorption effects and knows the general energy dependence of the detector efficiency, it is not necessary to calibrate every element of interest as the dependence with atomic number varies in a systematic way over most of the periodic chart. A typical PIXEA sensitivity curve is shown in Fig. 6.4 for two different absorbers across the Si(Li) detector face (0.4-mm thick Mylar and a 0.8-mm thick polyethylene absorber). These data were obtained using point deposits from 10 μl of calibrated solutions. The linearity of PIXEA systems have been verified for a variety of elements from 6 ng to above 2 μg by using such point deposits [6]. For thin-layer samples, such as air filter membranes, which are larger in area than the PIXEA beam area, one merely converts the number of

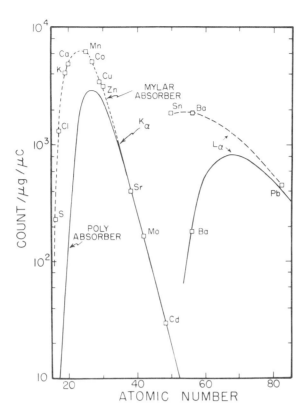

FIG. 6.4 Overall sensitivity of PIXEA system for $K\alpha$ and $L\alpha$ characteristic X rays for elements from S to Pb.

counts in the peak for the respective $K\alpha$ or $L\alpha$ X-ray line into $\mu g/cm^2$ for the filter. If the area of the deposit is less than the beam area the total amount of mass in the irradiated area can be obtained immediately from the values of a sensitivity curve like that in Fig. 6.4.

The aspect of attenuation of the emitted X rays within the sample has been avoided by assuming the samples are thin enough that the absorption of the X rays of interest is negligible. Likewise, the decrease in characteristic X-ray production as the particle beam loses energy in penetrating the sample is also negligible for thin layers. How thin does this mean the samples have to be?

Because of the interrelation between these two effects, it will simplify matters to discuss the latter one first. Here we will limit the description to a 3-MeV proton beam; other energies or types of particles can be handled in a similar fashion. From standard energy-loss tables [11], one can determine that a 3-MeV proton will drop in energy to about 2.87 MeV in passing through a 1-mg/cm^2 thick organic sample comprised predominantly of CH_2. From measurements [12] of characteristic X-ray production in the 3-MeV region, it is known that the yield of X rays for a 2.87-MeV beam is about 0.94 of the 3-MeV yield. Thus the averaged yield for a 1 mg/cm^2 sample would be down by about 3% from that for an extremely thin layer. Therefore, to within 3%, a membrane as thick as a standard 1-mg/cm^2 Nuclepore filter can be considered to be a thin specimen. In such an organic matrix, the average absorption [13] of a 6-keV X ray (~Mn Kα) will be less than 0.5%, and for a 3-keV X ray (~Ar Kα) it will be ~3.5%. So, for elements heavier than Mn, the correction for Kα absorption will be negligible in a 1-mg/cm^2 organic matrix and for elements below Ar, a correction needs to be applied if one is to attempt to achieve an accuracy of better than 3%. If the target is thicker than 1 mg/cm^2, one can continue to use the tables already referenced to calculate the necessary correction factors, either in an iterative manner (which can be done fairly accurately) or else by integration using some functional dependence for the proton-energy loss curves and for the X-ray yield function. For the case of a thick target for which the proton beam stops inside a substance, Willis [14] has prepared a table of corrections for an organic matrix. A diagram of the processes in a thick target for which the target is aligned at the angle θ with respect to the proton-beam axis is shown in Fig. 6.5. The results obtained by Willis for $\theta = 35°$ are listed in Table 6.1. Using these corrections, one can convert the measured yield for a thick organic matrix to the amount of an element present in the sample. Of course, the correction factor assumes a homogeneous distribution of the element throughout the organic matrix. Shabason and Cohen [10] have made similar calculations for ^{16}O beams.

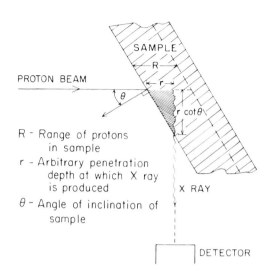

FIG. 6.5 Schematic diagram of X-ray production by protons incident on a thick target.

TABLE 6.1

Proton-Energy Loss and X-Ray Absorption Correction Factors for a Thick Sample of CH_2 for $\theta = 35°$

Element	Correction factors		Integrated total	Effective[a] mass (mg)
	X-Ray absorption	Proton-energy loss		
Cl	2.15	1.48	2.63	2.09
K	1.54	1.62	2.22	2.48
Ca	1.37	1.67	2.10	2.62
Ti	1.20	1.78	2.00	2.74
Cr	1.11	1.87	1.99	2.76
Fe	1.07	1.96	2.06	2.66
Cu	1.03	2.06	2.10	2.62
Zn	1.03	2.15	2.10	2.62
As	1.02	2.15	2.17	2.63
Br	1.01	2.19	2.21	2.49
Sr	1.008	2.28	2.32	2.36
Mo	1.004	2.34	2.34	2.34

[a] Using 12.0 mg/cm^2 for proton range in CH_2 and 0.46 cm^2 for irradiated area at $\theta = 35°$, gives 5.5 mg for irradiated mass.

For complex matrices, one must have some knowledge of the dominant abundances, e.g., Al, Si, and S in fly ash, or else the thick-target correction factors must be generated in some other manner.

One other method for correcting the relative yields from thick targets is to use a homogeneous dope comprised of several elements. This was done for lyophilized placenta to generate a calibration curve for one of our studies. The results are exhibited in Table 6.2 for several thick pellets for vanadium and nickel. Here the amount-measured columns show values obtained merely by dividing the X-ray yield by the sensitivity values in Fig. 6.4 and by a factor of 2.4 (mg) to account for the cross-sectional area A of the proton beam and for the effective yield parameter P for this matrix. The parameter P is important to consider here as it does not vary much for most environmental samples, and it is the crux to making approximate elemental abundance assignments to such samples without carrying out the mathematical integration just described. Let the actual yield of characteristic X rays produced when a 3-MeV proton

TABLE 6.2

Doped Placenta

Element	Amount measured (ppm dry weight)					mdl
	Run 1	Run 2	Run 3	Run 4	Run 5	
K	10,800	12,200	11,000	12,600	14,000	300
Ca	620	530	480	760	700	60
V	0.0	68	53	76	5	7
Fe	650	660	620	640	700	3
Ni	89	0.7	108	0.9	4.2	1.6
Cu	8	9	10	11	11	1.3
Zn	60	60	58	62	63	1.1
Br	120	40	44	44	72	0.3
Rb	16	13	19	19	17	0.3
Dope values						
V	0	68	52	85	0	
Ni	92	0	111	0	0	

stops in a thick sample, be given by Y_0. Assuming that there is no energy loss for the 3-MeV protons when they penetrate the sample and no X-ray absorption, one can calculate a thickness T that would give the same yield Y_0 as in the actual case. Such a calculation gives T = 5.2 mg/cm^2 for an organic matrix. Thus, we can speak of an effective irradiated mass being the product of the density and the effective volume which equals AT. For our beam of cross-sectional area of 0.46 cm^2, the effective mass m = PAT is 2.4 mg. The convenience in having this number is that for an organic matrix, one can use the 3-MeV sensitivity curve in Fig. 6.4 to divide into the yield to obtain the number of µg per 2.4 mg of sample.

How can one test this concept further? Measurements can be made on other doped substances or on standard reference materials (SRM). The effective masses obtained for the NBS SRM of orchard leaf and coal were 2.4 mg and 2.1 mg, respectively, when the mass of the measured Fe was divided by the published NBS relative abundances for Fe.

Quantitative analyses for liquids have been carried out by the usual technique of adding an internal standard before deposition. In this way it is not necessary to insure that the beam entirely encompasses the deposit, although if it does, one can make a consistency check if a known volume was originally deposited. When using a porous Nuclepore substrate, we relied upon reproducibly depositing 20 µl aliquots from a calibrated micropipette, making only periodic checks with internal standards. Although one is working with residues, it is necessary to carefully investigate X-ray absorption and proton-energy loss effects when the residue forms a thick layer (i.e., greater than 1 mg/cm^2 of organic material), when heavy drying rings are present and when depositing material of high Z as one obtains in ashing procedures.

One method used to check that the calculated X-ray attenuation corrections are valid is to dope the sample with a standard having high Z, like barium which produces Kα and Lα X rays of energies 32 and 4.4 keV, respectively. Knowing the ratio of the intensities of these lines for a thin layer of barium, a measure of the ratio of a Ba-doped solution or thick target will give the size of the attenuation correction for a 4.4-keV X ray since for the proton penetration depths involved in PIXEA there will be no noticeable absorption of a 32-keV X ray.

6.4.2 Beam-Heating Effects and Volatilization

Except for the He-filled chamber method described in Ref. 7, all other systems currently operate with the samples located in a hard vacuum. Therefore, these systems cannot measure the presence of highly volatile substances like some mercury compounds.

Also, one of the problems PIXEA analysts try to avoid is excessively heating the samples during irradiation. Two extremes have been employed in PIXEA studies in regard to this avoidance. Some groups have used extremely thin targets so that the proton-energy loss is down. Then, in order to compensate for the loss in counting rate, these people increase the beam currents to the range from 0.5 to 1.0 µA. Campbell et al. [8] prefer wet-ashing the samples to increase the thin-target counting rates. Jolly and White [15] introduced a thin rotating target which ideally should withstand even higher beam currents than the static targets. Unfortunately, as the samples become thinner, the significance of the impurities in the substrate becomes greater. The high current backing is typically a thin carbon foil, but such foils have had variable amounts of noticeable impurities.

At the other extreme, one can employ thick targets mounted on backings which the proton beam need not reach. Then the problem of impurities in the backing completely disappears. Furthermore, one has a gain in counting rate of 10- to a 100-fold due to the increased sample amount. However, one must now worry about the dissipation of the power delivered by the particle beam. As mentioned earlier, the steady state temperature for 100 nA of 3-MeV protons in a thick target is 300°C, assuming little heat conduction and insignificant X-ray emission. Clearly highly volatile elements or compounds in the sample cannot survive a steady beam of 1 µA in this case. The temperature of 300° is certainly high enough to drive off some organic compounds, and perhaps some metallic compounds. In order to check the effect of heating, numerous tests have been made. One is merely to observe the time profile of the elemental concentrations during successive irradiation periods. Another is to increase the beam intensity until clear evidence of some form of evaporation occurs. The results of two of the tests related to data to be described later are shown in Tables 6.3 and 6.4. Here, it is shown that for two long runs for leaves, and four long runs for crab muscle, there is no evidence for evaporation outside of the statistical errors or the detectability limit, whichever is greater. This is true even for such volatile elements as bromine and arsenic. In actual data runs, the samples are only subjected to 30-75% of the beam intensities used in these tests. To test if some evaporation occurs in the first few seconds, poorer statistic runs have

TABLE 6.3

Evaporation Test with 50 nA of 3-MeV Protons Incident on a Thick-Leaf Sample[a]

Element	Run 1	Run 2	mdl
Mn	90	90	3.7
Fe	232	227	2.9
Ni	2.6	2.8	2.0
Cu	33	33	1.6
Zn	60	65	1.3
As	0.0	0.7	2.6
Se	26	31	3.1
Br	6440	6540	3.4
Sr	37	36	4.1
Pb	112	103	11.0
Acc. µC[b]	10	20	
Time lapse[c]	0	5	

[a] Values given in ng/0.4 cm^2.
[b] Total accumulated charge in µC at the beginning of run.
[c] Total time in minutes that beam was on target before beginning run.

TABLE 6.4

Evaporation Test with 80 nA of 3-MeV
Protons Incident on a Thick Section of Crab Muscle[a]

Element	Run 1	Run 2	Run 3	Run 4	mdl
K	9,200	9,900	10,600	10,200	70
Ca	1,000	1,070	1,090	1,050	20
Ti	3	5	0	2	5
V	0.6	4.7	0.0	2.8	4
Mn	3	2	2	1	1.5
Fe	34	31	31	31	1.2
Ni	0.1	1.8	1.3	1.2	1.0
Cu	22	23	24	23	1.0
Zn	333	352	358	357	0.8
As	24.8	19.9	16.6	16.0	0.8
Se	4.7	2.8	1.9	2.2	0.8
Br	37.3	40.1	39.6	41.4	1.0
Rb	5.3	4.3	5.4	5.2	1.0
Sr	24.6	27.0	22.4	23.2	1.0
Pb	0.0	0.5	0.3	0.0	3.0
Acc. µC[b]	0	50	100	150	
Time lapse[c]	0	10	20	30	

[a] Values given in ppm (dry weight).

[b] Total accumulated charge in µC at the beginning of run.

[c] Total time in minutes that beam was on target before beginning run.

been made in about 10-sec intervals and again no evidence of evaporation was seen for the more dominant peaks in the spectrum. In fact, if the beam current is excessively high, one will observe an apparent buildup of the heavy elements as the organic material is gradually vaporized.

6.5 DATA ANALYSIS

6.5.1 Typical Spectra

In this section a brief discussion of gross structure in typical PIXEA spectra will be presented. If one studies several spectra, it is possible to note the elements which typically give a strong signal above the bremsstrahlung background and those elements which are not observed at all. It is convenient to display all of the data on a log scale which unfortunately gives the appearance of broad peaks and poor signal-to-noise ratios. In this section we also include some comments about choices of beam energy, filters, and sample types.

Spectra obtained with a 30-μm section of kidney are shown in Fig. 6.6 for two different absorbers positioned over the Si(Li) detector. The bremsstrahlung produced in the sample by the knocked-out electrons, as mentioned earlier, causes a large background at low X-ray energies, which corresponds to the lower channels. Because the absorbers selectively attenuate the lowest energy X rays, and because the Si(Li) detector efficiency drops at low energies, the observed background appears to roll off below channel 80. The product of the absorption effect and the bremsstrahlung shape causes the broad hump around channels 80 and 60 for the polyethylene and Mylar absorbers, respectively. The proton bremsstrahlung becomes significant around channel 200 and is partially responsible for the leveling off of the counts in the channels above 250. The solid curve drawn through the data is merely a guide to the eye in Fig. 6.6.

The purpose of the absorbers over the Si(Li) detector is to avoid unnecessary pulse pileup in the electronics of the analyzer system caused when count rates become excessive. Routinely, we will irradiate each specimen first with the Mylar absorber in place, using about 1-10 nA of beam. Then, the polyethylene absorber is inserted and a spectrum is recorded for a longer period of time with the beam increased to about 20-60 nA. If the counting rate is still too high, i.e., if it exceeds ∼3000 count/sec, we will make the measurement a third time using an aluminum absorber and the higher beam currents. Such a series of spectra is illustrated in Fig. 6.7. From the Mylar, polyethylene, and aluminum spectra we extract abundances for the elements from S → Fe, Mn → Sr, and Cu → Mo, respectively. The elements around mercury and lead are obtained from the last two spectra. Only on rare cases will we observe cadmium and without moving the detector into the vacuum chamber, it is difficult

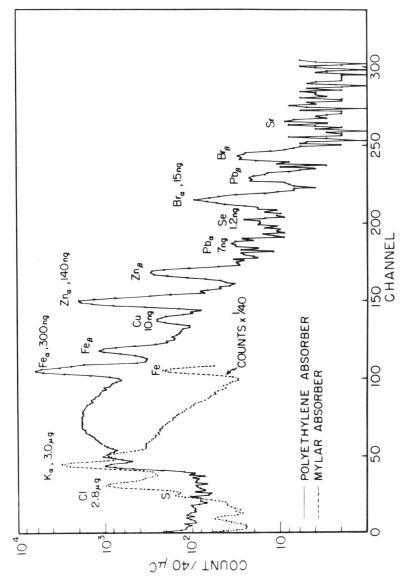

FIG. 6.6 Spectra obtained with PIXEA for a 30-μm human kidney section (proton beam area 0.5 cm^2, 3.0 MeV) for a Mylar and polyethylene absorber over the Si(Li) detector.

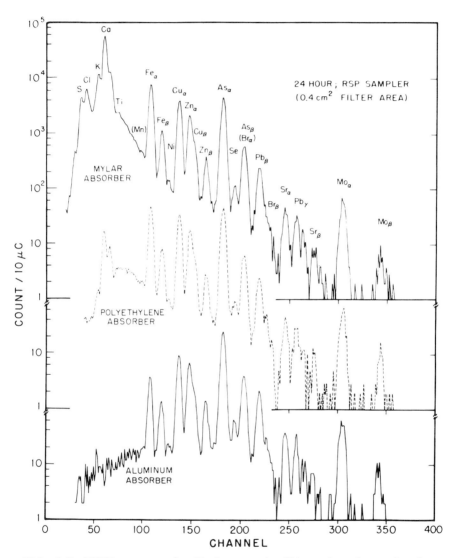

FIG. 6.7 PIXEA spectra for Nuclepore air filter for three absorbers.

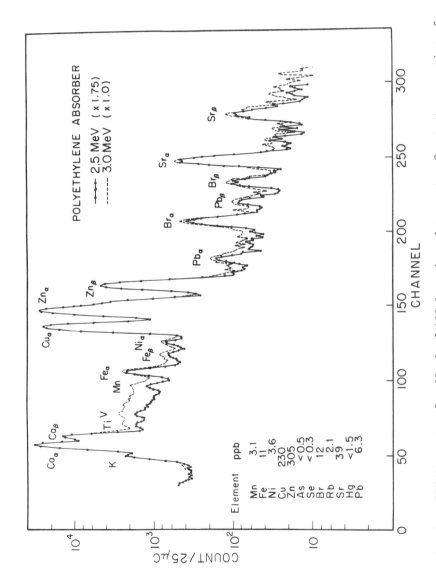

FIG. 6.8 PIXEA spectra for 80 μl of 100:1-condensed tap water for proton energies of 2.5 and 3.0 MeV.

PROTON AND ALPHA EXCITATION   149

to extract meaningful values for anything lighter than sulfur.

The effect of lowering the beam energy from 3.0 to 2.5 MeV is illustrated in Fig. 6.8. These spectra are for 80 μl of tap water (condensed 100:1) deposited on a Nuclepore membrane of 1 $\mu g/cm^2$ in thickness. The X-ray yield for the 2.5-MeV beam was multiplied by 1.75 to emphasize the improvement in signal-to-noise ratio gained by lowering the proton energy.

For these particularly thin samples where beam-heating effects are not a serious problem, it is clear that one makes an appreciable gain in minimum detectability limits (mdl) by lowering the energy if the beam intensity can be increased by 1.75 to compensate for the lower cross section for the production of characteristic X rays. This increase in beam is necessary because X-ray analysts are usually concerned with the mdl that can be achieved in some fixed irradiation time interval. For thick samples in which beam heating is a problem, probably the 3-MeV beam will give the better mdl for systems comparable to ours.

6.5.2 Computer Fitting Codes

In order to operate a PIXEA system economically, one must be able to process a large amount of spectra and eventually convert these spectra into elemental abundances. There are a number of ways of generating such abundances either on-line or off-line. One method is to "strip down" a spectrum, using previously measured information about background shapes and line shapes generated with calibrated standards. Another method is to make a least squares fit to the entire spectrum using functions to represent the background and the peaks, and then convert the peak area to elemental abundances using a calibration curve like the one in Fig. 6.4. It is not our purpose here to discuss the pros and cons of these methods and various combinations thereof. However, we will present a brief description of the technique used at our laboratory to document one of these methods. The code is called TRACE, and is an off-line version of GAUSS-N, and has been under development for over two years at our laboratory. It still operates in a semiautomatic mode, i.e., with interactive operator input.

In order to minimize computer time, the code fits only about a third of the spectrum at one time. The operator chooses the region, the initial guess for the background level, the elements to be searched for, and the number of iterations before jumping out to a cathode-ray tube (CRT) display loop or to a line printer for listing the elemental abundances. After an initial background fit to a five-parameter polynomial, the code determines initial peak-height guesses for the elements of interest, having been given an initial energy-calibration guess. In a few seconds the code generates a functional representation of the data based on the initial background fit, the energy- and width-calibration guesses, a gaussian resolution function, and fixed $K\alpha:K\beta$ and $L\alpha:L\beta:L\gamma$ ratios. This calculation can be displayed for comparison to the data if the operator chooses, or else the program can proceed to make a least squares fit to the data. In the latter case, if desired, all of the amplitudes representing the elements of interest, two energy parameters, two width parameters, and the five background parameters are searched upon. Typically we do not search on the background parameters as the initial background fits are quite suitable. After the prearranged number of iterations has been reached or if chi squared (χ^2) (see Sec. 4.6, p. 84) does not improve more than a specified amount, the computer lists the χ^2, the peak areas and their standard deviation, the backgrounds, the abundances in ng or ppm, and the 3σ detectability limit based on the background under the peak. Three overlapping slices of the spectra are successively fit and listed in a total time of about 50 sec. One such fit is represented by the solid curve in Fig. 6.9, where the data were obtained for a thick section of crab muscle. The vertical bars on the data represent the statistical (standard) deviation associated with each datum point. Clearly, the representation of the background and the peaks by the functions employed is sufficient for this spectrum. The abundances shown give the extracted ppm values on a dry weight basis.

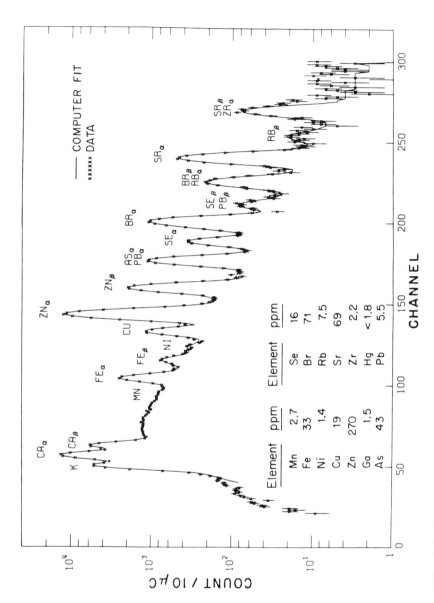

FIG. 6.9 PIXEA data (crosses) for thick section of crab muscle. Solid curve is computer-fit to data.

(There is probably evidence in the spectrum for elements heavier than zirconium, i.e., above channel 290, but the computer search in this instance was directed to only include peaks out to channel 275 and a background to channel 300.)

Two comments about the present code should be given here. The initial calculation which uses the peak heights to estimate the amplitudes is very close to the final value in most (environmental) samples. Thus, with care, a rapid scan of elemental abundances for a series of samples can be made in a few seconds each using only initial parameters. This also means that for all practical purposes, after only one iteration the fit has converged to its final value in most cases. Second, in order to avoid the cost of a skilled X-ray analyst operating an off-line computer, the code should be made completely automatic with sufficient flags to alert the analyst before he reports the results. Such an automatic procedure is employed by Cahill's PIXEA group who process 1000 samples in 36 hr. The group at Florida State University has made a preliminary report on their fitting code, which also should be automated soon [16].

6.6 APPLICATIONS OF PARTICLE-INDUCED X-RAY EMISSION ANALYSIS (PIXEA)

As the two methods XRF and PIXEA are so similar, it is not necessary to give details about many of the applications that can be approached with both techniques. For that reason, it should be sufficient here to itemize many of the sample types that have been irradiated and the scientific fields to which such studies related. A description of air filter analyses will also be omitted here but we will give a description of several other interesting applications. Lastly, we will mention some of the advantages and disadvantages of PIXEA as compared to XRF.

One can group or divide irradiated samples into a number of arbitrary categories. Our choice is the following:

1. Biological and agricultural: plant tissue, soil, soil extracts, tree bores, water, aerosols, animal tissue, animal foods.

2. Environmental: drinking and river water, marine life and sea food, air and water filters, coal, fuel oil, fly ash, gasoline, ion-exchange membranes, human and animal lung cells, roadside plants, sediment.
3. Biochemical: proteins and enzymes, "pure" water and chemicals, animal tissue.
4. Industrial and municipal: air filters from inside factories, cloth, aerosol sprays, drinking water, water at sewage outfalls, downstream air filters, alloys.
5. Geochemical: soil, minerals, soil extracts, and plants from mining sites.
6. Medical and clinical: human organs and tissue, serum, blood, urine, hair, solid and liquid food.

One of the important aspects of PIXEA is the potential to make miniscans practical. The proton beams are typically 0.5 cm in diameter upon incidence on the samples although reducing the beam down to 1 mm^2 in area presents no mechanical problems. However, to use such a beam one must recognize the power dissipation problem and cannot assume that the heating effects of a 100 nA beam deposited in a 1 mm^2 area of sample will be the same for 100 nA deposited over the 20-mm^2 area corresponding to a diameter of 0.5 cm. Very few elemental analyses data have been reported using beams in the neighborhood of 1 mm^2 to date although the air filter devices at Florida State University and the University of California at Davis (see Sec. 6.3.2) are intended to be used in conjunction with rectangular beams about 2 × 4 mm. The results of a larger area scan intended to study the amount of nonuniformity of the deposit across a 37-mm diameter Nuclepore membrane from a simple EPA air filter system are shown in Fig. 6.10. The membrane was studied at five positions along a diameter as shown in the inset. A falloff in concentration from 4 to 1 from the center to the edge is seen for some of the elements which probably compose the heavier particulates, and a falloff of 2.5 to 1 for other elements which compose lighter particulates. Such information is significant if these filters are to be analyzed in XRF systems which have a nonuniform exciting beam or in analysis systems

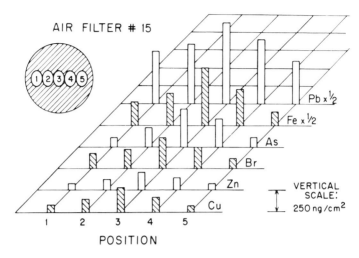

FIG. 6.10 Results for six elements from a PIXEA scan at five positions. Inset represents manner of scan.

which do not utilize the entire filter area. In our laboratory we have used the scanning ability of PIXEA to investigate (1) the uniformity of thin films and deposited solutions that are commercially available for calibration of XRF systems, (2) the variation of elements along hair, (3) the variations for different regions of single leaves, (4) the uniformity of surface deposits across a 20 × 25-cm fiber-glass (air) filter, and (5) the variation of elements along lengths of tissue, e.g., lengths of fish muscle.

Another study for which PIXEA has supplied useful data concerns metal uptake in plants obtained at mining sites, reclamation problem areas, and roadside and control areas. For some of these studies, PIXEA data were also obtained for whole soil samples and acid extracts from the same soils. One project concerns genetic effects on plants growing under highly toxic conditions. A portion of the results of a particular study involving the falloff of elemental abundances in plants and soil as a function of distance away from a heavily traveled road is shown in Figs. 6.11, 6.12, and 6.13. One of the plants studied was ribwort plantain, *plantago lanceolata*. Three plants were collected from each of five sites. Leaves from the plants

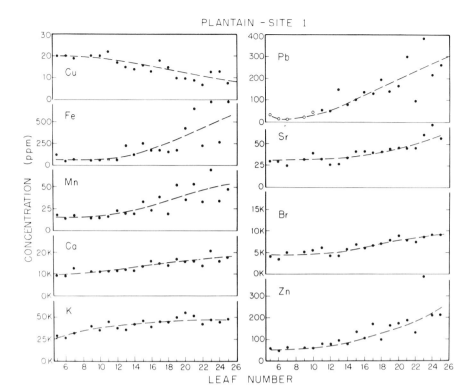

FIG. 6.11 Elemental abundances for 20 plantain leaves as function of position along stem. The leaf ages increase with increasing number (roadside site).

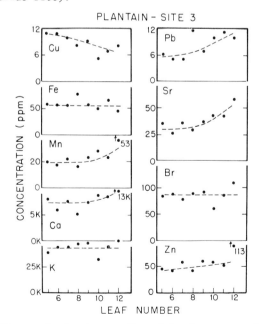

FIG. 6.12 Elemental abundances for eight leaves as function of age (see Fig. 6.11). Collected from site located 75 m from roadside.

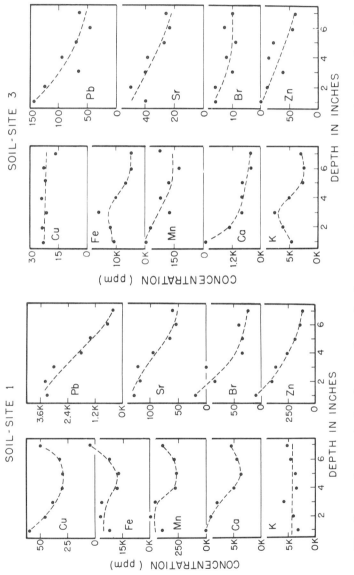

FIG. 6.13 Elemental abundances for soil from two sites. Abundances given as function of depth below surface.

were removed, washed, dried, and mounted according to their position along the stem. The leaf number in Figs. 6.11 and 6.12 corresponds to this ordering which assigns the lowest number to the newest or youngest leaves. Figure 6.11 shows a large buildup of particular elements with age for one of the roadside plants, whereas Fig. 6.12 provides a comparison for the elemental abundances and the buildup pattern for another plant from site E3A which is 75 m from roadside. Although adjacent leaves showed large differences in abundances, the systematic patterns are obvious.

For the soil study, soil bores were taken to a depth of 7 in. and samples for each 1-in. depth were homogenized, finely ground, meshed, and pelletized. The results for soil from the two sites from which the plants were collected are exhibited in Fig. 6.13 for the various depths. Again, the systematic trends and the relative abundances for the two sites are obvious.

Earlier, in Fig. 6.8, results were shown for an analysis of the residue from evaporation of clean drinking water. Even though the substrate was relatively thick (that is, 1 mg/cm^2) sensitivities down to the ppb level of the original 8 ml of H_2O were observed. If a thinner backing was used for the deposition, it should be possible to reduce the detectability limit another factor of 2 to 4. This reduction is based on the lowering of the bremsstrahlung background level which is primarily due to the substrate in Fig. 6.8. At some point a limit will be reached. This is where the bremsstrahlung from the dominant residue elements, for example, $CaCO_3$ or NaCl, exceeds that due to the backing. In fact, for sea water, the detectability level is greater than that implied in Fig. 6.8 because of the NaCl content, whereas for condensed, distilled H_2O, the detectability levels presumably would be several orders of magnitude lower. For fluids like urine and serum, the organic content causes a deterioration of the detectability limit. For these solutions, ashing may be necessary to achieve required sensitivities.

The final results to be shown here relate to a study of the uptake of metals in marine life that are being poisoned by some still unknown mechanism from nearby localized sources of heavy pollution.

In this study muscle and gonad were obtained from crab, and muscle from Dover sole from both the contaminated region and a control region. The gonad specimens were smeared onto a Mylar substrate for analysis and whole muscle chunks were attached to aluminized Mylar film. Large variations in metal abundances were seen from specimen to specimen, in both the contaminated and the control samples. In Fig. 6.9 a spectrum for a thick piece of crab muscle already has been shown. The results for Dover sole muscle from the contaminated region are exhibited in Fig. 6.14. Here, the median ppm level is given by a horizontal bar near the center of vertical boxes. The top and bottom edges of the boxes represent the upper and lower values observed for the samples studied. Several interesting results came from these data, one being that the same plot for the muscle from the control region is nearly identical to the one shown in Fig. 6.14 for the contaminated region. The detectability limits

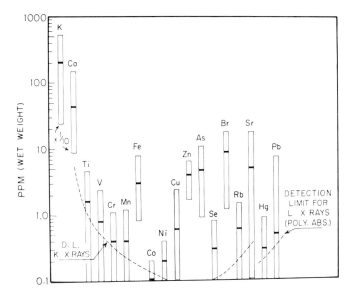

FIG. 6.14 Concentration of elements in muscle of Dover sole collected from polluted area. Median concentration represented by horizontal bar. Upper and lower ends of rectangles represent the highest and lowest concentrations observed.

shown by the dashed curves represent interference-free limits for about 20 μC of 3-MeV protons and for a polyethylene absorber.

Very little has been said or implied in this text to indicate the definite aspects of XRF analysis which make it superior to PIXEA and vice versa. This issue is difficult to handle because both methods are undergoing rapid improvements and because *concentrated* PIXEA efforts are a relatively new happening. Furthermore the boundary conditions of budget, available equipment, available accelerator time, and ability of the personnel are necessary input for any comparison. Even if one starts with the assumption that both types of facilities are perfectly outfitted for X-ray analyses, one must consider the following questions:

What kind of diverse sample load will need to be handled?
What will be the maximum turnaround time that is permissible?
What will be the annual load?
What other responsibilities, e.g., sample prepping, will be required by equipment operators?

Also particle accelerators are versatile and can be used in parallel to make ion-scattering analyses (ISA) for identifying light elements between hydrogen and fluorine. So the issue has many facets.

Nevertheless, it is possible to make some statements based on limited observations. It is our feeling that for routine aerosol analyses of large diameter air filters, i.e., 37 mm diameter, modern X-ray fluorescence units give sufficient sensitivity (compared to PIXEA which may have lower detectability limits at the present time) to provide the X-ray information necessary for aerosol study programs. In this case XRF systems should be used because of the ease of operation, i.e., they can "run alone," and the ease of maintenance of the basic system. If on the other hand, one devises low-volume air sampling devices, such as portable personnel monitors, or seven-day "streaker" devices [9], then the deposit pattern might be as small as 1 mm^2 and might require a narrow beam for the exciting radiation. In this case only a particle beam may suffice.

For tissue analyses, PIXEA is limited to identifying positive signals typically for only about 10 elements heavier than potassium (see Fig. 6.13). The reason here is that the other heavy elements are only present in normal tissue at levels under the 200 ppb (wet weight) level. Without ashing of samples or long, expensive irradiation periods, it is hard to improve on these limits. XRF systems on the other hand can work with larger tissue samples, but then the limitation might be counting rates and attenuation corrections. At the present time, one pulsed XRF system is capable of measurements under 100 ppb (wet weight) in 30 mg/cm^2 of tissue in 100 sec [17].

PIXEA requires that the targets be placed in vacuum to achieve the maximum sensitivity, but this is also true for XRF for somewhat similar reasons.

For a routine operation, the cost of analyses using PIXEA and XRF will be under $5 for an already equipped facility, and as such, the specimen collection fee, sample preparation time, and the assimilation of the elemental abundance information will be the most expensive items. Hence, the cost of the analysis technique probably will not be the decisive factor in selecting the analytical system for many of the larger projects.

It is still our feeling that for many elements, PIXEA has lower detectability limits for *absolute* amounts deposited on thin substrates than XRF. Basically this is because large amounts of focused proton beam are available with very low X-ray backgrounds.

Finally, XRF systems with multichannel wavelength-dispersive analyzers are becoming more practical and more sensitive. Such systems have a potential for X-ray analyses of lighter elements than Si(Li) or energy-dispersive systems are capable of studying. These systems also can avoid some of the problems yet unanswered for Si(Li) systems--problems of interferences in the region of the Kα and Kβ lines for titanium, vanadium, and chromium which overlap the Ba Lα, Lβ, and Lγ lines, for instance. PIXEA presently does not work effectively with crystal analyzers because of low counting rates; however some proposals to overcome this difficulty are presently surfacing.

NOTE ADDED IN PROOF

Due to the typical delays in preparing a text of this nature, the time interval between submission of most manuscripts and the publication date is nearly 24 months. Since this chapter was submitted for publication (November 1975) numerous advances have been reported in the methods described here, and in many places we could now replace "will be" or "should be" to "has been." The largest collection of new articles, *Nuclear Instruments and Methods, Volume 142* (1977), is a summary of the 1st International Conference on Particle Induced X-Ray Emission and Its Analytical Applications held in Lund, Sweden. Also, *X-Ray Fluorescence Analysis of Environmental Samples* edited by T. G. Dzubay (Ann Arbor Science Publishers, Ann Arbor, Michigan, 1977) is comprised of a set of papers contributed by experts in the XRF and PIXEA field. Lastly, at the Lund Conference, for almost arbitrary reasons, the body agreed to use the acronym PIXE over the others--PIXEA and PIX--in future writing. The present chapter ends our use of PIXEA.

REFERENCES

1. L. S. Birks, *Anal. Chem.*, 44 (5), 557R-562R (1972).

2. T. B. Johansson, R. Akselsson, and S. A. E. Johansson, *Advan. X-Ray Anal.*, 15, 373-387 (1972); *Nucl. Instrum. Methods*, 84, 141 (1970).

3. Thomas A. Cahill, Report to the California Air Resources Board and Project Clean Air, University of California, Davis, Rep. No. UCD-CNL 162, October 1972. See also Thomas A. Cahill, *Proc. 3rd Conf. on Applications of Small Accelerators*, J. L. Duggan, ed., Denton, Texas, 1974, p. 184.

4. F. C. Jundt, K. H. Purser, H. Kubo, and E. A. Schenk, *J. Histochem. Soc.*, 22, 1 (1974).

5. F. Folkmann, C. Gaarde, T. Huus, and K. Kemp, *Nucl. Instrum. Methods*, 116, 487 (1974); F. Folkmann, J. Borggren, and A. K. Jeldgaard, *Nucl. Instrum. Methods*, 119, 117 (1974).

6. R. L. Walter, R. D. Willis, W. F. Gutknecht, and J. M. Joyce, *Anal. Chem.*, 46, 843 (1974).

7. G. G. Seaman and K. C. Shane, *Nucl. Instrum. Methods*, 126, 473 (1975).

8. J. L. Campbell, B. H. Orr, A. W. Herman, L. A. McNelles, J. A. Thomson, and W. B. Cook, *Anal. Chem.*, 47, 1542 (1975).

9. T. B. Johansson, R. E. Van Greiken, J. W. Nelson, and J. W. Winchester, *Anal. Chem.*, 46, 843 (1975); J. W. Nelson, J. W. Winchester, and R. Akselsson, in *Proc. 3rd Conf. on Applications of Small Accelerators*, J. L. Duggan, ed., 1974, p. 139.

10. L. Shabason and B. L. Cohen, *Anal. Chem.*, *45*, 284 (1973).

11. Hans Bichsel, *Amer. Inst. Phys. Handbook*, McGraw-Hill Book Co., New York, 1972, pp. 8-142.

12. C. J. Umbarger, R. C. Bearse, D. A. Close, and J. J. Malinify, *Advan. X-Ray Anal.*, *16*, 102 (1973); R. Akselsson and T. B. Johansson, *Z. Phys.*, *266*, 245 (1974); C. E. Busch, A. B. Baskin, P. H. Nettles, and S. M. Shafroth, *Phys. Rev.*, *A7*, 1601 (1973).

13. William J. Veigele, *Atomic Data*, *5*, 51 (1973).

14. R. D. Willis (unpublished); see also L. Shabason, B. L. Cohen, G. H. Wedberg, and K. C. Chan, *J. Appl. Phys.*, *44*, 408 (1973).

15. R. K. Jolly and H. B. White, *Nucl. Instrum. Methods*, *97*, 103 (1971).

16. H. C. Kaufmann and R. Akselsson, *Advan. X-Ray Anal.*, *18*, 353 (1970).

17. J. M. Jaklevic, D. A. Landis, and F. S. Goulding, LBL Report 4248; to be published in *Advan. X-Ray Spectrometry*, 1975.

Chapter 7

ELECTRON PROBE MICROANALYZERS

Kurt F. J. Heinrich

Analytical Chemistry Division
National Bureau of Standards
Washington, D. C.

Harvey Yakowitz

Institute for Materials Research
National Bureau of Standards
Washington, D. C.

7.1	Introduction	164
	7.1.1 Historical Development	164
	7.1.2 Present Instrumentation	166
	7.1.3 Recent Trends and Projections	170
7.2	Sample Requirements	171
7.3	Qualitative and Semiquantitative Analysis	173
	7.3.1 Wavelength Spectrometry	174
	7.3.2 Energy Spectrometry	176
7.4	X-Ray Area-Scanning Images	178
7.5	Quantitative Electron Probe Microanalysis	181
	7.5.1 Empirical Model for the Calibration Function	183
	7.5.2 Theoretical Approaches to the Calibration Function	185
	7.5.3 Generation Correction (Atomic Number Correction)	187
	7.5.4 Absorption Correction	188
	7.5.5 Characteristic Fluorescent Correction	189

	7.5.6 Continuum Fluorescence Correction	190
	7.5.7 Iteration and Computation	190
7.6	Analysis Involving Special Specimen Configurations	192
7.7	Standards	195
7.8	Applications of Electron Probe Microanalysis (EPMA)	196
	7.8.1 Selected Applications Involving Quantitative Analysis	198
	7.8.2 Sources of Information on Applications	199
References		201

7.1 INTRODUCTION

7.1.1 Historical Development

The use of focused electron beams for micrographic examination dates back to the 1930s. The first scanning electron microscope (SEM) was built by Ardenne [1], and an instrument which is the forerunner of the present scanning electron microscope was described by Zworykin et al. in 1942 [2]. The first electron probe microanalyzer was constructed in 1948 by Castaing [3], who demonstrated that qualitative and quantitative analyses of regions of solids containing approximately 10^{-13} g of material could be obtained. Castaing's investigations, and those of other early users of electron probe microanalyzers [4-7], received wide attention among material scientists, and soon electron probe microanalyzers were being developed and marketed by commercial manufacturers. Today there are more than 2,000 electron-beam instruments with X-ray spectrometers in operation throughout the world.

Castaing's instrument was based on an electron-optical column that could focus accelerated electrons to a diameter of about 1 μm at the plane of the specimen. An optical microscope was mounted coaxially with the electron-optical column. Thus, the region of interest on the specimen surface could be observed visually and bombarded with the electron beam at the same time. A single curved-crystal spectrometer of the Johann [8] type was used for the detection and measurement of the X-ray emission. No provisions were made for lateral displacement (scanning) of the electron beam.

In 1956, Duncumb [9] reported on an electron probe microanalyzer provided with beam scanning. The beam diameter and thus the resolution of the scanning electron micrographs were of the order of 1 μm, comparable to the resolution of the optical microscope which had been inserted into Castaing's fixed-beam instrument, and commensurate with the dimensions of the region emitting primary (electron-excited) X rays at conventional operating voltages (8-25 kV). The X-ray spectrometer was semifocusing rather than fully focusing as was Castaing's. For this reason, the X-ray signal was less sensitive to small displacements of the point of X-ray generation; the optical microscope, which served in Castaing's instrument to define the correct position of the specimen plane, could therefore be dispensed with.

Duncumb also was the first to use an energy-dispersive X-ray spectrometer [10]. For this purpose he employed a flow-proportional detector. Dolby refined the technique of pulse-height analysis with single-channel analyzers, and showed that X rays emitted from elements of atomic number less than 10 could be observed and displayed in scanning images [11,12]. Shortly thereafter, curved Blodgett-Langmuir pseudocrystals [13] were introduced as diffracting devices, and the X-ray emission of elements of low atomic number could be routinely monitored.

The development of lithium-drifted energy-dispersive detectors for X rays in the main wavelength range of interest to microanalysis (1-10 Å) further stimulated the use of pulse-height analysis [14], both for rapid qualitative tests and for quantitative application [15]. Such detectors are also used in the scanning electron microscope, which, after being further developed by Smith [16], Oatley [17], and by other investigators [18,19], were used for a wide range of applications. Since, at the same time, the electron optics of conventional electron probe microanalyzers were improved to provide spatial resolution equal to that of most scanning electron microscopes, the difference between the two instruments tends to disappear. Electron transmission techniques were also introduced in scanning electron microscopes [20], and X-ray spectrometers, as well as scanning coils, were attached to transmission electron microscopes. Therefore, many techniques developed for the electron probe microanalyzer can be applied with other electron-beam instruments, and vice versa.

7.1.2 Present Instrumentation

The essential components of an electron probe microanalyzer are (Fig. 7.1):

1. A source of accelerated electrons (electron gun)
2. Electron optics which focus these electrons into a fine beam
3. A specimen manipulator which orients the specimen region of interest towards the electron beam

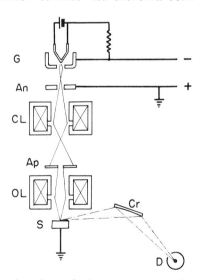

FIG. 7.1 Schematic view of the electron probe microanalyzer. G: electron gun, An: anode plate, CL: condenser lens, OL: objective lens, Ap: beam-defining aperture, S: specimen, Cr: X-ray analyzer crystal, D: X-ray detector.

4. An X-ray spectrometer or set of X-ray spectrometers

Among the ancillary components, the following are the most important (Fig. 7.2):

1. A pumping system to maintain within the instrument the vacuum required for the formation of the electron beam and for the reception of some of the signals
2. A high-voltage power supply to produce the potential required for the electron-beam acceleration
3. An optical microscope for visual observation of the specimen
4. Provisions for the scanning of the electron beam (scanning coils and associated electronics)
5. An electronic network for the amplification, signal processing, and data output

Typical parameters for the more important components of the electron probe microanalyzer (EPMA) and the scanning electron microscope (SEM) are listed in Table 7.1. The main differences between the two instruments are the achievable minimum beam diameter, which is smaller by a factor of 20-30 for the scanning electron microscope, and the existence of a light microscope and crystal spectrometers in the electron probe microanalyzer. These differences reflect the purposes for which the instruments were originally designed.

The *electron probe microanalyzer* serves mainly to provide qualitative and quantitative elemental analysis of a small region of a solid specimen through the measurement of characteristic X rays. At the operating potentials required for such an analysis, the X-ray emitting region is of one to several micrometers in diameter, due to the diffusion of the exciting electrons within the target and the transparency of the target to the emerging X rays. If the analysis is to be quantitative, the target configuration (surface flatness, electron-beam incidence angle, and X-ray emergence angle) must be well defined, so that a rigid specimen state system is advantageous. Wavelength-dispersive (Bragg) X-ray spectrometers are preferred to energy-dispersive units since they are superior in spectral

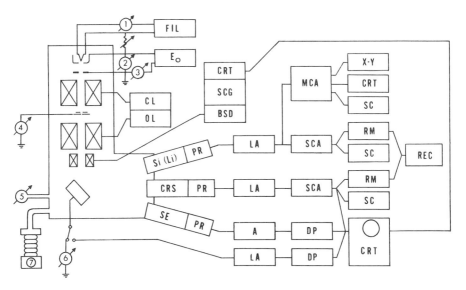

FIG. 7.2 Electron probe control and readout system. 1: Filament current meter, 2: operating potential meter, 3: gun emission current meter, 4: monitor current meter, 5: vacuum meter, 6: specimen current meter, 7: vacuum pump, FIL: filament power supply, E_0: high-voltage power supply, CL: condenser lens, OL: objective lens, CRT: cathode-ray tube, LA: linear amplifier, A: amplifier, PR: preamplifier, BSD: beam scan driver, SCA: single-channel analyzer, RM: rate meter, X-Y: X-Y recorder, REC: strip-chart recorder, SE: secondary electron detector, Si(Li): energy-dispersive X-ray detector, CRS: crystal spectrometer, SC: scaler, DP: data processor, SCG: scan generator.

resolution, typically by a factor from 10 to 50. The high stability of the thermal emission gun makes it a good source of electrons for comparative X-ray measurements. However, the combination of the low brightness of this gun with the low efficiency of wavelength-dispersive spectrometers limits the achievable minimum electron-beam diameter; at beam currents of 10^{-8} to 10^{-7} A, which are required to produce strong X-ray signals, an electron-optical column with a thermionic emission gun does not usually permit the formation of an electron beam of diameter below 0.1 μm. For the proper operation of conventional vertical focusing crystal spectrometers, the position of the specimen in the direction of the beam (elevation) must be

TABLE 7.1

Comparison of the Electron Probe Microanalyzer
and the Scanning Electron Microscope

Parameter	EPMA	SEM
Voltage range	5-30 kV	0.5-30 kV[a]
Beam current range	10^{-6}-10^{-9} A	10^{-7}-10^{-12} A
Beam diameter	0.3 μm	0.01 μm
Electron-beam incidence angle	Usually 90°	Variable
Optical microscope	Yes	No
Wavelength-dispersive spectrometers	2-4	Optional
Energy-dispersive spectrometer	Optional	Optional
Secondary electron detection	Optional	Yes
Specimen current detection	Yes	Optional
Column vacuum	10^{-5} torr[b]	10^{-5}-10^{-6} torr[b]
Signal processing:		
a. Nonlinear amplification (gamma)	No	Yes
b. Differential amplification	Yes	Yes
c. Derivative processor	Optional	Optional

[a] Scanning transmission: up to 200 kV.
[b] 10^{-5} torr ≃ 1.333 × 10^{-3} Pa.

well controlled. Such control is best achieved with the aid of an optical microscope.

The *scanning electron microscope*, in turn, was constructed to provide images with topographic or structural information. The efficiency of production of secondary electrons permits the use of beam currents of the order of 10^{-12} A; the generation close to the specimen surface of signals, such as secondary or low energy-loss backscattered electrons [21], provides a potential for high spatial resolution (10 nm or less), and the stability requirements can be reduced so that nonthermionic guns, such as field-emission cathodes, become attractive. In view of the superb imaging capability of the scanning electron microscope, the optical microscope is usually not of interest as a component. The specimens, whose surface topography is to be studied, frequently have uneven surfaces, and flexible stages which can provide a variety of specimen orientations are desirable.

With the arrival of lithium-drifted silicon X-ray detectors, elemental analysis can also be performed with scanning electron microscopes. On the other hand, the advantages of high-resolution microscopy for electron probe microanalyzers were rapidly recognized, and instruments having the characteristics of both devices are now available. Insofar as the conditions of flatness of the specimen and the spectral resolution of the silicon detectors permit, the procedures of quantitative electron probe analysis can also be performed in SEM's, so that a further distinction is unnecessary.

7.1.3 Recent Trends and Projections

In recent years, the scanning electron microscope has found rapidly increasing acceptance while little development has been seen in the design and use of the conventional electron probe microanalyzer. Moreover, other variants of the electron beam instrument, such as the scanning transmission microscope and the scanning Auger electron microscope, have shown great promise [22]. The shifts in emphasis are in part related to the increasing use of the silicon X-ray detector.

Present trends indicate that the characterization of microstructures requires a variety of tools besides the X-ray spectrometer. Furthermore, the spatial resolution of conventional microprobe analysis and of structural determinations by the Kossel X-ray diffraction technique is insufficient for many purposes. There is also less interest in the minor points of quantitation and more in on-line analysis because, in many areas of technological investigation, speed of analysis is more important than a fully quantitative analysis. In such a situation, a scanning electron microscope with a silicon detector may be the most effective tool. On the other hand, even in conventional microprobe analysis, the high resolution of the scanning electron microscope is very useful in the selection of analysis sites, and we expect that all future instruments will have this capability. Most of them will offer, on an optional basis, more auxiliary techniques, such as the observation of electron channeling and magnetic domain patterns, Auger-electron analysis and cathodoluminescence spectrography [23]. Such diversity will also require more efficient high-vacuum systems, probably based on ion and turbomolecular pumps. Significant improvements can also be expected in the handling of instrumental output. These will be mentioned following a discussion of the analytical methods.

7.2 SAMPLE REQUIREMENTS

For a successful quantitative electron probe analysis, the specimen must have the following properties:

1. Its surface must be prepared so as to permit the selection in the specimen of the region of analytical interest (good visibility, without preferential removal of specimen constituents).
2. The geometrical configuration and orientation of the specimen surface must be well defined and known, particularly in terms of flatness and X-ray emergence angle.
3. The specimen must conduct the heat and electric charges injected by the electron beam without compositional changes.
4. The prepared region of the specimen must have, and maintain, a composition representative of that of the original specimen.

The problems of specimen preparation become more difficult as the depth of the analyzed region diminishes. For this reason and because of the need for conduction of heat and electricity, the specimen conditions for electron probe analysis are more stringent than in X-ray fluorescence spectrometry, for example.

Clearly, it is not possible to completely satisfy all four requirements for all specimens. We cannot, for instance, render conductive an electric insulator without changing its surface composition in some way. Also, many specimens subjected to EPMA have been prepared originally for metallography; yet, the purposes and procedures of metallography differ from those of analysis. Observation by a light microscope is aided by procedures which selectively remove certain materials such as the disordered contents close to grain boundaries. But in etching techniques, the surface composition may be altered by selective dissolution and even by redeposition of components. Such procedures must, therefore, be avoided. At the same time, attention must be given to the possibility of artifacts in the analysis due to smearing or imbedding of polishing materials.

Nonconductive specimens usually must be coated with carbon or some metal such as aluminum or gold, to avoid electric charging or heating. The coating is chosen so as to avoid interfering X-ray emission, and from this point of view, carbon is attractive. Carbon also has the advantage of high optical transparency at the thicknesses usually applied (200-400 Å). The coating can interfere with the X-ray emission from the specimen in two ways: by absorbing part of the energy of the electron beam and by absorbing emergent X-ray photons. The electron absorption is serious when low-excitation voltages (<10 kV) are used. The X-ray absorption is usually less troublesome except for long-wavelength lines. Therefore, and in view of the toxicity of beryllium compounds, beryllium is not recommended as a coating. Regardless of the choice of material, it is good practice to apply conductive coatings simultaneously to the specimen and the standards used for quantitation since the analytical errors due to coatings tend to cancel when the same thickness is applied to specimen and standard.

The handling of particulate material [24] and of organic specimens [25] offers special problems. Particles (such as air pollutants and industrial ashes) must be affixed to a substrate which should emit little X-ray background. Beryllium, if used for this purpose, must be high (>99.9%) purity. Vitreous carbon, although it emits more continuous radiation, is a useful alternative. The particles must be carefully coated to avoid heating or electrostatic repulsion from the substrate. The coating must cover all sides of the particle and a firm bridge to the substrate must be formed to facilitate the flow of heat and current. Special rotating evaporator stages or coaters based on plasma sputtering are useful in this respect.

Biological tissues can either be used as thick (electron-opaque) specimens or in thin sections. The usefulness of thick specimens is, however, limited, at least for soft tissue, by the deep penetration (>10 µm) of electrons in such low-density materials, and by the effects of continuum fluorescence. Furthermore, the preparation of biological tissue is complicated by the high mobility of some elements of interest, notably sodium, potassium, and chlorine. Due to this mobility, the traditional histological preparation procedures are unsuitable, and the most promising prospects presently entertained involve the use of sections of tissue which were rapidly frozen to very low temperatures without dehydration [26]. For detailed discussions of the analysis of biological tissues, see Ref. 27.

7.3 QUALITATIVE AND SEMIQUANTITATIVE ANALYSIS

The wavelengths of the characteristic lines provide information as to which emitting elements are present, but the investigation always involves some degree of estimate of the relative abundance of the specimen constituents derived from the intensities of the observed characteristic lines. We will distinguish three modes of operation:

1. In a *qualitative analysis*, the concentrations, expressed in mass fractions of the components, are roughly estimated by visually comparing the line intensities emitted from different elements. In this case, the accuracy of the estimate depends on the familiarity of the analyst with the

X-ray yield (emitted photons/electron) and with the efficiency of the spectrometer system. Both vary extensively as a function of the atomic number of the emitter. Hence, the accuracy of the estimate depends on the experience of the operator, the specimen composition, and the equipment used. Errors by a factor of 10 are not uncommon; it is customary to classify detected elements as "major, minor, or trace."

2. A higher accuracy is obtained in *semiquantitative analysis* in which the observed X-ray intensities are compared with those obtained from pure elements or other available standards. No special precautions (e.g., dead-time corrections) are taken to assure high accuracy, and matrix effects (nonlinearity of the analytical calibration curves) are ignored. With judicious selection of standards and operating conditions, mainly avoiding measurement of soft X-ray lines ($\lambda > 3$ Å) at operating voltages above 15 kV, the errors can be less than 10-20%. Such procedures are particularly indicated where irregularities in the specimen surface would preclude a fully quantitative analysis.

3. For a *fully quantitative analysis*, the X-ray measurements on flat, electron-opaque specimens and known standards are performed with high precision, and dead-time corrections and background corrections are included; the ratio of characteristic X-ray intensities from specimen and standard is used in a "corrective procedure," either empirical [28] or based on the physical laws of X-ray generation [29], to obtain an estimate of the specimen composition. An accuracy of 1-3% can usually be attained.

7.3.1 Wavelength Spectrometry

When wavelength-dispersive spectrometers are used in a qualitative mode, the X-ray intensities, observed by means of a rate meter, are recorded as a function of time while the spectrometer is mechanically

ELECTRON PROBE MICROANALYZERS   175

driven through its wavelength range. Such an operation typically
consumes 20-40 min for the entire spectrometer range. As noted in
Chap. 1 (e.g., Figs 1.2 to 1.6), any given crystal covers only part
of the useful spectral range (usually 1-50 Å). A complete wavelength scan covering the lines of all elements of atomic number above
five or six thus requires the use of several crystals. But, in most
microprobe investigations, the composition of many points on the
specimen is of interest. Therefore, the duration required for a
complete qualitative wavelength scan is a significant limitation.
Several approaches have been used to minimize this inconvenience:

 1. Scan simultaneously with several [2-4] spectrometers.
 2. Program the scanning speed so that regions containing no
    lines of interest are scanned rapidly.
 3. Use a large number of spectrometers of fixed wavelength
    tuned to the elements of interest.
 4. Use energy-dispersive detectors (see Sec. 7.3.2).

Solutions 2 and 3 have the disadvantage that the presence of an unsuspected element may be overlooked. This holds even if a rapid scan
covers all wavelength regions, when the unsuspected element is present
at low concentrations, since small peaks may be hidden in statistical
noise.

The interpretation of wavelength-dispersive spectra is considerably complicated by the appearance of high-order reflections from
the diffracting crystal. The elimination of the peaks produced by
these reflections by electronic means is possible but complicated.
A systematic procedure for the interpretation of complex spectra is
therefore a useful time saver. It is based on a listing of all lines
and reflections observable in practice. After checking off all lines
emitted by each identified major constituent, the number of remaining
lines is greatly reduced, and misinterpretations of higher reflections
can be avoided [30].

The curved-crystal spectrometers of the Johann or Johansson
type require that the source of radiation be on a line normal to the
plane of the focal circle (and thus represented as a focal "point"

in Fig. 1.8). When the focal circle lies in a plane which contains the electron beam (vertical spectrometer), the focal condition is fulfilled only at one point along the electron beam. Hence, the elevation of the specimen must be carefully controlled. The control is facilitated by the optical microscope which is in focus only when the specimen is at the correct position. Regardless of the spatial orientation of the spectrometer, the focusing property of the spectrometer limits the lateral deflection permissible in area scanning without significant loss of signal intensity.

7.3.2 Energy Spectrometry

The time required for the qualitative characterization of a specimen region can be substantially reduced when a lithium-drifted silicon detector is used to obtain a photon-energy spectrum. This economy is due to the simultaneous observation of all lines within the spectrometer range and to the fact that the spectrum can be observed during data collection, so that the time of observation can be varied according to needs. A further simplification arises from the absence in the observed spectrum of the lines in higher orders of reflection found in wavelength spectrometry.

The use of the energy-dispersive spectrometer also allows a greater range of specimen elevation than the wavelength spectrometer. For this reason, energy-dispersive spectrometers are preferable to crystal spectrometers in scanning electron microscopes, which are not normally provided with optical microscopes. The wide angular signal acceptance of the solid state detector should, however, be restricted by careful collimation so as to exclude the detection of X rays generated from diverse instrument components by backscattered electrons.

Energy-dispersive spectra are not entirely free of artifacts [14, 31] (see also Chap. 2, Sec. 2.3.1). The main peak generated by a line emission is considerably wider than that obtained with a crystal spectrometer (Fig. 7.3). The more limited resolution causes the line-to-background ratio to be lower by a factor of about 10; hence, under

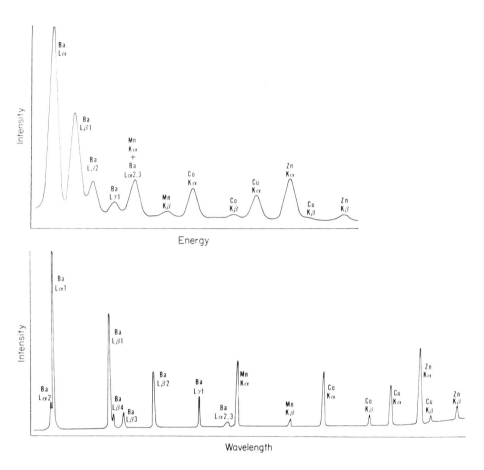

FIG. 7.3 Energy- and wavelength-dispersive spectra of an experimental glass. Upper spectrum: Li-drifted silicon detector; resolution: 162 eV full width half maximum. Shaping time: 10 μsec at Mn Kα. Lower spectrum: LiF crystal, 10 cm radius. To simplify the comparison, the upper spectrum is plotted with energies increasing from left to right, and the lower spectrum with wavelengths increasing from right to left.

comparable circumstances, the limits of detection attainable with crystal spectrometers are usually lower. Although the efficiency of the solid state spectrometer is high, there is a limitation to the usable count rate since the entire spectrum is received by the detector, and, at high rates, the dead-time corrections become large and the line resolution deteriorates. Moreover, at such count rates, artifacts due to pulse pileup in the detector become significant [31, 32]. For this reason, the high efficiency of the solid state detector is only advantageous when the total X-ray emission from the specimen is low (e.g., in the analysis of small particles).

The reduced spectral resolution of energy-dispersive systems causes line interferences to occur which are usually fully resolved with crystal spectrometers. For instance, it is impossible to separate, with the Si(Li) detector, the Kα emission of sulfur from the L$\alpha_{1,2}$ emission of lead. Interference situations also arise from the superposition of Kα and Kβ lines of various elements, as well as among lines at the L level of neighboring elements. Hence, there are analytical situations which only wavelength-dispersive spectrometers can resolve.

7.4 X-RAY AREA-SCANNING IMAGES

The technique of X-ray area scanning provides the investigator with a two-dimensional representation of the distribution of one or more elements on the specimen surface. The representation is a map or image of the specimen surface portion covered by a rectangular raster scan of the electron beam. The gray values of the image are related to the intensity of the X-ray signal emitted at the corresponding locations of the specimen surface. In practice, they are obtained by photographing the surface of a cathode-ray tube (CRT), the beam of which scans in synchronism with the electron probe. The scanning pattern can be a set of parallel lines or a grid of points. The gray levels of the image are obtained by modulating the CRT brightness with a signal derived from the emission of a characteristic X-ray line (X-ray scan) obtained with a crystal spectrometer or an energy-dispersive spectrometer. For topographic orientation, scanning electron micrographs of the same scanned area can be obtained; in this case, secondary or backscattered electrons, or the current flowing from the target to ground, can be used as signals.

ELECTRON PROBE MICROANALYZERS 179

The scanning images obtained from electron signals look strikingly like macroscopic objects under a source of illumination and observed visually (Fig. 7.4). It is tempting to conclude that there

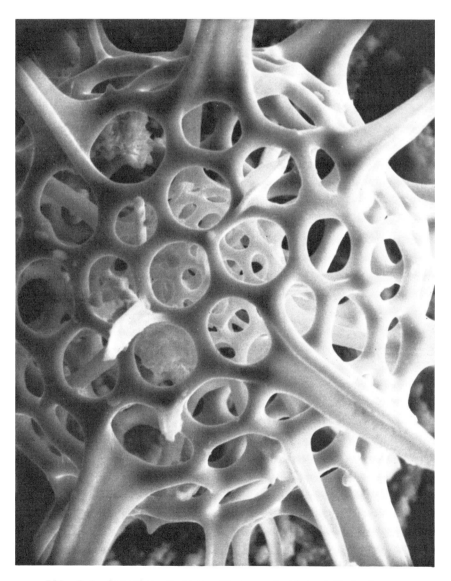

FIG. 7.4 Scanning electron micrograph of a radiolarian.

is a relation such that the electron source is analogous to a light source, and the detector is analogous to the observing eye. However, the configuration of the image (dimensions, foreshortening) depends not on the position of the detector but on the position of the virtual electron source. The distribution of gray levels (lights, shadows, and shades), in turn, does not depend primarily on the position of the electron source, but rather on that of the detector. (Although the angle of electron incidence codetermines the efficiency of emission of secondary and backscattered electrons, the detector position determines which surfaces emit electrons that reach the detector.) Hence, the virtual source of electrons is analogous to the eye of the observer, and the detector is analogous to the source of light.

In the procedure commonly followed for X-ray images, the pulses produced by the detector (after amplification) are applied to generate dots of light on an otherwise dark CRT screen. This standard X-ray area scan is illustrated in Fig. 7.5.

The magnification of the scanning image is given by the ratio of the scanning excursion on the specimen to that on the CRT. The range of useful magnifications is determined by the CRT resolution, the effective size of the signal source, and the maximum permissible excursion of the electron beam on the specimen surface [33]. For a CRT with a screen width of 8 cm, if a lateral signal dispersion of 2 µm is assumed, sharp images cannot be obtained at magnifications of 500 (one picture element is 1 mm wide) or above. On the other hand, the angular acceptance of the X-ray spectrometer--particularly of the fully focusing crystal spectrometer--limits the range of useful magnifications towards the lower side. With typical crystal spectrometers, the signal loss becomes substantial for area scans larger than 200 µm in side length. For such conditions, the useful magnification is limited to 400 or above. It can be appreciated that for a fully focusing crystal spectrometer, the useful range is narrow.

In practice, X-ray scanning images of poor resolution are frequently paired with scanning electron micrographs of the same area, which show smaller details due to the smaller size of the effective source for secondary or backscattered electrons or for target current.

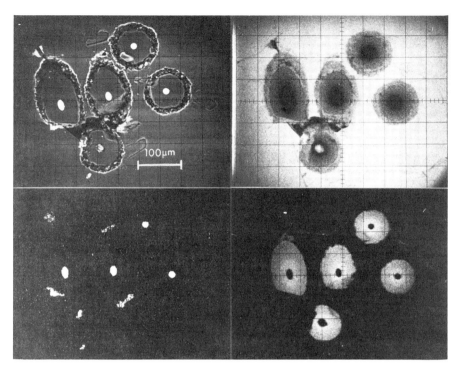

FIG. 7.5 Scanning images of an alumina-silica composite. Upper left: specimen-current image; upper right: Al Kα image (white regions have high aluminum levels); lower left: W Lα image; lower right: Si Kα image.

7.5 QUANTITATIVE ELECTRON PROBE MICROANALYSIS

The intensities of emitted characteristic X-ray lines can usually be measured to better than 1% in reasonable observation times; they grow with the concentrations of the emitting element. Therefore, the X-ray emission of a line from a target containing two elements can be expressed by a calibration curve (Fig. 7.6), or by an equivalent mathematical function (calibration function). For historical reasons, concentrations are often expressed as mass fractions (or weight fractions), and it is sometimes assumed, without good reason other than analogy from chemical analysis, that an "ideal" calibration curve should be linear. Curves obtained in practice are continuous and monotonic. For a given emitting element and X-ray line,

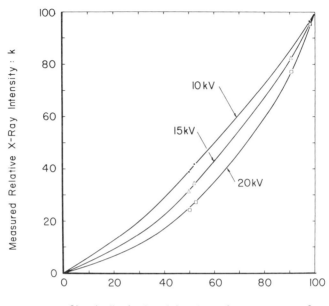

FIG. 7.6 Calibration curves at three operation potentials (marked in kV) for Al Kα in binary Al-Mg alloys.

the shape of the curve depends on operating conditions, such as the X-ray emergence angle and the operating voltage, and on the concentration of *all* elements present (matrix dependence or interelement effect). For this reason, the graphic expression of calibration functions for ternary or more complex specimens is impractical. The observation of characteristic X-ray intensities is complicated, in practice, by the background due to continuous radiation and by the limited speed of recovery of a detector and electronics after detection of a pulse (dead time [34]) (see also Chap. 2, Sec. 2.3.2).

Electron probe microanalysis is based on relative measurements. The ratio of the count rate for a line from the specimen to that from the pure element, for the same conditions, is called the relative X-ray intensity k. Where the pure element cannot be observed under the electron beam (e.g., chlorine, calcium, sulfur, phosphorus), the count rate for the pure element can be deduced from that obtained with a multielement standard or compound standard such as an oxide or silicate. Due to chemical line shifts for lines of long wavelength [35], compound standards are also preferred for the determination of elements of low atomic numbers (7-15) in oxidic specimens [36]. If the corresponding analytical curve, or an equivalent algorithm (i.e., the analytical calibration function), is available, the relative intensity to be obtained from the standard (k^S) can be determined. If the X-ray intensities emitted from the specimen and the standard are I'^* and I'^S, respectively, the relative intensity from the unknown specimen k^* can be obtained by:

$$k^* = \frac{I'^*}{I'^{elem}} = \frac{I'^*}{I'^S} k^S \tag{7.1}$$

If the analytical function is only imperfectly known, the systematic error in the analytical result decreases with decreasing difference in composition between specimen and standard. Moreover, when the specimen is similar in composition to the standard, errors in the measurement of intensities such as those due to chemical shifts, dead time, or the background also tend to cancel.

One may conclude from the above that the use of matching standards is always preferable. The advantages of such standards must, however, be weighed against the risks inherent in the use of inhomogeneous or poorly characterized materials. In many cases, matching standards of the required quality are unavailable, and the analyst must be prepared to use pure elements or other available, but not matching, standards. For this reason, it is important to have accurate models for the analytical calibration function which are applicable to a wide range of materials.

7.5.1 Empirical Model for the Calibration Function

Experience indicates that many binary systems can be characterized with good approximation by the equation

$$\frac{1-k}{k} = \alpha \frac{1-C}{C} \tag{7.2}$$

where C is the concentration of the element being determined, and α is a constant independent of specimen composition. The model can be extended to the determination of element a in multielement specimens as follows:

$$k_a = \frac{\alpha_a C_a}{\Sigma \alpha_i C_i} \tag{7.3}$$

Equation (7.2) is that of a hyperbola; it can be transformed as follows:

$$\frac{k}{C} = \frac{1}{\alpha} + (1 - \frac{1}{\alpha})k \tag{7.4}$$

Hence, k/C is a linear function of k (Fig. 7.7). The hyperbolic model was first proposed to describe the generation of X rays within the specimens by Castaing and Descamps [37], and applied to emerging X-ray intensities by Ziebold and Ogilvie [28]. A successful adaptation of this model to the analysis of oxidic minerals was developed by Bence and Albee [38]. This procedure has the advantage that it can be applied to the data reduction for on-line analysis by means of small computers. It is particularly efficient when the main differences between specimen and standard are in the absorption

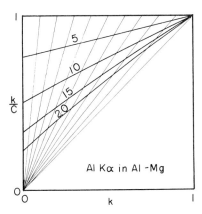

FIG. 7.7 Plot of k/C as a function of k, for the same system shown in Fig. 7.6. Operating potentials are noted in kV. The lines concurring at the point (0,0) connect points corresponding to the same aluminum mass fraction (c), in steps of 10%.

ELECTRON PROBE MICROANALYZERS

of the X rays on their path towards the specimen surface, and less so when they are due to large differences in the average atomic number or to the emission of indirectly excited (fluorescent) radiations.

7.5.2 Theoretical Approaches to the Calibration Function

Castaing, in his doctoral thesis [3], laid the foundations for the theoretical treatment of the analytical calibration function. Although many other investigators have greatly contributed to further the art of electron probe quantitation, the principles given by Castaing are still valid.

The characteristic radiation emitted by a target contains contributions of direct (electron-excited) and indirectly excited (fluorescent) radition. The latter can be produced by the action of the continuous spectrum (continuum fluorescence) or, under appropriate conditions, by that of one or more characteristic lines (characteristic fluorescence). Fluorescent excitation requires that the exciting photons have an energy larger than the minimum ionization energy for the shell giving rise to the indirect emission. Secondary emission due to the continuum fluorescence is always present in varying proportions to the primary production (Fig. 7.8), while characteristic fluorescence only occurs in some combinations of elements.

Both primary and fluorescent emission suffer attenuation on the path towards the specimen surface. The absorption factor f_p [or $f(\chi)$] denotes the probability of emergence of a photon generated within the specimen under defined experimental conditions and propagating within the solid angle of acceptance of the X-ray spectrometer. We call I_p, I_f, and I_c the generated intensities (photon/electron) of primary emission, fluorescence due to characteristic lines, and fluorescence due to the continuum, respectively. We call f_p, f_f, and f_c the respective absorption factors. We will also denote parameters related to the specimen by the asterisk (*) and those related to the standard by (s). The relative emitted intensity can be obtained by:

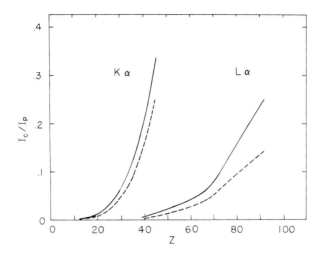

FIG. 7.8 Ratios of generated (continuous lines) and emitted (broken lines) continuum fluorescence intensities to the respective primary intensities for elements at an operating potential of 20 kV.

$$\frac{I'^*}{I'^s} = \frac{k^*}{k^2} = \frac{I_p^* f_p^* + \Sigma\, I_f^* f_f^* + I_c^* f_c^*}{I_p^s f_p^s + \Sigma\, I_f^s f_f^s + I_c^s f_c^s} \qquad (7.5)$$

The fluorescence due to the continuum is usually neglected in the use of Eq. (7.5). Calculations by means of a computer program [39], which include the correction for continuum effects, indicate that the errors in omitting these effects seldom exceed 2% relative.

If a pure element is used as a standard, the fluorescent excitation by characteristic lines in the standard is also negligible:

$$\frac{I'^*}{I'^s} = \frac{I_p^* f_p^* + \Sigma\, I_f^* f_f^*}{I_p^s f_p^s} \qquad (7.6)$$

This equation is traditionally presented in the form of a product of various factors:

$$\frac{I'^*}{I'^s} = C^* \left(\frac{1}{C^*} \frac{I_p^*}{I_p^s} \right) \left(\frac{f_p^*}{f_p^s} \right) \left(1 + \frac{\Sigma\, I_f^* f_f^*}{I_p^s f_p^s} \right) \qquad (7.7)$$

or

ELECTRON PROBE MICROANALYZERS

$$C^* = k k_Z k_A k_F$$

where

$$k = \frac{I'^*}{I'^s} \quad k_Z = \frac{1}{C^*} \frac{I_p^s}{I_p^*} \quad k_A = \frac{f_p^s}{f_p^*} \quad k_F = \frac{1}{1 + (\Sigma\, I_f^* f_f^* / I_p^* f_p^*)}$$

The coefficients k_Z, k_A, and k_F are called, respectively, the atomic number, absorption, and fluorescent corrections, and this simplified correction procedure is called the ZAF correction method [29].

7.5.3 Generation Correction (Atomic Number Correction)

The generation of primary X-ray photons in a specimen depends on three processes: the deceleration of the incident electrons within the target, the ionization of the target atoms by electrons of a given (instantaneous) energy E, and the loss of electrons which are reemitted from the target after scattering acts (backscatter loss). The energy loss of the electrons within the target is usually assumed to be continuous and expressible by means of a "stopping power" S:

$$S = \frac{-dE}{dx} \tag{7.8}$$

which indicates the average loss of energy E (in eV) as a function of the length of path x, measured in g/cm^2 of the target material. The ionization cross section Q, for a particular inner shell is defined by the equation:

$$n = QN\frac{i}{a} \tag{7.9}$$

in which a is the area of a layer of N atoms crossed normally by i electrons. The symbol n denotes the number of ionizations of the atomic shell under consideration.

Bethe has developed algebraic models for the stopping power [40] and for the ionization cross section, both of which depend on the instantaneous electron energy. The choice of parameters to be used in Bethe's equations is still a matter of controversy [41].

The backscatter factor R provides a correction for the loss of ionization due to the exit through the specimen surface of backscattered electrons. An algorithm for this backscatter correction, which depends on target composition and operating voltage, was given by Duncumb and Reed [42].

The number of generated primary photons of a given X-ray line of element a per incident electron can be obtained from these parameters as follows:

$$I_{pa} = RN_A \omega p \frac{m_a}{\Sigma \, m_i A_i} \int_{E_o}^{E_q} \frac{Q}{-S} \, dE \tag{7.10}$$

where N_A is Avogadro's number, ω the fluorescent yield of the excited shell q, p the relative intensity of the line with respect to the sum of all lines of this shell, m_i the molar fraction of element i in the specimen, and A_i its atomic weight [42]. E_o is the operating potential, and E_q the critical excitation potential for the emitting shell.

The ratio of the X-ray generation probabilities from specimen and standard provides the basis for the "atomic number correction term," k_Z. The integration in Eq. (7.10) can be performed numerically or analytically [43], or approximated by a summation [42].

7.5.4 Absorption Correction

If an electron beam impinges normally upon the flat surface of a thick sample, the absorption correction for the X rays generated within the specimen depends upon the energy of the incident electrons, the X-ray emergency angle ψ, and the mass absorption coefficient $\mu/\rho \, (cm^2 \, g^{-1})$ of the radiation in the specimen. The last two parameters can be combined in the parameter $\chi = \mu/\rho \, (*\lambda) \csc \psi$. The symbol $\mu/\rho \, (*\lambda)$ denotes the mass absorption coefficient of the specimen at the wavelength λ.

The absorption correction factor, commonly called $f(\chi)$, denotes the probability that a photon generated within the specimen will survive absorption on its way to the specimen surface in the direction towards the X-ray spectrometer. When the radiation of interest is primary (electron-generated), the specific notation f_p is preferable. Philibert [44] developed a model for f_p from theoretical considerations, simplifications, and adjustment to experiments:

$$f_p = \frac{1}{(1 + \chi/\sigma)[(1 + h/(1 + h)(\chi/\sigma)]} \tag{7.11}$$

with $h = \frac{1.2A}{Z^2}$

where Z is the atomic number of the target, A the atomic weight, and σ an "electron absorption" parameter.

We have shown [45] that the dependence on specimen composition of f_p is negligible, and that the following model is simpler and more accurate than Eq. (7.11).

$$f_p = [1 + 1.2 \times 10^{-6}(E_o^{1.65} - E_q^{1.65})\chi]^{-2} \tag{7.12}$$

Here, E_o is the operating voltage (kV), and E_q the critical excitation voltage for the observed radiation. This simple model is used in the computer program FRAME for the ZAF correction [46].

7.5.5 Characteristic Fluorescence Correction

A rigorous treatment of the correction for fluorescence excited by characteristic lines was given by Henoc et al. [47]; the resulting equations are complex, but are easily programmed for computer reduction [39]. A simplified procedure derived by Reed [48] is widely used in practice.

Significant secondary excitation occurs only if the energy of the exciting line exceeds (by a factor not larger than 1.5) that of the critical excitation energy of the excited line. In multielement materials such as complex alloys or minerals, it is not a trivial task to sort all edges and lines and, therefore, the decision as to which secondary excitations occur is best arrived at by a computer routine [39,46].

7.5.6 Continuum Fluorescence Correction

The mechanism of fluorescence due to the continuum is, in principle, identical to that of characteristic fluorescence with the following complications:

1. The corresponding correction requires a model for the generation of continuous radiation. The accuracy of such models is still a subject of controversy.

2. Since the continuum is polychromatic, its effects must be integrated with respect to wavelength, and separate regions must be set for the integration bounded by the absorption edges of all elements in the specimen between the operating voltage and the critical excitation voltages for the line of interest. In a multielement specimen, this requirement leads to complex sorting routines which consume time and computer memory, and which cannot be substantially reduced by simplification of the correction model.

3. All targets, including elements, which emit primary X rays, also exhibit, though to varying degrees, X-ray fluorescence. The inclusion of continuum fluorescence considerably increases the size of computations for data reduction, and for this reason many programs do not include this correction, the effects of which rarely exceed 2-3% relative in the accuracy of the final result (see Fig. 7.9).

7.5.7 Iteration and Computation

Many parameters entering the correction calculation, such as mass absorption coefficients, stopping powers, and fluorescent intensities, are composition-dependent and, therefore, the correction factors for a specimen of unknown composition cannot be determined accurately. In the ZAF procedure (see below), none of the correction factors can be calculated without an estimate of specimen composition. The problem is solved in practice by an iterative procedure in which successively improving concentration estimates are used to calculate the correction factors with increasing accuracy until the

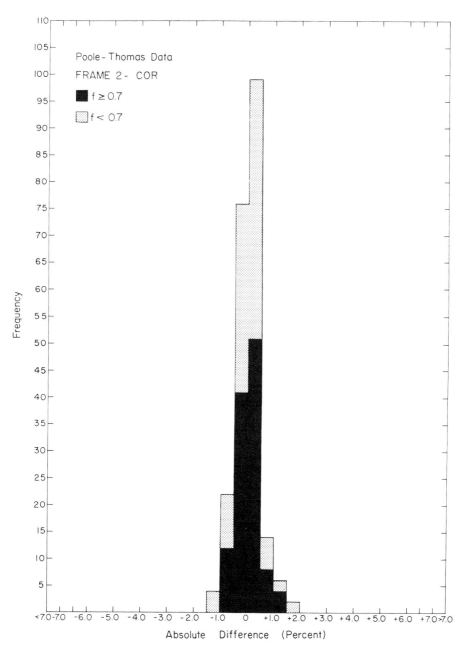

FIG. 7.9 Differences between results of FRAME2 (without continuum fluorescence correction) and COR (with continuum fluorescence correction) for 236 analyses of a wide variety of specimens.

estimates are essentially unchanged in successive iterations. A review of a large number of correction data reduction programs for electron probe microanalysis was written by Beaman and Isasi [49].

The complexity of the calculations precludes a manual computation for practically all cases. Moreover, until a few years ago the limitations of available computation devices forced the analysts to introduce considerable simplifications in their procedure. The program COR presents an attempt to obtain high accuracy by the avoidance of several simplifying assumptions and, particularly, by a full treatment of the continuum fluorescence correction [39]. A critical comparison with a carefully simplified procedure of much lesser length (ZAF [46]) indicates that in most cases the simplified program is sufficiently accurate, particularly in view of the uncertainties which still persist in the complete procedure (Fig. 7.9).

7.6 ANALYSIS INVOLVING SPECIAL SPECIMEN CONFIGURATIONS

The procedures we have discussed apply to flat, electron-opaque samples struck normally by the electron beam. However, many specimens received by the analyst do not conform to this situation, as illustrated in Fig. 7.10. The case represented by (a) is very important since most SEM specimen-X-ray detector configurations are included in (a). Experience indicates that if sample and standard are tilted to the same angle and are flat, serious analytical errors do not occur [50,51]. Nevertheless, further study is required to establish more rigorously the effect of specimen tilt on the analytical result.

Case (b), that of a thin, homogeneous unsupported film, is mainly of theoretical interest, but case (e), of a film of variable mass thickness, is often encountered in biological studies. Methods for dealing with such specimens usually presume that the specimen is so thin that electron scattering within the sample is negligible. Then the influence of other elements on the measured element is negligible, and the correction equation relating the observed intensity to the mass cross section of the element is linear [52,53]. The question is, to what specimen thickness does linearity hold? Russ

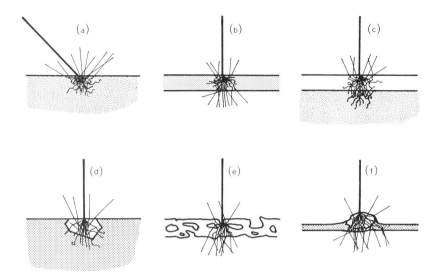

FIG. 7.10 Special specimen configuration: (a) inclined electron beam, (b) thin unsupported foil, (c) thin coating, (d) small inclusion, (e) biological tissue preparation of irregular mass thickness, and (f) particle mounted on thin foil.

has discussed this point in detail [54]. In biological samples of soft tissues, a thickness of over 1 μm may be tolerated. Hall and coworkers have developed methods in which cases (b) and (e) can be handled by examining the ratio of characteristic line intensity to continuum X-ray intensity [55-57]. Analytical methods for treating case (b) with only one pure element standard have been proposed [58,59]. None of these methods require that the film thickness be known.

A thin coating on a thick substrate [case (c)] presents a more complicated situation because of the effect of electron backscattering on X-ray production, which varies with the composition of the substrate. There are empirical or semiempirical treatments of this case [60-63], but more work is needed. Kyser and Murata have used the Monte Carlo technique to analyze Bi-Mn and Co-Pt alloy films approximately 50 μg/cm^2 in thickness [64]. The results agreed well with those obtained by nuclear backscatter analysis.

The Monte Carlo calculation, in which the electron is made to follow a random walk within the specimen, can be a valuable method for treating complex specimen shapes. In the Monte Carlo procedure, a scattering model for the electron is selected, and the electron trajectory simulated by random number substitution into this model. A sufficient number of trajectories must be calculated so that statistical validity is achieved; usually 1000-5000 trajectories are computed. The application of the Monte Carlo method to X-ray microanalysis has been reviewed in detail recently [65]. The Monte Carlo method is not a panacea; the results depend upon virtually the same parameters as in the ZAF method. However, the Monte Carlo procedure is an important (perhaps the only) means of predicting X-ray emission from complex geometrical shapes, such as those shown in Fig. 7.10.

The composition of particulate matter [case (f)], such as airborne dust, is of great practical interest. Several publications (for example, Refs. 66,67) have reported results of electron probe microanalysis. The increasing attention given to the pollution of air and water will extend further the range of application.

The accurate quantitative analysis of particles in the micrometer or submicrometer range is difficult. The particles are often of irregular shape, and their density and, hence, the total mass is unknown. It is difficult to accurately determine the back- and sidescattering and the transmission of electrons and, hence, the generation of X rays. Furthermore, most specimens contain low-atomic number elements which cannot be determined directly. Procedures for the quantitative analysis of small particles have been proposed by Hoffmann et al. [68] and by Armstrong and Buseck [69].

The case (d), which deals with the analysis of small particles imbedded in a matrix of different composition, is probably the most difficult situation found in quantitative electron probe microanalysis. If an element sought within the particle is also present in the matrix, the resulting interference is a formidable obstacle which sometimes renders an accurate analysis impossible.

7.7 STANDARDS

The basic requirement for standards is that they be homogeneous at the microscopic levels of spatial resolution, stable with respect to time and environment, properly prepared for use in the instrument, and, if not pure elements or compounds known to be stoichiometric, carefully analyzed by independent chemical methods.

A number of materials have been proposed as standards. These include intermetallic compounds, fully ordered alloys, supposedly stoichiometric inorganic compounds, natural and synthetic minerals, organometallics, and homogenized solid solutions. Materials such as intermetallic compounds may not be homogeneous; a single vertical line on a phase diagram is not a reliable criterion for homogeneity. All prospective standards should be carefully checked for homogeneity by measurements. The requirement for homogeneity is particularly important; yet, reliable homogeneity tests are tedious and time consuming [70].

The National Bureau of Standards has certified some metal alloy systems suitable for use as microanalytical standards. At present, the standards available are a low-alloy steel [71], gold-silver and gold-copper alloys [72], a tungsten-20% molybdenum alloy [73], an iron-3.22% silicon alloy [74], cartridge brass [75], and an iron-chromium-nickel alloy [76].

An effort is presently underway at the National Bureau of Standards to establish the usefulness of glass standards for the analysis of oxidic materials. Glasses can be prepared with the requisite degree of homogeneity, and a wide range of elements can be incorporated. However, in view of the mobility of the alkalies in glasses irradiated by electron beams, the presence of these elements should be avoided where possible.

The elemental standards frequently used in electron probe microanalysis are not difficult to obtain since high purity (>99.99%) is not required. If specimens must be covered with a conductive layer, it is desirable that the corresponding standard be coated in the same operation, particularly for analysis with X rays of long wavelength.

7.8 APPLICATIONS OF ELECTRON PROBE MICROANALYSIS (EPMA)

The chief advantage of electron probe microanalysis is that it can be applied to virtually any solid, e.g., metals, bone, oxides, minerals, glasses, soft tissue, paper, art objects, composite and particulate materials, thin films on substrates. [The list of information sources (Sec. 7.8.2) contains examples of all these and more.] For many cases, especially for inorganic materials, the techniques of electron probe microanalysis are well developed and the applications are straightforward. Difficulties are experienced in applications to particles, and to medical and biological specimens, especially thin, soft-tissue sections. Nevertheless, many investigations have been carried out on such samples as well.

What follows are a few examples of applied X-ray microanalysis, meant to illustrate the kind of information obtainable from the technique.

Qualitative determinations of elements in secondary cracks from the failed Point Pleasant (West Virginia) bridge were instrumental in postulating the failure mechanism. The bridge collapsed on December 15, 1967; preliminary investigation showed that the source of failure was one of the 55-ft-long eyebars connecting the superstructure. Examination of secondary cracks in this eyebar showed a definite sulfur gradient from the origin of the root of the crack. Near the origin, most of the sulfur was not combined as manganese sulfide, although inclusions of manganese sulfide are present throughout the bridge steel. Particles high in lead and calcium were also found near the origin. Elements of atomic number less than 11 were absent, except for oxygen. The appearance of the crack in the SEM and the sulfur distribution suggested the possibility of sulfide-driven stress corrosion as a mechanism of crack propagation. (The calcium- and lead-rich regions are minute paint particles which fell into the crack.) Tests confirmed that the eyebar steel fails because of stress corrosion in a hydrogen atmosphere [77].

Ruff has identified elements in wear particles of steel recovered from a ball-bearing bench test. Elements such as silicon, potassium, and zinc, in addition to those usually present in steel, were found in oil residues; the particles are iron oxides which have absorbed oil [78].

Kiessling and Lange have performed an exhaustive study of the inclusions which can be found in irons and steels. Hundreds of types of inclusions have been classified in this way. The results have been combined into an atlas for inclusion identification [79].

Läuchli recently reviewed applications of X-ray microanalysis in botany [80]. A specific example is the determination of K^+- and Cl^--ion distributions in frozen barley root specimens by means of energy-dispersive analysis. The concentration of K^+ was found to increase from the root epidermis up to the xylem parenchyma; the Cl^- gradient was in the opposite direction.

Another study shows that K^+ is specifically involved in the process whereby stomata open when exposed to light and close in the dark [80].

X-Ray microanalysis was used to show that rhizoids of the alga chara contain statolith vacuoles filled with biocrystallites of barium sulfate. This work appears to support the theory that statoliths are effective through mass and not due to chemical effects, since barium sulfate is inert. This also appears to be the first evidence for any physiological function of barium in living organisms [80].

Sjösström and coworkers have studied inclusions in muscle tissues occurring under a variety of pathological conditions [81]. For major element determinations, sections of plastic embedded tissue have been found to be satisfactory. For example, the presence of calcium has been confirmed in Liesegang rings in hyperthalamic periodic paralysis.

In investigations of hard tissue, such as bones, teeth, and pathological calcifications, the ratio of calcium to phosphorus was used to determine the progress of mineral growth toward the fully developed stage [82].

7.8.1 Selected Applications Involving Quantitative Analysis

A large number of metallurgical investigations has been based on the quantitative capabilities of the electron probe microanalyzer. A typical example, in which microprobe results are used with other techniques for the accurate determination of binary metal phase diagrams is contained in Ref. 83.

The discovery of the nonterrestrial mineral sinoite (Si_2N_2O) represents an instance in which the determination of light elements was possible. The mineral was found in the study of the Jajh deh kot Lalu meteorite (an enstatite chondrite). The standards used were pure silicon, SiO_2, $MgSiO_3$, and BN for silicon, oxygen, and nitrogen, respectively. The occurrence of sinoite indicates that there was insufficient oxygen available to bind all of the excess silicon as SiO_2 when the meteorite formed [84].

The analysis of thin films of MnBi and CoPt on silica substrates was performed by Kyser and Murata with the aid of a Monte Carlo calibration [64]. This computation was used to prepare calibration curves of the intensity ratio from films of various thicknesses and compositions to pure bulk standards. Agreement with nuclear backscatter analysis was good (e.g., 25.5 wt % manganese by X-ray analysis, and 26.6 wt % manganese by nuclear backscatter analysis).

Coleman [25] has reviewed methods of quantitation in biological specimens as well as the effect of specimen preparation on the results. In mineralized portions of 1-µm thick sections of 3-day old chick calvarium (a region of the top of the skull), analysis for calcium varied from 18.9 to 25.0 wt %, depending on the preparative technique; for phosphorus, the range was 10.6-12.4 wt %. Even more striking were the results for potassium and sodium, for which the ranges were 0.02-0.14 and 0.11-0.55 wt %, respectively.

Hall has summed up the present situation for quantitative analysis of biological material with the following statement [27].

"We do not lack methods of analyzing quantitatively the specimen which is "seen" by the probe. The pressing problem is to ensure that this specimen, as it exists during the microanalysis, is sufficiently similar to the original material that we want to analyze. We have to control the changes occurring during specimen preparation, and we have to control beam damage and other environmental effects within the microprobe instrument."

7.8.2 Sources of Information on Applications

The literature contains many examples of applications of EPMA to a wide variety of fields. We list here suggested references citing such examples.

> Use of Electron Probe Microanalysis in Physical Metallurgy,
>   D. A. Melford, *J. Inst. Metals*, *90*, 217 (1962).
> *Symposium on Fifty Years of Progress in Metallographic Techniques*, ASTM Spec. Tech. Publ. *430* (1968).

The pertinent section is entitled Progress in Electron Probe Microanalysis (K. F. J. Heinrich, organizer), pp. 291-408. Papers on scanning (Heinrich) and applications (J. D. Brown) contain many examples of analytical applications.

> *Electron Probe Microanalysis*, Advances in Electronics and Electron Physics, Suppl. 6 (A. J. Tousimis and L. Marton, eds.), Academic, New York and London, 1969.
> *The Electron Microprobe* (T. D. McKinley, K. F. J. Heinrich, and D. B. Wittry, eds.), Wiley, New York, 1966.
> *Microprobe Analysis* (C. A. Andersen, ed.), Wiley, New York, 1973.

This book contains chapters on applications in geology (K. Keil), ceramics and glass technology (W. T. Kane), biology (W. L. Robison), particulate analysis (M. Bayard), and solid state electronics (D. B. Wittry), among others.

> *Proceedings of Apollo 11 Lunar Science Conference*, 1970.
> *Proceedings of Second Lunar Science Conference*, MIT Press, Boston, 1971.

In these volumes the reader will find many examples of microanalysis applied to selenological problems.

Microprobe Analysis as Applied to Cells and Tissues (T. Hall, P. Echlin, and R. Kaufmann, eds.), Academic Press, New York and London, 1974.

This book contains many papers on specimen preparation and analysis of biological material.

Energy-Dispersive X-Ray Analysis (J. C. Russ, ed.), ASTM Spec. Tech. Publ. *485* (1971).

The series of volumes from the international conferences on microanalysis contains many applications papers:

X-Ray Microscopy and Microradiography (V. E. Cosslett, A. Engström, and H. H. Pattee, Jr., eds.), Academic Press, New York, 1957.

X-Ray Microscopy and X-Ray Microanalysis (A. Engström, V. E. Cosslett, and H. Pattee, eds.), Elsevier, New York, 1960.

X-Ray Optics and X-Ray Microanalysis (H. H. Pattee, V. E. Cosslett, and A. Engström, eds.), Academic Press, New York and London, 1963.

X-Ray Optics and Microanalysis (R. Castaing, P. Deschamps, and J. Philibert, eds.), Hermann, Paris, 1966.

Fifth International Congress on X-Ray Optics and Microanalysis (G. Möllenstedt and K. H. Gaukler, eds.), Springer Verlag, Berlin, 1969.

Proceedings of the Sixth International Conference on X-Ray Optics and Microanalysis (G. Shinoda, K. Kohra, and T. Ichinokawa, eds.), University of Tokyo Press, Tokyo, 1972.

A joint Franco-British workshop on techniques and applications of biological microanalysis was held in Creteil, France in March 1975. The proceedings are published as a special issue of the Journal de Microscopie et de Biologie Cellulaire, Vol. 22 (P. Favard, P. Galle, and P. Echlin, eds.), pp. 121-520.

The Proceedings of the National Conferences on Electron Probe Analysis (first through tenth) contain many other examples of a wide variety of applications. These proceedings are available from Kurt F. J. Heinrich, at the Nat. Bureau of Standards, Washington, D. C. 20234.

REFERENCES

1. M. von Ardenne, Z. Techn. Phys., 19, 407 (1938).
2. V. K. Zworykin, J. Hillier, and R. L. Snyder, ASTM Bull., 117, 15 (1942).
3. R. Castaing, Thesis, University of Paris, 1951.
4. D. B. Wittry, Thesis, Cal. Tech., 1957.
5. I. B. Borovskii and N. P. Ilin, Dokl. Akad. Nauk SSSR, 106, 655 (1953).
6. L. S. Birks and E. J. Brooks, Rev. Sci. Instrum., 28, 709 (1957).
7. M. E. Haine and T. Mulvey, J. Sci. Instrum., 26, 350 (1959).
8. H. H. Johann, Z. Phys., 69, 185 (1931).
9. P. Duncumb, Thesis, Cambridge University, 1956.
10. P. Duncumb, Proc. 1st Int. Conf. X-Ray Optics and Microanalysis (V. E. Cosslett, A. Engström, and H. H. Pattee, Jr., eds.), Academic Press, New York, 1957, p. 617.
11. R. M. Dolby, Brit. J. Appl. Phys., 11, 64 (1960).
12. R. M. Dolby, Proc. 3rd Int. Conf. X-Ray Optics and Microanalysis (H. H. Pattee, V. E. Cosslett, and A. Engström, eds.), Academic Press, New York, 1963, p. 483.
13. B. L. Henke, Advan. X-Ray Anal., 7, 460 (1964).
14. R. Fitzgerald, in Microprobe Analysis (C. A. Andersen, ed.), Wiley, New York, 1973, p. 1.
15. E. Lifshin, M. F. Ciccarelli, and R. B. Bolon, in Practical Scanning Electron Microscopy (J. I. Goldstein and H. Yakowitz, eds.), Plenum Press, New York, 1975, p. 263.
16. K. C. A. Smith, Thesis, Cambridge University, 1956.
17. C. W. Oatley, The Scanning Electron Microscope, Part I: The Instrument, University Press, Cambridge, 1972.
18. T. E. Everhart and R. F. M. Thornley, J. Sci. Instrum., 37, 246 (1960).
19. R. F. W. Pease and W. C. Nixon, J. Sci. Instrum., 42, 81 (1965).
20. G. R. Booker, D. C. Joy, J. P. Spencer, and H. Graf von Harrach, SEM/74 (O. Johari, ed.), IITRI, Chicago, 1974, p. 225.

21. O. C. Wells, SEM/74 (O. Johari, ed.), IITRI, Chicago, 1974, p. 1.
22. H. Yakowitz, SEM/75 (O. Johari, ed.), IITRI, Chicago, 1975, p. 1.
23. M. D. Muir and D. B. Holt, SEM/74 (O. Johari, ed.), IITRI, Chicago, 1974, p. 135.
24. M. Bayard, in *Microprobe Analysis* (C. A. Andersen, ed.), Wiley, New York, 1973, p. 323.
25. J. R. Coleman, in *Practical Scanning Electron Microscopy* (J. I. Goldstein and H. Yakowitz, eds.), Plenum, New York, 1975, p. 491.
26. A. Saubermann and P. Echlin, *J. Microscopy*, *104*, 3 (1975).
27. T. Hall, P. Echlin, and R. Kaufmann, eds., *Microprobe Analysis as Applied to Cells and Tissues*, Academic Press, New York, 1974.
28. T. O. Ziebold and R. E. Ogilvie, *Anal. Chem.*, *36*, 322 (1964).
29. K. F. J. Heinrich, *NBS Tech. Note*, *521*, 13 (1969).
30. K. F. J. Heinrich and M. A. Giles, *NBS Tech. Note*, *406*, 53 (1967).
31. S. J. B. Reed, *J. Phys. E. Sci. Instrum.*, *5*, 997 (1972).
32. R. Woldseth, *X-Ray Energy Spectrometry*, Kevex, Burlingame, Calif. (1973).
33. K. F. J. Heinrich, *NBS Tech. Note*, *278* (1967).
34. K. F. J. Heinrich, D. L. Vieth, and H. Yakowitz, *Advan. X-Ray Anal.*, *9*, 208 (1966).
35. E. W. White, in *Microprobe Analysis* (C. A. Andersen, ed.), Wiley, New York, 1973, p. 349.
36. K. Keil, in *Microprobe Analysis* (C. A. Andersen, ed.), Wiley, New York, 1973, p. 189.
37. R. Castaing and J. Descamps, *J. Phys. Radium*, *16*, 304 (1955).
38. A. E. Bence and A. Albee, *J. Geol.*, *76*, 382 (1968).
39. J. Henoc, K. F. J. Heinrich, and R. L. Myklebust, *NBS Tech. Note*, *769* (1973).
40. H. A. Bethe, *Ann. Phys.* (Leipzig), *5*, 325 (1930).
41. K. F. J. Heinrich and H. Yakowitz, *Mikrochim. Acta*, 123 (1970).
42. P. Duncumb and S. J. B. Reed, in *Quantitative Electron Probe Microanalysis* (K. F. J. Heinrich, ed.), *NBS Spec. Publ.*, *298*, 133 (1968).
43. J. Philibert and R. Tixier, in *Quantitative Electron Probe Microanalysis* (K. F. J. Heinrich, ed.), *NBS Spec. Publ.*, *298*, 13 (1968).

44. J. Philibert, *Proc. 3rd Int. Conf. X-Ray Optics and Microanalysis* (H. H. Pattee, V. E. Cosslett, and A. Engström, eds.), Academic Press, New York, 1963, p. 379.
45. K. F. J. Heinrich and H. Yakowitz, *Anal. Chem.*, *47*, 2408 (1975).
46. H. Yakowitz, R. L. Myklebust, and K. F. J. Heinrich, *NBS Tech. Note*, *796*, 46 (1973).
47. J. Henoc, F. Maurice, and A. Zemskoff, *Proc. 5th Int. Conf. X-Ray Optics and Microanalysis* (G. Möllenstedt and K. H. Gaukler, eds.), Springer, Berlin, 1969, p. 187.
48. S. J. B. Reed, *Brit. J. Appl. Phys.*, *16*, 913 (1965).
49. D. R. Beaman and J. A. Isasi, *Anal. Chem.*, *42*, 1540 (1970).
50. R. B. Bolon and E. Lifshin, *Proc. 8th Int. Conf. Electron Probe Anal.*, MAS, New Orleans, paper 31 (1973).
51. J. W. Colby, D. R. Wonsidler, and D. K. Conley, *Proc. 4th Nat. Conf. Electron Probe Anal.*, MAS, Pasadena, Calif., paper 9 (1969).
52. M. H. Jacobs and J. Baborovska, *Proc. 5th Eur. Conf. Elect. Micros. Inst. Phys.*, London, 1972, p. 136.
53. G. Cliff and G. W. Lorimer, *Proc. 5th Eur. Conf. Elect. Micros. Inst. Phys.*, London, 1972, p. 140.
54. J. C. Russ, in *Microprobe Analysis as Applied to Cells and Tissues* (T. Hall, P. Echlin, and R. Kaufmann, eds.), Academic Press, New York, 1974, p. 269.
55. T. A. Hall, in *Quantitative Electron Probe Analysis* (K. F. J. Heinrich, ed.), *NBS Spec. Publ.*, *298*, 269 (1968).
56. T. A. Hall and P. D. Peters, in *Microprobe Analysis as Applied to Cells and Tissues* (T. Hall, P. Echlin, and R. Kaufmann, eds.), Academic Press, New York, 1974, p. 229.
57. T. A. Hall, in *Electron Microscopy and Analysis, Conf. Ser. #10*, Inst. Phys., London, 1971, p. 146.
58. J. C. Russ, *Proc. 9th Nat. Conf. Electron Probe Anal.*, MAS, Ottawa, Canada, paper 22 (1974).
59. H. Yakowitz, *J. Microscopie Biol. Cell.*(Paris), *22*, 137 (1975).
60. W. E. Sweeney, R. E. Seebold, and L. S. Birks, *J. Appl. Phys.*, *31*, 1061 (1960).
61. J. W. Colby, *Advan. X-Ray Anal.*, *11*, 287 (1968).
62. Y. Oda and K. Nakajima, *J. Jap. Inst. Met.*, *37*, 673 (1973).
63. H. Yakowitz and D. E. Newbury, SEM/76 (O. Johari, ed.), IITRI, Chicago (in press).

64. D. F. Kyser and K. Murata, *IBM J. Res. Devel.*, *18* (4), 352 (1974).

65. K. F. J. Heinrich, H. Yakowitz, and D. E. Newbury, eds.,*Monte Carlo Calculations in Electron Probe Microanalysis and Scanning Electron Microscopy*, NBS Spec. Publ. (in press).

66. J. Sisefsky, *Nature*, *203*, 708 (1964).

67. T. Mamuso, *Atompraxis*, *14*, 1 (1968).

68. H. J. Hoffmann, J. H. Weihrauch, and H. Fechtig, *X-Ray Optics and Microanalysis* (R. Castaing, P. Deschamps, and J. Philibert, eds.), Hermann, Paris, 1966, p. 166.

69. J. T. Armstrong and P. R. Buseck, *Anal. Chem.*, *47*, 2178 (1975).

70. J. I. Goldstein and J. W. Colby, in *Practical Scanning Electron Microscopy* (J. I. Goldstein and H. Yakowitz, eds.), Plenum Press, New York, 1975, p. 435.

71. R. E. Michaelis, H. Yakowitz, and G. A. Moore, *J. Res. NBS*, *68A*, 343 (1964).

72. K. F. J. Heinrich, R. L. Myklebust, S. D. Rasberry, and R. E. Michaelis, *NBS Spec. Publ.*, *260*, 28 (1971).

73. H. Yakowitz, R. E. Michaelis, and D. L. Vieth, *Advan. X-Ray Anal.*, *12*, 418 (1969).

74. H. Yakowitz, C. E. Fiori, and R. E. Michaelis, *NBS Spec. Publ.*, *260*, 22 (1971).

75. H. Yakowitz, D. L. Vieth, K. F. J. Heinrich, and R. E. Michaelis, *Advan. X-Ray Anal.*, *9*, 289 (1966).

76. H. Yakowitz, A. W. Ruff, Jr., and R. E. Michaelis, *NBS Spec. Publ.*, *260*, 43 (1972).

77. D. B. Ballard and H. Yakowitz, SEM/70 (O. Johari, ed.), IITRI, Chicago, 1970, p. 321.

78. A. W. Ruff, Jr., NBS Report NBSIR 74-474, p. 15 (1974).

79. R. Kiessling and N. Lange, *Nonmetallic Inclusions in Steel*, Iron and Steel Inst. London, Spec. Rept. 90 (Part I), 1964, and Publication 100 (Part II), 1966.

80. A. Läuchli, *J. Microscopie Biol. Cell.*, *22*, 433 (1975).

81. M. Sjösstrom, *J. Microscopie Biol. Cell.*, *22*, 415 (1975).

82. H. J. Hohling and W. A. P. Nicholson, *J. Microscopie Biol. Cell.*, *22*, 185 (1975).

83. R. M. Waterstrat and R. C. Manuszewski,, NBS Report 10571, 1971; NBS Report 73-415, 1975.

84. K. Keil and C. A. Andersen, *Geochim. Cosmochim. Acta*, *29*, 621 (1965).

Chapter 8

BONDING AND ELECTRON
SPECTROSCOPY FOR CHEMICAL ANALYSIS

H. K. Herglotz

Engineering Department
E. I. du Pont de Nemours and Company
Wilmington, Delaware

8-1 Introduction and Definition	205
8-2 X-Ray Spectroscopy	208
8-3 Electron Spectroscopy for Chemical Analysis (ESCA)	217
References	221

8-1 INTRODUCTION AND DEFINITION

X-Ray spectra are very attractive for compositional analysis because of their atomic nature which makes the spectra independent of the chemical combinations of the atoms. Thus, the analytical chemist is not burdened with the task of unraveling complex spectra to arrive at his results. The paucity of lines, exhibited in Fig. 1.1, and the simple relationship between wavelength and atomic number of Eq. (1.1) have assured X-ray spectroscopy its secure place in the analytical laboratory and an expanding usefulness.

This independence of X-ray spectra from the bonding state of the emitting atoms, however, is true only as long as the "resolution" of the instruments in wavelength Δλ or energy ΔE is smaller than the energy changes of atomic levels due to chemical combination, which is in the 1 to 10-eV range. Nearly all wavelength-dispersive and all energy-dispersive instruments in the analytical laboratory fall into this low-resolution category.

It was R. Swinne [1] who drew attention to the existence of a bonding effect in the early days of X-ray spectroscopy (1916), while Bergengren [2] is credited with the first experimental confirmation of the predicted, very small effect, namely, by measuring wavelength shift in the absorption spectrum of phosphorus due to chemical bonding. Since then, a side branch of X-ray spectroscopy has evolved which uses this phenomenon to

Extract information about the valence state of elements
Aid in the determination of the chemical structure of compounds
Shed light on the electron distribution in conduction bands of metals and semiconductors

Since its inception, interest in these applications has varied during the decades, as reflected by the number of papers appearing in the literature that deal with the subject.

Recently, information on bonding has been achieved with another, X-ray-related method: electron spectroscopy for chemical analysis (ESCA). Figure 8.1 illustrates how closely related the two methods are. ESCA and X-ray fluorescence analysis deal with the consequences of a *deep ionization*: an X-ray quantum absorbed by an atom generates a vacancy in a deep (core) electron level; a photoelectron is emitted which carries away kinetic energy E_k equal to the difference between the energy of the ionizing X-ray quantum $h\nu_i$ and the binding energy E_b of the core electron.

$$E_k = h\nu_i - E_b \tag{8.1}$$

The measurement of E_k with an electron spectrometer furnishes information about the binding energies characteristic of the atom and its bonding state.

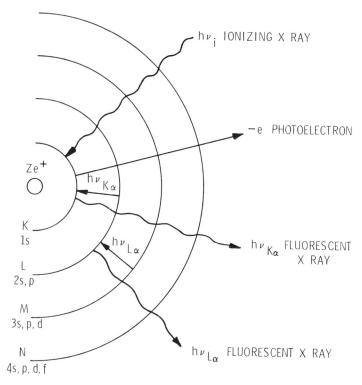

FIG. 8.1 Photoelectron emission and X-ray fluorescence in the (simplified) Rutherford-Bohr model of the atom.

The fluorescent X-ray quantum (essence of all X-ray fluorescence analysis) is a consequence of the resultant deep ionization introduced in the atom. Relaxation is accomplished by emission of the fluorescent X-ray quantum $h\nu_f$, which furnishes information about the two electronic levels involved in the core excitation and relaxation. Information from ESCA requires monochromatic X-ray excitation; interpretation of spectra for bonding studies is simpler because it involves only one atomic level (Fig. 8.1). In contrast, X-ray spectra can be excited by bremsspectrum (see Chaps. 1 and 2); the information carried by the fluorescent X-ray quantum involves two electronic levels, its interpretation is not always straightforward. However, ESCA has not completely displaced X-ray spectroscopy as a source of bonding information. Instead, the two methods often supplement each other. In some cases, ESCA's nature as a shallow surface method precludes its use, as in the case of the valence state of iron in an iron ore. A long exposure of the surface to an oxidizing environment reveals only trivalent iron by ESCA analysis. No information from the bulk is obtained.

In the remainder of this chapter instrumental arrangements required by both methods to probe the chemical state and a few typical examples of applications are presented. The chapter has been kept short in order to maintain the proper ratio of the size of this branch of X-ray spectroscopy to the compositional applications described in the rest of the book. Readers seeking more detailed information about characterization of bonding are referred to the summarizing articles, both historical and modern, of Refs. 3-11.

8.2 X-RAY SPECTROSCOPY

All early demonstrations of bonding effects were in the light-element range where the wavelengths λ of X-ray lines and their chemical shifts $\Delta\lambda$ are both large, according to the relationship

$$\Delta\lambda = -\frac{\lambda^2}{hc} \Delta E \tag{8.2}$$

where h = Planck's constant, c = velocity of light, and ΔE = energy change due to bonding.

The advantage of a large $\Delta\lambda$ more than compensates for the experimental inconveniences of measurements at the longer wavelengths of light elements such as the need for vacuum and the lower fluorescent yield (see Chaps. 1 and 2). Chemical shifts for the K spectra of the elements of the third row of the periodic table are in the order of magnitude 0.001 to 0.01 Å (at wavelengths in the ~5 to 10-Å range) which is easily measurable [12] with relatively unsophisticated X-ray spectrographs; for the K series of elements of the fourth row (potassium to bromine), where λ = 1-3 Å, $\Delta\lambda \sim$ 0.0001-0.003 Å [12], and this measurement requires far more instrumental skill. In the "ultrasoft" region of the K series of the lightest elements and L and M series of higher Z elements, Δλ becomes very large, but the measurements require instrumental arrangements quite different from those for the 1 to 3-Å region [13-17].

The high instrumental resolution Δλ or ΔE needed in this branch of X-ray spectroscopy can be obtained in various ways by using

1. Nearly perfect crystal and a large distance between crystal and detector [18]
2. Large glancing angle θ to increase Δθ [19]

$$\Delta\theta = \frac{\Delta\lambda}{\lambda} \tan\theta \qquad (8.3)$$

(See also Chap. 1)

3. Long-period organic crystals for very long wavelengths where nearly ideal crystals do not exist [13-16]
4. Ruled gratings for the same wavelength range [17,20,21]

Combinations of these various routes sometimes are more effective than a "pure" approach. No general recipe for success exists, because knowledge of the underlying concepts and resourcefulness are the basic requirements.

Let us support this statement and demonstrate strategy with a few examples which are chosen because they closely approximate reality, both with respect to the tasks and to the approaches taken.

1. The X-ray spectroscopist is asked whether a mineral contains sulfur in the form of sulfide or sulfate. The wavelength of sulfur $K\alpha_{1,2}$ is listed in Ref. 22 at 5.3613 Å, and Kβ at 5.020 Å. He could choose the $K\alpha_{1,2}$ doublet for his analysis, but would probably select the Kβ line because

> It is a single line and not a poorly resolved doublet.
> It comes from a transition of an outer electron (M shell) and should show bonding effects to a higher degree.

Looking up the crystals listed in Chap. 1, he can calculate Bragg angles for the Kβ line of sulfur, which are listed in Table 8.1.

TABLE 8.1

Common Spectrometer Crystals Suitable
for the Sulfur Kβ Line at ~5.02 Å

Crystal	2d (Å)	Bragg angle θ
LiF (220)	2.84	
LiF (200)	4.02	
$CaCO_3$ (200)	6.06	55.96°
Quartz (10$\bar{1}$1)	6.70	48.53°
Quartz (10$\bar{1}$0)	8.50	36.20°
ADP (110)	10.64	28.15°
ADP 2nd order (220)	5.32	70.67°

ADP diffracts sulfur Kβ radiation in second order at the favorable Bragg angle θ = 70.67°. He can calculate the dispersion $\Delta\theta/\Delta\lambda = (1/\lambda) \tan\theta$ using Eq. (8.3) and finds it equal to 0.57 radian/Å.

For chemical shifts of ~0.005 Å which he can expect for an element of the second row of the periodic table, the displacement Δθ of the Kβ line of sulfate vs. sulfide should be approximately 0.16°; this should be measurable by most laboratory spectrographs.

After preparing the sample as advised in Chaps. 10-12, he performs wavelength-dispersive analysis, ignoring the intensity of lines, since the amount of sulfur in his sample is not his concern at the moment. His calibration would consist of recording the sulfur Kβ line of a sulfide and a sulfate in the purest form available and of comparing the results of his unknown with those of the calibration materials.

If he can afford the additional time, he should record the $K\alpha_{1,2}$ doublet as well; and also check the literature (in this case the review articles referenced at the end of the chapter and in the chemical abstracts). There, he discovers that A. Faessler [23] has measured carefully both the Kβ and $K\alpha_{1,2}$ of sulfides and sulfates and his findings are shown in Table 8.2.

TABLE 8.2

Wavelength for K Lines
from Sulfides and Sulfates [23]

	$K\beta_1$	$K\alpha_1$ [a]	$K\alpha_2$ [a]
CaS	5.0176	5.3612	5.3640
CaSO$_4$	5.0150	5.3580	5.3608
Δλ	0.0026	0.0032	0.0032

[a] From Ref. 23, averages of several sulfides and sulfates.

Our spectroscopist, of course, should have studied the literature before plunging into the problem. Then he would have learned that in this case the Kα doublet shows a larger chemical shift than the Kβ line. Experience tells that the course taken by our fictitious spectroscopist is closer to reality.

Faessler used a calcite crystal. Data vital to our spectroscopist's objective can be easily calculated from the 2d spacing of 6.06 Å and are given in Table 8.3. Calcite crystals can be obtained in nearly perfect form and are therefore more suitable for this precision work, more so, perhaps, than the ADP crystal first suggested.

2. In the second example chosen, another spectroscopist is expected to investigate the valence state of iron in a biological sample. He can expect shifts only in the order of magnitude of

TABLE 8.3

Bragg Angle θ, Dispersion Δθ/Δλ for Sulfur Lines Diffracted on Calcite, CaCO$_3$ 2d = 6.06 Å

	λ [22]	θ	Δθ/Δλ (degrees/0.03 Å)
SKβ$_1$	5.0163	55.87	0.05
α$_1$	5.3596	62.18	
α$_2$	5.3624	62.24	0.06

0.0001 Å and therefore can accept this task only if

His spectrometer has the necessary angular precision, i.e., backlash of the angular drive is smaller than 0.03° (see below how he arrived at this figure)

His instrument permits registration of 2θ values close to 180°, where $\Delta\theta/\Delta\lambda = (1/\lambda) \tan\theta$ is large

A search of the literature for a crystal with θ near 90°, primarily from Table 1.1, leads to the following results:

Bragg Angles

		LiF (420) 2d = 1.79	LiF (220) 2d = 2.84	CaF$_2$ (220) 2d = 3.86	Quartz (13$\bar{4}$0) [24] 2d = 2.356
α_1	1.9323		42.87	30.04	55.10
α_2	1.9365		43.00	30.11	55.28
β_1	1.7530	78.33	38.11	27.00	48.08
$2\alpha_1$	3.8646				
$2\alpha_2$	3.8730				
$2\beta_1$	3.5060			65.27	

He has relatively little choice and therefore would select lithium fluoride (420). He would test the crystal quality by recording the $K\beta_1$ line of iron to see that it is not heavily distorted by crystal defects. Reflections at high Bragg angles are very sensitive to distortions of the lattice. At the Bragg angle of θ = 78.33° for the $K\beta_1$ line, his dispersion $\Delta\theta/\Delta\lambda = (1/\lambda)\tan\theta$ is calculated to 158.34°/Å; he can expect a shift from the valence state of iron of not more than 0.0001-0.0003 Å, and therefore a shift of the Bragg angle of ∼0.03°. Calibration with both a pure bivalent and trivalent iron standard tells him whether his instrument is capable of resolving the shift in wavelength due to the difference in valence level.

The method has been applied for practical purposes: the state of cobalt in vitamin B_{12} was determined as trivalent by Boehm et al. [24]. These researchers used the absorption edge of cobalt instead of an emission line.

Another way to approach this task would be by using a double-crystal spectrometer, shown in Fig. 8.2. The two nearly perfect crystals can be placed in a parallel (a) or antiparallel (b) arrangement. In the first case, the dispersion is zero, while in the second case it is twice that of a common spectrometer.

The parallel arrangement is very useful for obtaining information about the perfection of its two crystals and this is a frequent application [25]. For extremely precise wavelength information, the width of the reflection curve (the so-called rocking curve) of the analyzing crystals is determined first with the parallel arrangement; then, the second crystal is switched to the antiparallel or dispersive position. The previously determined "crystal width" can be eliminated by deconvolution and the natural line shape of an X-ray line, that is, the function intensity vs. λ can be determined. Small chemical shifts can be measured in the same way.

Usually, the double-crystal spectrometer is not part of the equipment of the regular X-ray laboratory. Therefore, it is not discussed in greater detail here. For more information, reference is made to the literature [26,27].

Very minute chemical shifts for elements of the fourth row of the periodic table, in the wavelength range 1-3 Å, have also been

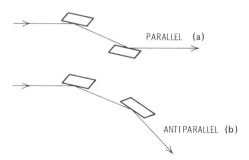

FIG. 8.2 Concept of double crystal spectrometer (a) in parallel position $\Delta\theta/\Delta\lambda = 0$ and (b) in antiparallel position $\Delta\theta/\Delta\lambda \approx (2/\lambda)$ tan θ (Ref. 18, p. 129).

measured in an unconventional way with small diffraction cameras which use a diffracting polycrystalline medium instead of the conventional crystal. The very high dispersion $\Delta\theta/\Delta\lambda$ is achieved by the choice of a polycrystalline material diffracting at θ very close to 90°, where $\tan \theta$ and therefore $\Delta\theta/\Delta\lambda$ [see Eq. (8.3)] are very large [19]. This approach was developed for low-budget laboratories which could not afford a large X-ray spectrograph. This situation is rather infrequent today and this experimental approach will therefore not be discussed at length in this chapter.

3. As the last arbitrarily selected though representative sample, let us talk about the shift $\Delta\lambda$ due to mechanical deformation of an aluminum alloy [19]. Since the energy remaining in the metal after deformation is small, it is best to resort to very long wavelengths where, from Eq. (8.3), we might expect a measurable $\Delta\lambda$, even for a minute ΔE. The wavelength of the aluminum L line was determined in Ref. 19 with λ = 180 Å, which is the centroid, while the actual line extends from 170 to 200 Å (Fig. 8.3). A change in

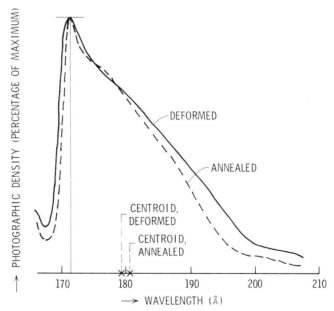

FIG. 8.3 Intensity distribution of Al $L_{2,3}$ band of a hardenable aluminum alloy: deformed ———; and annealed ----.

energy ΔE = 1 eV causes, under these conditions, a Δλ = 2.6 Å [Eq. (8.2)]. The 0.8-Å shift measured in Ref. 19 means that the transition M → L of the aluminum atom in the deformed lattice represents, on the average, an energy 0.3 eV higher than in the undeformed lattice.

This measurement was accomplished with a ruled grating and utilizes the same diffraction effect as an optical grating spectrograph. However, there are no mirrors for X rays comparable to optical reflectors.

In the X-ray region, the phenomenon of total reflection is substituted for it. X rays have a refractive index

$$\mu = \frac{v_{vacuum}}{v_{solid}} < 1 \quad (v = \text{phase velocity})$$

and can therefore be totally reflected, but since μ is nearly equal to 1, total reflection occurs only at grazing angles. (See Compton and Allison [26], pp. 40-42.) With an arrangement described in Fig. 8.4, one can disperse and record wavelengths without moving or rocking the grating. The resolving power equals: $\lambda/\Delta\lambda$ = nN, where n is the order of the diffraction and N the number of grooves participating. Good gratings exhibit $\lambda/\Delta\lambda$ of ~1000; at a wavelength of 180 Å, Δλ is therefore 0.18 Å. To increase the luminosity of the spectrograph and thus to shorten exposure times, curved gratings are used [28]. More about X-ray spectroscopy with gratings can be found in Refs. 18 and 26. Ruled gratings have been used extensively for measurements of bonding shifts [17].

There exists another approach that has been equally successful in the X-ray spectroscopy of very long wavelengths, where no well-grown single crystals exist. Layers of fatty acids or soaps deposited on a flat substrate can act as substitute for crystals (see, e.g., Table 1.1, last item).

Not only the dispersing crystal, but also the detector and the exciting X-ray tube require unique features in this X-ray range of very long wavelengths that overlap with the short wavelengths of UV.

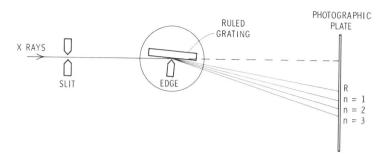

FIG. 8.4 Experimental arrangement for X-ray spectroscopy with a ruled grating: R = reflected X-rays; n = 1, 2, 3 (three orders of diffracted X rays).

Flow-proportional counters with extremely thin windows are needed since this "ultrasoft" X-radiation penetrates poorly and requires good vacuum in the spectrograph. Also, the poor fluorescent yield requires very powerful exciting sources. Procedures and some of the interesting results for this ultrasoft region are summarized by B. L. Henke [13]. Bonding effects become very strong in this X-ray region, manifested in wavelength and intensity changes. One of the major applications is the determination of electron distributions in the outer shells in metals and semiconductors. The intensity distribution I(E) (E = energy) of the aluminum L line of Fig. 8.3 contains the electron distribution N(E) in the M shell (conduction band) but not in a straightforward form. To extract N(E) from I(E) is not always easy. References 29 and 30 are devoted to this subject.

We have concerned ourselves almost exclusively with bonding effects in the emission spectra, although the first experimental demonstration took place in the absorption spectrum [2], as mentioned previously. Absorption spectra show interesting effects besides wavelength shifts, effects traceable to atomic or lattice structure. Applications in the X-ray laboratory are rather rare and confined to highly specialized research laboratories; sample thickness is of critical influence on the results. References 4-6, 33, and 34 should be consulted if there is intention to enter the field. Recently, very intense polychromatic radiation, which is required for absorption spectroscopy, has become available from electron-synchrotrons, and has made elegant experiments possible [35]. Very few researchers, of course, have the opportunity to use electron synchrotrons as X-ray sources, because only a few exist worldwide. Therefore, this experimental branch of X-ray spectroscopy has received only scant coverage in this book.

X-Ray absorption spectroscopy is also hardly ever applied for compositional analysis. Therefore, it has not been covered in this book, which by charter deals with practical spectroscopy. However, recently a commercial X-ray absorption analyzer for lead in gasoline and sulfur in fuel oil has been advertised (American Laboratory, January 1975, p. 47).

We return after this brief excursion into compositional analysis, to the theme of the chapter, which is bonding analysis. Lately, valuable help has come to the researcher in the field from ESCA, as mentioned previously. Therefore, a few words about instruments and procedures of this relatively young member of practical spectroscopy seem in order.

8.3 ELECTRON SPECTROSCOPY FOR CHEMICAL ANALYSIS (ESCA)

It is nearly as old as X-ray spectroscopy itself, but only recent instrumental developments have converted it from a scientific demonstration into a tool of the analytical laboratory [7-10]. We follow here the nomenclature of Siegbahn et al. because he and his collaborators have contributed most to its transition to an analytical method. The name ESCA also makes reference to the method's objective. We restrict ourselves to the option (among several variations) that deals with X-ray excitation and measurement of the kinetic energy of the photoelectron, as defined in Fig. 8.1. Tables 8.4 and 8.5 are incomplete summaries of various names and branches of electron spectroscopical methods.

In spite of its close relationship with X-ray fluorescence spectroscopy (XRFS), ESCA has distinct features of its own that fill gaps in XRFS, namely

   Sensitivity to all elements (hydrogen only under special
     conditions)

TABLE 8.4

Nomenclature

ESCA	Electron spectroscopy for chemical analysis
IEES	Induced electron-energy spectroscopy
SES	Secondary electron spectroscopy
PES	Photoelectron spectroscopy
XRPS, XPS	X-Ray photoelectron spectroscopy
UPES, UPS	UV-Photoelectron spectroscopy
AES	Auger-electron spectroscopy
ELS	Energy-loss spectroscopy
ISS	Ion-scattering spectroscopy

TABLE 8.5

Electron Spectroscopical Modes of Excitation, Secondary Events, and Major Analytical Methods

	Secondary events		
Excitation	(1) Ionization	(2) X-Ray emission	(3) Auger electron
X Rays	Photoelectron spectroscopy	Fluorescent X-ray spectrometry	Auger-electron spectroscopy
Electrons	Energy-loss spectroscopy	X-ray spectrometry	Auger-electron spectroscopy
		Soft X-ray appearance potential spectrometry	

Free from the affliction of nearly vanishing fluorescent yields for light elements

Clearer bonding information (see Fig. 8.5) than from XRFS

Information about the composition and bonding at the surface apart from the bulk since the ESCA signal comes from a shallow depth of less than 100 Å

ESCA shares with XRFS some of the difficulties the latter experiences in the long-wavelength region: need to measure the sample in a vacuum; exclusion of liquids and substances with high vapor pressure.

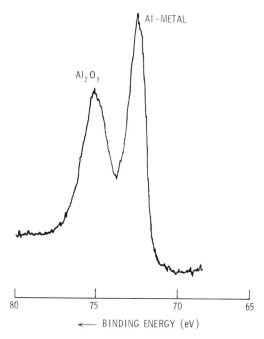

FIG. 8.5 Aluminum $2p_{1/2,3/2}$ line (= ionization of $L_{II,III}$ levels of aluminum) from aluminum coated by oxide layer.

The instrumental needs and experimental conditions of the two methods are very similar as is best demonstrated by Ref. 31, which describes a combined ESCA/XRF instrument, illustrated in Fig. 8.6. The sample on the tip of a piston is introduced through a vacuum lock into the instrument proper, where it is excited by X rays from an annular target surrounding it. Photoelectrons from the sample are analyzed by the electron-energy analyzer described in detail in Ref. 32. By moving the electron analyzer sideways (without breaking the vacuum) an energy dispersive analyzer can measure the fluorescent X rays generated simultaneously with the photoelectrons by the X-ray excitation. Results obtained with the instrument are typified by Fig. 8.7, which demonstrates the composite nature of the information that is obtained: ESCA reveals the presence of carbon and fluorine in the sample, a Teflon-coated galvanized steel sheet, but not zinc and iron, which therefore must be absent in the surface.

The energy-dispersive X-ray analyzer of this combined instrument, with a resolution of roughly 150 eV, does not yield, of course,

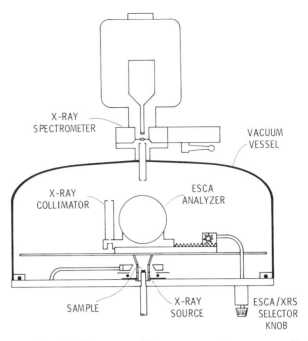

FIG. 8.6 A combined ESCA X-ray fluorescence Instrument [31].

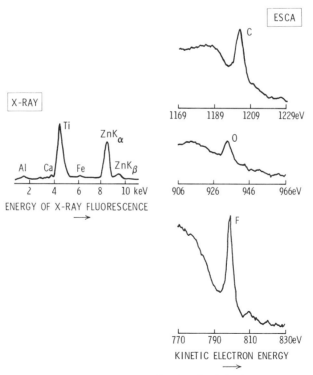

FIG. 8.7 X-Ray and ESCA results obtained with instrument of Fig. 8.6 from a galvanized steel, coated with a fluoropolymer, pigmented with titanium dioxide.

bonding information. It was chosen because it fills the most frequent need of the laboratory for which it was developed. It could be replaced, if the need arises, by a high-resolution wavelength-dispersive X-ray spectrograph. Then, bonding information would be available from both ESCA (surface) and XRF (bulk, up to escape depth of fluorescent X rays).

Worthy of mention are the electron-excited Auger spectra because of their significance to bonding studies in metals and semiconductors. The Auger process is another relaxation process of the deeply ionized atom, competing with the emission of fluorescent X-ray quanta. The interested reader is referred to one of the many informative review articles of Auger spectroscopy, e.g., Refs. 36 and 37.

It would be beyond the scope of this chapter to describe all variations of electron spectroscopy and all instrumental developments. The reader who intends to adopt ESCA into his program is advised to resort to one or more of the ESCA publications [7-11] where he finds detailed information and references guiding him to answers of specific questions.

REFERENCES

1. R. Swinne, Zum Ursprung der γ-Strahlenspektren und Röntgenstrahlenserien, *Physikalische Zeitschrift*, 17, 481 (1916).
2. J. Bergengren, Über die Röntgenabsorption des Phosphors, *Zeitschrift für Physik*, 3, 247 (1920).
3. M. Siegbahn, *Spektroskopie der Röntgenstrahlen*, Springer, Berlin, 1931.
4. H. K. Herglotz, Einflüsse der Bindung auf das Röntgenspektrum, *Abhandlungen des Dokumentationszentrums für Technik und Wirtschaft*, No. 13, Vienna, 1952.
5. A. Faessler, Röntgenspektrum und Bindungszustand, *Landolt-Börnstein*, 1, (4), 769 (1955).
6. D. J. Nagel and W. L. Baun, Bonding Effects in X-Ray Spectra, in *X-Ray Spectroscopy* (L. V. Azároff, ed.), McGraw-Hill, New York, 1974.

7. K. Siegbahn, C. Nordling, A. Fahlman, R. Nordberg, K. Hamrin, J. Hedman, G. Johansson, T. Bergmark, S. E. Karlsson, I. Lindgren, and B. Kindbert, *ESCA, Atomic, Molecular and Solid State Structure Studied by Means of Electron Spectroscopy*, Almqvist and Wiksells, Uppsala, 1967.

8. K. Siegbahn, C. Nordling, G. Johansson, J. Hedman, P. F. Hedén, K. Hamrin, U. Gelius, T. Bergmark, L. O. Werme, R. Manne, and Y. Baer, *ESCA Applied to Free Molecules*, North Holland, Amsterdam, 1971.

9. A. D. Baker and D. Betteridge, *Photoelectron Spectroscopy; Chemical and Analytical Aspects*, Pergamon Press, Oxford and New York, 1972.

10. H. K. Herglotz and H. L. Suchan, ESCA - A New Tool for Surface Research, *Advances in Colloid and Interface Science*, 5, 79-103 (1975).

11. S. B. M. Hagström and C. S. Fadley, X-Ray Photoelectron Spectroscopy, in *X-Ray Spectroscopy* (L. V. Azároff, ed.), McGraw-Hill, New York, 1974.

12. A. Faessler and M. Goehring, Das Kα-Dublett von Schwefel in Metallsulfiden, *Zeitschrift für Physik*, 142, 558-574 (1955).

13. B. L. Henke, Application of Multilayer Analyzers to 15-150 Å Fluorescence Spectroscopy for Chemical and Valence Band Analysis, *Advan. X-Ray Anal.*, 9, 430-440 (1966); and B. L. Henke and M. A. Tester, Techniques of Low Energy X-Ray Spectroscopy, *Advan. X-Ray Anal.*, 18, 76-106 (1975).

14. D. W. Fischer and W. L. Baun, Experimental Techniques for Soft X-Ray Spectroscopy, *Technical Documentary Report No. RTD-TDR-63-42R2*, Office of Technical Services, U. S. Department of Commerce, Washington, D. C., 1964.

15. D. W. Fischer and W. L. Baun, Experimental Dispersing Devices and Detection Systems for Soft X-Rays, *Advan. X-Ray Anal.*, 7, 489-496 (1964).

16. W. L. Baun, Detection and Spectroscopy of Long Wavelength X-Rays, *Advan. X-Ray Anal.*, 13, 49-67 (1970).

17. J. E. Holliday, The Use of Soft X-Ray Spectroscopy as a Tool for Studying the Surface Region of Metals and Alloys, *Advan. X-Ray Anal.*, 16, 53-62 (1973).

18. M. Siegbahn, *Spektroskipie der Röntgenstrahlen*, 2nd ed., Springer, Berlin, 1931, pp. 99-102.

19. H. K. Herglotz and E. Schiel, Effects of Chemical State and of Mechanical Deformation on the Wavelength of X-Ray Emission Lines, *Nature* (London), 205, 1093-1095 (1965).

20. J. B. Nicholson and D. B. Wittry, A Comparison of the Performance of Gratings and Crystals in the 20-115 Å Region, *Advan. X-Ray Anal.*, 7, 497-511 (1964).

21. A. Franks, K. Lindsey, J. M. Bennett, and R. J. Speer, The Theory, Manufacture, Structure and Performance of N.P.L. X-Ray Gratings, *Philos. Trans. A*, 277, 503-543 (1974).

22. J. A. Bearden, X-Ray Wavelengths, U. S. Atomic Energy Commission, Division of Technical Information, *NYO-10586*, Oak Ridge, Tennessee, 1964.

23. A. Faessler, Röntgenspektrum, Elektronenstruktur und chemische Bindung, *Österreichische Chemiker-Zeitung*, 57, No. 1/2, 7-12 (1956).

24. G. Boehm, A. Faessler and G. Rittmayer, Röntgenspektroskopische Bestimmung der Wertigkeit des Kobalts im Vitamin B12, *Zeitschrift für Naturforschung*, 9b, 509-513 (1954).

25. A. D. Kurtz, S. A. Kulin, and B. L. Averbach, Effects of Growth Rate on Crystal Perfection and Lifetime in Germanium, *J. Appl. Phys.*, 27, 1287-1290 (1956).

26. A. H. Compton and S. K. Allison, *X-Rays in Theory and Experiment*, D. Van Nostrand, New York, 1935, p. 726.

27. J. S. Thomsen, High Precision X-Ray Spectroscopy, in *X-Ray Spectroscopy* (L. V. Azároff, ed.), McGraw-Hill, New York, 1974, pp. 26-132.

28. M. Siegbahn, Messungen langer Röntgenwellen mit Optischen Gittern, *Ergebnisse der exakten Naturwissenschaften*, 16, 104 (1937).

29. D. J. Fabian, ed., *Soft X-Ray Band Spectra and the Electronic Structure of Metals and Materials*, Academic Press, London and New York, 1968.

30. A. Faessler and G. Wiech, eds., *X-Ray Spectra and Electronic Structure of Matter*, Munich, 1973.

31. H. K. Herglotz and D. R. Lynch, A Combined Photoelectron/X-Ray Fluorescence Spectrometer, *Advan. X-Ray Anal.*, 17, 509-520 (1974).

32. J. D. Lee, A Nondispersive Electron Energy Analyzer for ESCA, *Rev. Sci. Instrum.*, 44, 893 (1973).

33. L. V. Azároff, Theory of Extended Fine Structure of X-Ray Absorption Edges, *Rev. Mod. Phys.*, 35, 1012 (1963).

34. F. W. Lytle, Determination of Interatomic Distances from X-Ray Absorption Fine Structure, *Advan. X-Ray Anal.*, 9, 398-409 (1966).

35. M. L. Perlman, E. M. Rowe, and R. E. Watson, Synchrotron Radiation-Light Fantastic, *Phys. Today*, 27, No. 7, 30-37 (1974).

36. W. C. Johnson, D. F. Stein, and A. Joshi, Auger Electron Spectorscopy--A Review, *Can. J. Spectrosc.*, 17, 88-92 (1972).

37. J. C. Riviere, Auger Electron Spectroscopy, *Contemp. Phys.*, 14, 513-539 (1973).

Chapter 9

SELECTION AND SAFE OPERATION OF X-RAY INSTRUMENTS*

Paden F. Dismore

Jackson Laboratory
Organic Chemicals Department
E. I. du Pont de Nemours and Company
Wilmington, Delaware

9.1	Introduction	225
9.2	Use of X-Ray Spectroscopic Equipment	226
	9.2.1 Objective of the Laboratory	226
9.3	Types of X-Ray Spectroscopy Equipment	229
9.4	X-Ray Safety	234
	9.4.1 Types of Hazards	235
	9.4.2 Legal Requirements for Radiation-Producing Equipment	236
	9.4.3 Safety Attitudes, Habits, Training, and Rules	237
References		239

9.1 INTRODUCTION

A general description of the equipment required for X-ray spectrometry is found in Chaps. 1 and 2 of this volume. Chapters 5, 6, and 7 provide descriptions of more specialized components. More detailed

---

*Research and Development Division Publication No. 534

descriptions of the equipment are readily available in the literature [1,2]. However, it is difficult for a prospective purchaser of X-ray spectrometric equipment to decide on the relative merits of commercially available equipment or even to decide which of the various types of equipment are best suited to his needs.

The rather significant number of accidents (one incident per 200 installations per year; Ref. 3), all of which are avoidable by proper attention to safety, indicates the need for thorough consideration of this aspect. This chapter will attempt to aid the X-ray spectroscopist in the solution of these two problems.

9.2 USE OF X-RAY SPECTROSCOPIC EQUIPMENT

9.2.1 Objective of the Laboratory

It may seem peculiar to begin a discussion of X-ray spectroscopic equipment with the apparently simple question of what are the objectives of the laboratory. The obvious answer to this question is to determine qualitatively and quantitatively what elements are present in the sample. This objective, however, has to be much better defined in order to decide which of the many types of equipment are best suited for a particular task.

There are several different tasks that an X-ray spectroscopy laboratory may be asked to perform. Rasberry [4] defined four broad types of application: (1) on-stream process control, (2) off-line quality assurance, (3) routine service analyses, and (4) general purpose analyses. To these might be added a fifth class of X-ray spectrometric research.

Each of these applications has different requirements. For example, a process-control application would require analysis of only a very few (frequently one) elements in an essentially constant matrix but would probably be used in the rather hostile environment of a manufacturing facility. This requires a relatively simple spectrometer tuned to the elements of interest. A simple go-or-no-go output is all that is required. On the other hand, the instrument

SELECTION AND SAFE OPERATION OF X-RAY INSTRUMENTS 227

must be extremely rugged to provide dependable service in the hostile environment; it must have extremely reliable safeguards to prevent accidents to the people using it, who generally are not trained in X-ray safety; the instrument may require rather elaborate, custom designed sample introduction facilities.

In contrast, a general purpose analytical laboratory may be asked to determine many elements in a sample and these elements may be anywhere in the periodic table. Any elements may also be in the matrix. There may be some routine, repetitive work, but many analytical requests will be unique. This laboratory will require a much more sophisticated spectrometer. However, it will probably be operated in the more sheltered environment of a laboratory, by people specifically trained in X-ray work. Thus it does not have to be as rugged as the equipment for process control work.

The X-ray research laboratory will have its own unique requirements. This work is extremely varied but since the people are highly skilled and have an extensive background in X-ray physics, there is no need for consideration in this book of equipment for such a laboratory.

Other laboratories have requirements between these extremes and will require different balances of simplicity, versatility, ruggedness, etc.

Another factor entering into the choice of X-ray equipment is the sample load. Again there will be wide variations between laboratories ranging from a few samples per week to a few hundred samples per day. Generally, a higher sample load makes a high-speed instrument more desirable.

However, laboratory management should thoroughly analyze the laboratory's problem before investing in a high-speed instrument. X-Ray spectroscopy has the reputation of being a very fast analytical technique. Unfortunately this is true only for those samples that (1) are received ready to run, (2) have a procedure (including calibration) developed, and (3) require a minimum of clerical work. This is the job of the process-control or quality-control laboratory. Other samples may require so much time for sample preparation, method development, computation, and clerical work that an increase in the speed of the spectrometer has essentially no effect on the total time involved.

The variation in sample types which as previously indicated, requires different kinds of equipment, also requires people with different training. The on-line process-control spectrometer will probably be operated by manufacturing people with essentially no knowledge of X rays. This machine must therefore be extremely simple to operate. A quality control machine will be operated by a laboratorian who probably has a general knowledge of laboratory procedures and some on the job X-ray training. He will probably work under the supervision of someone with at least a rudimentary knowledge of X rays. In this case a simple but fairly versatile spectrometer is required. Detailed instructions for each type of analysis will have to be provided. Operation of a general purpose laboratory requires a technician, highly trained in practical X-ray spectroscopy, working under the supervision of a technically trained person with broad knowledge of chemistry and physics. In this case, a very versatile spectrometer can be put to good use.

An X-ray spectrometer, like any other instrument, will require some maintenance (another factor for the prospective purchaser to consider). Again the amount and kind of maintenance required will depend on the type of instrument. Some of this is relatively simple and must be done by the laboratory personnel (e.g., keeping the cryostat on an energy-dispersive detector filled with liquid nitrogen), while other work is so complex that in most laboratories it must be done by the vendor. Between these extremes, some maintenance will be done by the laboratory and some by the vendor. It is very important that laboratory people doing maintenance work understand what they are doing and that they realize their limitations. X-Ray spectrometers present two major hazards that are not generally present in other equipment: (1) the radiation hazard and (2) high voltages (up to 50,000 V) (see Sec. 9.4.3). Maintenance people must be very careful not to injure themselves and to be sure that the instrument is left in a safe operating condition.

The choice of an X-ray spectrometer will also depend on what other facilities are available. Some of these considerations are so fundamental as to appear ridiculous; nevertheless, they are sometimes overlooked. Is there adequate space available? In addition to space for the spectrometer, space is needed for sample preparation, sample storage before and after analysis, desk space, and file space. Another consideration is adequate electric power of adequate quality. Some spectrometers require 5 kW while others only require a few watts. This author knows of at least one X-ray installation where extra equipment had to be installed to eliminate voltage spikes from the power line. Another fundamental consideration is cooling capacity. The high-power units require water cooling of the X-ray tube. Will a chilled water supply be needed? Is the air conditioning adequate? This author has observed a 0.2% change in count rate per degree Fahrenheit change in room temperature. Condensate has also been observed on the chilled water lines if the relative humidity rises much above 50%. The author has also found corrosion of the spectrometer caused by high humidity. This corrosion can cause failure of safety devices leading to possible radiation exposure.

Many manufacturers are now offering computer-controlled X-ray spectrometers which adds another complication. If the spectroscopist does not have a knowledge of the use of his particular computer, outside computer expertise may be desirable.

9.3 TYPES OF X-RAY SPECTROSCOPY EQUIPMENT

In this section, we discuss the advantages and disadvantages of the various components of an X-ray spectrometer.

Atoms can be excited to emit X rays by almost any radiation of adequate energy. However, because of practical limitations very few sources of radiation are commonly used. Thus positive-ion beams are usable only if an accelerator is available (see Chaps. 6 and 16).

Electron beams are more common, however, being used in such instruments as microprobes and electron microscopes (see Chaps. 5 and 7). Both positive-ion and electron beam excitation require that the sample be in a high vacuum; this, of course, is a severe limitation.

Because of these limitations, most X-ray spectrometers use either a radioactive source or an X-ray tube for excitation, each of which has advantages and disadvantages. A radioactive source does not require power so a smaller, simpler instrument can be built. On the other hand, radioactive sources cannot be turned off. This requires more elaborate safety devices to prevent stray radiation. Radioactive sources emit radiation of discrete energies, so efficient excitation can be obtained for most elements by the proper choice of isotope. This advantage is also a disadvantage in that several sources are required if many different elements are to be analyzed. For safety, radioactive sources have low power, but this is not necessarily a disadvantage since solid state detector systems can be made very efficient.

X-Ray tubes are available in a wide range of power ratings with the higher power tubes giving a more intense X-ray beam. Because of the difference in detector efficiency, the low-power tubes are generally used with energy-dispersive systems, while high-power tubes are required for wavelength-dispersive systems. Naturally, the high-powered systems require a larger and generally more complex power supply. They also require more cooling. These requirements increase both the size and the cost of the system.

An X-ray tube emits both continuous radiation, or *bremsstrahlung*, and radiation characteristic of the target. As discussed in Chap. 2, the most efficient excitation is by energies slightly greater than the emitted radiation. Thus an X-ray tube will be most efficient in exciting elements of atomic number slightly less than that of the target. However, the continuous radiation can be used to excite most other elements with reasonable efficiency.

There are several ways that the radiation from an X-ray tube can be modified. The most obvious is, of course, by varying the voltage on the tube, which can be done on most instruments.

A second, frequently used technique, is to replace the X-ray tube with another with a different target. At least one spectrometer is commercially available with a dual target tube, which makes such a change quite simple. Another way of modifying the emitted radiation is by the use of filters. As shown in Fig. 2.4, different materials have different absorption characteristics for various energies. Frequently, a filter of the same material as the X-ray tube target is used to get relatively pure characteristic radiation. A transmission target X-ray tube is commercially available which does the same thing without an additional filter. Still another technique is secondary emission in which an X-ray tube excites characteristic radiation of a secondary target. This secondary radiation is then used to excite the sample. A choice of monochromatic radiation is thus readily available. However, this technique is relatively inefficient so that, generally speaking, higher power X-ray tubes are required.

If the laboratory will require any of these ways of modifying the exciting radiations, facilities for the modification should be built into the instrument. Improvised modifications often do not work well and frequently create safety hazards.

There are two commercially available ways of separating or dispersing the radiation emitted by the sample: energy dispersion and wavelength dispersion. The differences between these are described in Chaps. 1 and 2. Some of these differences are reemphasized here as a help in the choice of dispersion type.

Wavelength dispersion has the advantage of much greater resolution. The Kα and Kβ lines of adjacent, low-atomic number elements cannot be resolved in an energy-dispersive system but are resolved with wavelength dispersion. To overcome this limitation, energy-dispersive systems are frequently equipped with a minicomputer and mathematical techniques (e.g., peak stripping) are used to separate the radiation from overlapping peaks. These methods are more or less effective depending on the relative amounts of the elements producing adjacent peaks and the particular algorithm that is used.

Any detector requires a finite time to process a radiation pulse. In a wavelength-dispersive system, the radiation pulses are counted after dispersion so that only a part of the radiant energy is counted at any one setting of the crystal. In the energy-dispersive system the detector is also the dispersive element, so that all of the radiant energy is counted at once. Thus, the time to process a pulse, or dead time, becomes much more important. Also, the dead time in an energy-dispersive detector is longer (4.5-17 μsec) than for a proportional counter (1-2 μsec) used in a wavelength system. Thus an energy-dispersive system is limited to about 20,000 total counts per second (see Fig. 2.14). If small concentrations of material must be accurately measured, an energy-dispersive system will require long counting times because most of the total counts will be coming from the matrix and are not providing useful information. However, this disadvantage turns out to be an advantage in other types of problems. If relatively large amounts of several elements have to be measured in a sample, the energy-dispersive system will determine all of them simultaneously with a saving of time. Multichannel wavelength spectrometers are available which also allow simultaneous measurement of several elements. However, these instruments are generally preset so that it is difficult to change the elements to be measured. Gilfrich et al. [5] provide an excellent comparison of the two dispersion techniques. Some manufacturers have increased the maximum counting rate for their energy-dispersive detectors by using more sophisticated amplifiers and by using pulsed X-ray tubes [6,7].

Maintenance problems will be different for the two types of dispersion in that wavelength dispersion uses a mechanical system while energy dispersion uses an electronic system.

Several systems of recording data are available. The simplest is a rate meter which is probably the most desirable for an on-line process-control instrument. Frequently, the rate meter is connected to a strip-chart recorder for scanning. For accurate quantitative work a scaler-timer system is required. With this, a mechanical printer is a big help. Energy-dispersive systems use a multichannel analyzer to record the data with the results displayed on a cathode-ray tube. A Polaroid camera can be used to provide a hard copy of the display.

With the great reduction in cost of minicomputers in recent years, many newer spectrometers are equipped with one of these. The minicomputer can replace such components as the scaler-timer and the multichannel analyzer.

Acoustical noise is a problem that is frequently overlooked in spectrometers. Instead of a mechanical printer for use with a scaler-timer or a mechanical teletypewriter for use with a minicomputer, one of the newer silent printers is strongly recommended.

As mentioned previously, the trend is now to equip spectrometers with minicomputers which can do much of the data manipulation previously done by hand. Along with the obvious advantages of the computer, some problems are introduced. With the reduced cost of computer hardware, the cost of developing programs is becoming more important. For this reason many of the available programs are considered proprietary information and the details of the program are not always available. The user must therefore accept these programs on faith and hope that they provide the correct answers to his problems [8]. There is still no general agreement as to the best procedure for correcting the interelement effects [9]. Thus, programs for all of the presently used techniques are desirable. As these techniques are improved, new programs will be required. A prospective purchaser should investigate the following: (1) Will the vendor provide these new programs and will there be a service charge for them? (2) Will these new programs fit into the size of the computer provided with the spectrometer? (3) How easily are the new programs introduced into the computer?

In the computers on some spectrometers, it is extremely difficult to change the programs. This type should be avoided unless the instrument is purchased only for analysis of materials that can be handled by the system without updating. Similarly, programs written in machine language should be avoided except for routine analyses. A spectrometer manufacturer may offer a choice of several sizes of computers and a variety of peripherals (e.g., disks and magnetic tape cassettes) with his equipment. However, not all of the programs the manufacturer provides will always fit in all of his computers. Close questioning of the salesman is suggested on this point.

A big advantage of a computer-controlled system is that it can operate unattended, freeing the operator for other tasks, such as sample preparation and recordkeeping.

Sample introduction should also be investigated prior to purchasing a spectrometer. This may cause the most trouble in on-line process-control equipment where a process line has to be sampled automatically and continuously. This sampling system may have to be custom made.

An automatic sample changer that can handle many samples is a big timesaver with a computer-controlled system. With this system, the operator can load the spectrometer and then go on to other work.

There are now many manufacturers of X-ray fluorescence equipment, but whether the market is large enough for all to survive is questionable. It would be wise for a prospective purchaser to assure himself that replacement parts and service will be available during the life of the equipment. Service may also be a problem if the vendor's service organization does not have an office close to the equipment. If the system consists of components from several manufacturers, will one service organization handle the whole system? If the vendor has an applications laboratory that will provide help when needed, this is a definite advantage.

9.4 X-RAY SAFETY

Although one does not often hear of an accident involving industrial X-ray equipment, the available statistics indicate that this equipment is potentially hazardous if not operated properly. Jenkins [3] has estimated that there is one accident per year per 200 X-ray installations. Thus, adequate safety precautions are absolutely essential in the operation of any X-ray equipment.

9.4.1 Types of Hazards

It is well known that radiation causes physiological changes in the human body, so that most people are aware that a danger exists. However, it is human nature to become careless when living in the presence of danger. Thus, many accidents occur to people with considerable experience [3]. On the other hand, X rays are very insidious in that the effects of radiation exposure are not observed until considerable time has passed; a person cannot tell when he is being irradiated. This along with the complexity of X-ray equipment makes it imperative that people using the instrument have thorough instructions in X-ray safety.

The unit of measurement of radiation is called the *roentgen* (R), which is defined as that quantity of radiation which will produce one electrostatic unit of quantity of electricity of either sign in 0.001293 grams of air. This is equivalent to about 84 ergs. The roentgen is not a particularly good unit for X-ray safety. For radiation to cause any effect (including physiological effects), it must be absorbed, and air has different absorption qualities than most other materials.

Thus the rad (roentgen-absorbed dose) was defined as the amount of radiation that produces 100 ergs of absorbed energy per gram of absorber. The rem (roentgen-equivalent-man) takes into account the relative biological effectiveness of human tissue. The rem is the unit that is used in X-ray safety. Practically, the roentgen, rem, and rad are all equivalent for the radiation used in X-ray spectroscopy [10]. However, this equivalence definitely is not true for radiation from neutrons or heavily charged particles.

Jenkins and Haas [3] give a good brief summary of the biological effects of radiation. At the time of exposure, there are no biological effects. After a period of time, which may vary from several hours to years, the biological effects appear. Many different effects are observed depending on the dose and the part of the body exposed. Some of these are lowering of the white blood cell count, skin reddening, deep burns, cancer, cataracts, sterilization, and even death. Because no effects are felt at the time of exposure, it is necessary to be extremely careful to prevent exposure (see Sec. 9.4.3).

X-Ray equipment requires several hundred-volt power supplies for the detectors and several thousand volts for the X-ray generator. Thus, special precautions must be used to prevent electrical accidents. High-powered X-ray tubes require water cooling and precautions must be taken to keep the water away from the electrical connections. The author has observed condensation on the outside of water lines on humid summer days and, on at least one occasion, this condensation caused arcing of the high voltage.

The high-voltage power supplies have electrical capacitors which can store energy, so that even after the equipment is shut off, high-voltage static charges may still be present. After shutting off the equipment and before doing any maintenance work in the power supplies, all parts of the equipment should be grounded with a heavy insulated wire to remove these static charges. Note that even a full wave generator (not constant potential) can generate these charges.

Another unusual electrical hazard is that most X-ray spectrometers have several electrical interlocks to prevent radiation leakage. These must be inspected regularly to be sure that they are in good working order. In some of these safety devices, lead is used for shielding. In high humidities, lead can corrode which may cause sticking of movable parts. These also need to be checked regularly.

9.4.2 Legal Requirements for Radiation-Producing Equipment

The United States Government has established minimum safety regulations for use of ionizing radiation [11]. In addition, many states have their own protection codes. Both employers and employees using radiation-producing equipment should be thoroughly familiar with these regulations.

Briefly, these regulations require that any area where an employee has a reasonable chance of receiving more than 0.5 rem per year be labeled as a radiation area. Note that this is equivalent to normal background level. In these radiation areas the employer must provide means for monitoring the amount of radiation received by his employees. With certain exceptions, the employee shall not receive more than 1-1/4 rems per calendar quarter in the critical parts of the body (head and trunk, blood-forming organs, lens of eye, or gonads), more than 18-3/4 rems per quarter on hands, forearms, feet and ankles, or more than 7-1/2 rems per quarter on the skin of the whole body. The employer must keep records of the radiation exposure of his employees.

Probably the best way of providing this monitoring is by the use of film badges. There are several commercial houses that supply film badges, develop the film, and provide records. A list is found in Ref. 3.

Adequate means of preventing unauthorized use of radiation-producing equipment must be provided. This can be done, for example, by keeping the equipment locked out except when it is in use under the supervision of the spectroscopist.

The owner of the equipment also has the obligation to report serious radiation incidents to the appropriate authorities. These reporting requirements as described in the federal regulations [11] and various state regulations vary with the severity of the incident

9.4.3 Safety Attitudes, Habits, Training, and Rules

In this author's opinion there are no substitutes for good safety attitudes, habits, and training. There is no such thing as an absolutely "fail-safe" instrument and the operators must know enough to realize when the equipment is not functioning properly and to take adequate safety measures to correct the problem. Everyone, including supervisors, must know the possible hazards, how to prevent them, how to find them, and what to do in emergency circumstances.

Jenkins and Haas [10] provide an excellent set of operating recommendations for X-ray diffraction and spectroscopy. Every installation should have a copy of these recommendations and follow them. The following provides additional emphasis on some of their recommendations.

Laboratory management (expression used generally for those in positions of responsibility, also in other than industrial organizations) has the responsibility of seeing that the equipment is operated safely. This is done by insisting that adequate safety rules, operating instructions, and emergency procedures are written and enforced. Monitoring and other safety devices must be provided. Record-keeping systems for personnel monitoring, equipment inspections, and equipment modifications must be maintained. Safety instruction in the particular problems of X-ray spectroscopy must be provided. Management must also investigate unusual incidents and comply with federal, state, and local regulations. These management functions are often carried out by the appointment of a radiation protection officer. One very important part of management's responsibilities, not mentioned by Jenkins and Haas, is to project an image of an interest in safety. Even if a radiation protection officer is appointed, management should frequently and informally discuss safety problems, check records, etc., with the operators.

Even though it is management's responsibility to provide a safe working environment, it is the operator who suffers pain in the event of an accident. Thus, he must accept responsibility for his actions. He should know where potential radiation hazards exist and should keep all parts of his body away from those areas. He should know when the equipment is operating properly and when it is not. When it is not operating properly he should ask for help to get the equipment in good working order before continuing to use it. Equipment should not be operated if there is any question as to its safety. The operator should be particularly aware of slight malfunctions that do not apparently affect the performance of the instrument but may provide an indication of a safety device malfunction. He should rigidly comply with all safety rules and regulations. These are for his benefit. He should be familiar with, and use regularly, all monitoring devices provided for checking stray radiation. He should ask for, and receive, adequate instruction in the safety aspects of his job.

One of the most important points in a safety program is to stimulate a continual thinking about safety. The suggestions, given above, of management frequently asking about safety procedures is one way to do this. Various safety acts, such as putting on film badges and writing down results of surveys, also serve this purpose. Any other ways of encouraging safety thinking should be used at every opportunity, especially in a small organization where management also operates the equipment and working procedures are less formal.

Jenkins and Haas [10] give limits for the amount of stray radiation that is considered safe. They point out, and it is reemphasized here, that these are maximum limits. With good modern X-ray equipment, there should be essentially no difference in radiation levels with the instrument turned either on or off, there should be essentially no stray radiation. This operating goal can be attained, and it provides an additional margin of safety.

REFERENCES

1. R. Woldseth, *X-Ray Energy Spectrometry*, Kevex Corp., Burlingame, California, 1973.
2. E. P. Bertin, *Principles and Practice of X-Ray Spectrometric Analysis*, 2nd ed., Plenum Press, New York, 1974.
3. R. Jenkins and D. J. Haas, *X-Ray Spectrometry*, 4, 33 (1975).
4. S. D. Rasberry, in *Advances in X-Ray Analysis* (K. F. J. Heinrich, C. S. Barrett, J. B. Newkirk, and C. O. Ruud, eds.), Vol. 16, Plenum Press, New York, 1972, p. 56.
5. J. V. Gilfrich, P. G. Burkmalter, and L. S. Birks, *Anal. Chem.*, 45, 2002 (1973).
6. J. E. Stewart, H. R. Zulliger, and W. E. Drumond, *Abstracts 24th Annual Denver X-Ray Conference*, in *Advances in X-Ray Analysis* (R. W. Gould, C. S. Barrett, J. B. Newkirk, and C. O. Ruud, eds.), Kendall/Hunt, Dubuque, Iowa, 1976.

7. J. M. Jaklevic, D. A. Landis, F. S. Goulding, *Advances in X-Ray Analysis* (R. W. Gould, C. S. Barrett, J. B. Newkirk, and C. O. Ruud, eds.), Vol. 19, Kendall/Hunt, Dubuque, Iowa, 1976, p. 253.

8. L. S. Birks and J. V. Gilfrich, *Anal. Chem.*, *46*, 360R (1974).

9. R. Jenkins, *Advances in X-Ray Analysis* (R. W. Gould, C. S. Barrett, J. B. Newkirk, and C. O. Ruud, eds.), Vol. 19, Kendall/Hunt, Dubuque, Iowa, 1976, p. 1.

10. R. Jenkins and D. J. Haas, *X-Ray Spectrometry*, *2*, 135 (1973).

11. *Occupational Safety and Health Act Standards, Code of Federal Regulations*, Title 29, Chapter XVII, Part 1910.96.

APPLICATIONS

Chapter 10

THE GENERAL SERVICE LABORATORY

Ron Jenkins

Engineering Department
Philips Electronic Instruments, Inc.
Mahwah, New Jersey

10.1	Features of X-Ray Spectrometric Technique	242
	10.1.1 General	242
	10.1.2 Comparison of X-Ray Fluorescence (XRF) and Other Instrumental Methods	242
10.2	Choice of the X-Ray Method	246
10.3	Types of Sample and Preparation of the Specimen for Analysis	250
10.4	Types of Commercial Instruments and Their Areas of Applications	253
10.5	Typical Application Areas of the X-Ray Spectrometer	256
	10.5.1 Analysis of Cements	257
	10.5.2 Analysis of Alloyed Steels	258
	10.5.3 Analysis of Air Particulates	259
	10.5.4 Analysis of Unused Oils	260
	10.5.5 Forensic-type Applications	261
10.6	Setting of the X-Ray Analytical Method	262
	10.6.1 Selection of Instrumental Parameters	262
	10.6.2 Selection of Calibration Standards	266
	10.6.3 Compare Replicates and Confirm Precision	268
	10.6.4 Evaluation of Matrix Effects	269

10.6.5 Selection of Correction Scheme 270
10.6.6 Long-Term Evaluation of the Method 275
References 275

10.1 FEATURES OF X-RAY SPECTROMETRIC TECHNIQUE

10.1.1 General

The present day analytical chemist has a wide variety of techniques and methods of elemental analysis available to him. In addition to a whole range of classical "wet" techniques, many instrumental techniques are commonly employed, including atomic absorption spectrophotometry, ultra-violet emission spectroscopy, solid-source mass spectrometry, neutron activation analysis, and, of course, X-ray spectrometry. Each of these instrumental methods has its own characteristics and idiosyncrasies, but without exception, each is able to offer high precision, sensitivity, and speed over a wide atomic-number range. These techniques differ mainly in the types of interelement (matrix) interferences and the methods by which these interferences are minimized. It is, in fact, in this area that the X-ray fluorescence technique is particularly powerful since, relative to the other instrumental methods cited above, matrix effects in X-ray fluorescence are readily predictable and fairly easily controllable over wide concentration ranges. The current state of the art in X-ray fluorescence spectrometry allows application of fundamental correction methods to an increasing number of analytical problems, without the need for repetitive recalibration and large numbers of primary calibration standards.

10.1.2 Comparison of X-Ray Fluorescence (XRF) and Other Instrumental Methods

In selecting an analytical technique, the analyst must weight carefully the various advantages and disadvantages which each technique has to offer. Such features as speed, accuracy, sensitivity, and range of elements covered are all important relative to the initial capital cost of the equipment in question. Although each of these factors is in general quite critical, other variables may effect the decision of a given laboratory to select one or more of the methods available. For example, these might include ease of operation of the equipment, breakdown history and serviceability, ability to do both qualitative and quantitative analysis, sample size and form required, the destructive or nondestructive nature of the method, and so on.

Table 10.1 lists the more important features of the major instrumental analytical techniques. In general, a given technique will ultimately stand or fall on its ability to produce a result of

TABLE 10.1

Comparison of Instrumental Techniques for Elemental Analysis

	XRF[a]	UVE[b]	AA[c]	NAA[d]	SMS[e]
Speed	Fast	Fast	Fast	Moderate	Slow
Minimum sample	mg	µg	Few hundredths µg/ml	ng	pg
Accuracy	Excellent	Good	Excellent	Fair	Fair
Sensitivity	Fair	Fair	Good	High	Very high
Range	$Z \geq 9$	Most elements	Most elements	Many elements	Most elements
Quantitation	Fairly easy	Difficult	Easy for single elements	Difficult	Time consuming
Price range ($\times 10^3$ dollars)	50-150	50-150	20-40	100-150	70-150

[a]XRF: X-ray fluorescence.
[b]UVE: ultraviolet emission.
[c]AA: atomic absorption.
[d]NAA: neutron activation
[e]SSMS: solid-source mass spectrometry.

the required accuracy within a given period of time. Of the five techniques compared in Table 10.1, three are commonly employed in the routine analytical laboratory, namely X-ray fluorescence (XRF), ultraviolet emission (UVE), and atomic absorption (AA). Each of these techniques is fast, precise, and potentially accurate. Ultraviolet emission spectrometers are typically multichannel machines capable of very high throughput and are able to produce elemental data on 20-30 elements in less than 1 min. The best specimen form is that of a solid disk or pellet, although techniques are available for the handling of powders or liquids. Electron transitions in the ultraviolet region involve partially filled atomic and molecular orbitals, and as such, characteristic line intensities and interelement influences are difficult to predict. Consequently, the intensity/concentration algorithms which are employed in this field are rather empirical and type standardization is commonly employed. Atomic-absorption spectrometers are typically single-channel instruments capable of exceedingly high throughput on single elements. The best specimen form in this instance is as a solution, although powders may be handled, but with some difficulty. Matrix effects are predictable to a certain extent and are controllable either by dilution or addition of stabilizing ions. Calibration is simple since standard solutions are readily prepared. Machines are now becoming available for the simultaneous analysis of several elements, but the development of multichannel (>10) units is unlikely. X-Ray spectrometers are typically sequential or simultaneous machines which rarely match the speed or sensitivities of the UVE and AA spectrometers. Against this, however, X-ray wavelengths arise mostly from inner orbital transitions, hence the values of these wavelengths are virtually independent of chemical state (see Chap. 8). Matrix problems are easily predictable and readily correctable, thus the X-ray technique is generally the most flexible of the three techniques for wide concentration range, multielement analyses.

In terms of the range of elements covered, each of the three methods is potentially able to cover the greater portion of the atomic table. Restricted elements in AA are generally those which are not readily taken into solution. In the cases of both XRF and UVE, problem elements are generally those where spectral interference by other elements is present. For instance, typical of such interferences in XRF are Pb(Lα) and As(Kα); Mo(Lα) and S(Kα); and Cr(Kβ) and Mn(Kα). The very low (Z < 9) atomic number elements are not easily measurable by the classical XRF spectrometer due mainly to the fact that for these elements, electron transitions producing characteristic lines arise from molecular rather than atomic orbitals. This results in valence-dependent *bands* rather than valence-independent *lines*, with consequent difficulty in accurate intensity measurement. Further to this low fluorescent yields and high absorption reduces the sensitivity of the spectrometer for the longer wavelength X rays. Absorption problems also lead to low penetration (<1 µm) of these longer wavelengths making specimen preparation extremely critical. Although sufficient sensitivity is probably available for carbon, oxygen, and perhaps nitrogen, attempts are rarely made to make quantitative determinations of these elements.

The X-ray fluorescence spectrometer is reasonably sensitive both in terms of its ability to determine low concentrations in specimens of unlimited quantity, as well as in limited quantities of material. In the former case, detection limits down to the low ppm range are common for all but the lowest atomic number elements and absolute sample sizes down to a few milligrams can be handled without too much of a problem, particularly where the specimen can be spread thinly over an area of several square-centimeters. This sensitivity is similar to that which can be obtained with the UVE spectrometer and approaches that of AA. None of these techniques, however, come close to the sensitivities obtainable with the neutron activation or mass spectrometric methods.

Two additional features, which are almost unique to the X-ray fluorescence method, are also of great importance to the general service laboratory. The first of these is the great ease with which qualitative or semiquantitative analyses may be performed by the sequential X-ray fluorescence spectrometer. The full range of elements from fluorine upwards can be scanned in about an hour with the data being presented in the form of an easily interpreted wavelength/intensity pattern. Semiquantitative information can be obtained by direct comparison of characteristic wavelength intensities from the chart, along with some prior knowledge of the wavelength/sensitivity relationship (usually a smooth and fairly continuous function) of the spectrometer. The second unique feature of the XRS method is the low probability of the less experienced operator in making a gross interpretation error. X-ray spectra are unique and relatively uncomplicated. Spectral overlap is uncommon and the characteristic line intensity/element concentration relationship, although certainly not linear, varies by factors of two or three, even over very wide concentration ranges, rather than by orders of magnitude as in some of the other instrumental techniques. Thus, serious qualitative interpretation errors are rare and excessive mistakes in quantitation are easily avoided.

10.2 CHOICE OF THE X-RAY METHOD

In any instrumental analytical method, four major steps are commonly encountered:

1. Preparation of the specimen to be analyzed from the sample submitted for analysis
2. Setting up of the instrumental conditions to give optimum response and sensitivity for the parameter to be measured
3. Collection of instrumental data of sufficient precision
4. Conversion of the instrumental data to quantitative elemental information with the required accuracy

When choosing or setting up an analytical scheme where accuracy and/or time are of importance, it is necessary to establish which of the four stages is the "rate-determining-step" in the cycle. For example, a high-cost, high-speed, computer-controlled spectrometer giving count/concentration data in less than a minute is hardly justified where specimen throughput is low, or where specimen preparation time severely limits throughput. (See also Chap. 9.)

Figure 10.1 shows the typical stages required in the setting up of an X-ray method. Initially, a selection must be made of the type of spectrometer to be employed, and where the types of problems to be handled are varied, some compromise in speed, sensitivity, and cost will have to be reached. This will be discussed in later sections. Since the spectrometer must yield both sufficient sensitivity and adequate precision, some guide lines and rules of thumb are useful. The concentration of a given element should be related to the *net* intensity of the characteristic line in question. Measurements may have to be made at the peak-maximum position (I_p) as well as at the background (I_b). The net counting error S_n is given by

$$S_n = \frac{100}{\sqrt{t}} \frac{1}{\sqrt{I_p} - \sqrt{I_b}} \qquad (10.1)$$

where t is the total counting time spent on both peak (t_p) and background (t_b) positions. Where t is limited, it may be useful to divide this optimally between peak and background such that

$$\frac{t_p}{t_b} = \sqrt{\frac{I_p}{I_b}} \qquad (10.2)$$

(see also Chap. 4).

Equation (10.1) tells us two things: firstly, that where t is limited, S_n will be at a minimum where $\sqrt{I_p} - \sqrt{I_b}$ is maximum, and for this reason this factor may be used as a "figure of merit" for establishing optimum equipment parameters. Secondly, as I_p/I_b increases, $(\sqrt{I_p} - \sqrt{I_b}) \to \sqrt{I_p}$, in other words, the influence of the background on S_n becomes less important as the peak-to-background ratio increases. For most practical purposes, the background can be ignored where the peak/background ratio is greater than 10.

Most quantitative methods will require calibration standards for establishing equipment sensitivity for an analyte line, for

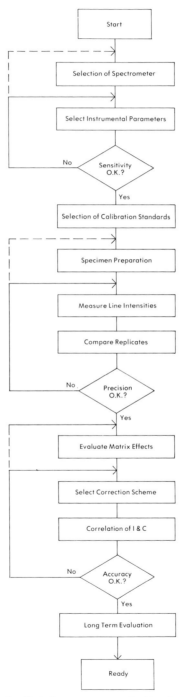

FIG. 10.1 Setting up an analytical method.

establishing the form of the intensity/concentration relationship and to overcome long-term instrumental drifts. Careful choice of standards is mandatory since it is vital that they should truly reflect the likely concentration ranges of *all* matrix elements.

Specimen preparation techniques are dealt with in more detail later; at this time suffice it to say that the specimen must be both representative and homogeneous over its analyzed volume.

Replicate determinations are the best measure of the precision of the analytical method, and where sufficient counts ($>10^6$) are accumulated, this precision should approach the potential precision of the spectrometer which is typically in the range 0.05-0.25%. When (and only when) sufficient precision has been demonstrated, attempts can be made to relate the characteristic line intensity with analyte concentration. In the first instance, this may be a preliminary study of absorption-edge and mass-absorption coefficient data to establish likely interferences. Secondly, this will probably involve some graphical plot of intensity vs. composition, or perhaps a regression analysis, to establish the form and magnitude of any deviation from linearity. Finally, some matrix correction scheme must be selected and tried. Table 10.2 lists some of the commonly employed matrix correction procedures for multielement and single-element analyses. Examples of the application of these methods will be given at the end of this chapter.

The ultimate test of the applicability and success of a given matrix correction procedure is to relate the error associated with the total analytical scheme with the precision of the measurements. For example, if a series of measurements has been made on a number of specimens and the standard error between X-ray and chemical results has been established, the successful matrix correction model will be that in which this standard error approaches the precision of the measurement, i.e., typically a few tenths of one percent. Experience indicates that typical standard errors S_T follow the form

$$S_T = K\sqrt{C + 0.01} \qquad (10.3)$$

where C is the average concentration of the analyte and K is a constant

TABLE 10.2

Matrix Correction Procedures

---

1. Multielement techniques
   a. Fundamental-parameter techniques: require few standards but a fair degree of computation; correction factors may be calculated; typical computer size, >32K.[a]
   b. Semiempirical methods: require more standards but less computation; correction factors determined by multilinear regression analysis or graphical methods; typical computer size, ≈8K.
   c. Type standardization: requires large numbers of standards but little or no computation.
2. Single-element techniques
   a. Internal standards: requires one internal standard per element; time consuming; best applied as a "spiking" method, where more of the analyte element itself is added to the specimen used as the standard.
   b. Use of scattered tube lines: simple to apply but rather empirical; the intensity of a scattered source line is roughly inversely proportional to the matrix absorption.

---

[a]1K means usually 1000 words of computer memory. Most computers use 16-bit words, but this is not universally true.

ranging between 0.005 and 0.05. "Good" XRF analysis would be that with a $K \leq 0.01$ and "poor" XRF analysis that with a $K > 0.05$.

The final stage in the setting up of the method is long-term evaluation of the precision and accuracy of the method.

10.3 TYPES OF SAMPLE AND PREPARATION OF THE SPECIMEN FOR ANALYSIS

The form of the sample which is submitted for analysis may range from an irregularly shaped solid piece, a heterogeneous powder mixture, a solution containing floating solid matter, to a single-phase solution. In nearly all cases, some preparation of the sample will be required, usually to render it homogeneous and representative over the relatively shallow penetration of the X-ray beam. It is frequently assumed that such preparation does nothing more than homogenize the sample, but such an assumption may be dangerous. Contamination of a powder sample in a grinding mill and loss or gain of water and/or carbon dioxide are not uncommon. Similarly, surface preparation of a bulk metal ingot may cause surface contamination from the grinding paste or by selective smearing of softer constituents of the alloy over the surface. Although it is to be hoped that such problems be avoided, or at least minimized, in sample preparation, the analyst must be aware of potential problem areas, and to this extent, it is useful to differentiate between the sample (the original material submitted for analysis) and the specimen (the prepared sample actually placed in the spectrometer). Thus, a well-prepared specimen is not only homogeneous to the penetration depth of the X-ray beam, but also its analyzed volume is identical in composition to the average composition of the sample.

Figure 10.2 illustrates the more important specimen preparation technique as applied to bulk samples, powders, and solutions. Bulk samples such as cast billets, sections of pipe, investment castings, and turnings, all need to be transformed into a disk, typically 3-5 cm in diameter and 0.5-2 cm thick. The penetration of X rays is roughly inversely proportional to the mass absorption coefficient of the specimen and such penetrations are typically in the 1-50 µm range, with the longer wavelengths (lower atomic numbers) having the smaller penetration. Thus, specimen preparation is most critical in the determination of low-Z elements. Where the sample is to be cast, for example, in taking a sample from a steel furnace, a mold of approximately the right shape can be employed, then the cast sample can be cut, and the surface ground. Metal turnings and irregularly shaped objects can be melted and cast into a similar ingot or taken into solution and analyzed as a single-phase liquid.

Solutions are, in general, ideal specimens for analysis since they are homogeneous and generally have low absorption coefficients, hence have large penetration depths. Dilution of the specimen also

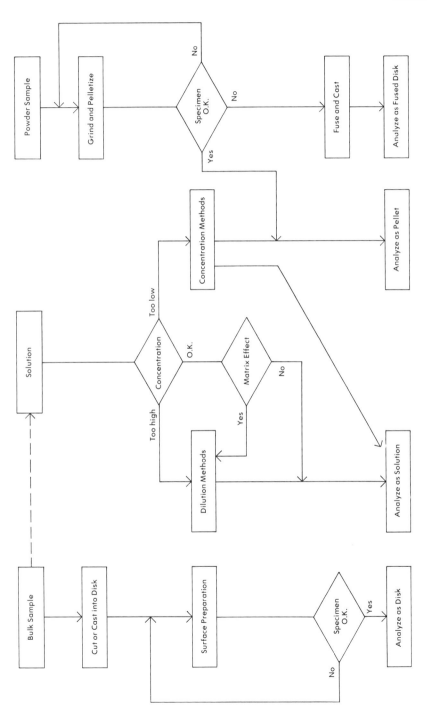

FIG. 10.2 Specimen preparation techniques.

tends to eliminate matrix effects because the nonanalyzed solvent tends to eliminate the total absorption of the specimen. Too high a concentration can cause loss of sensitivity (due to excessive self-absorption) and/or large matrix effects. In either case, dilution of the specimen with the inert solvent will reduce the problem. The major difficulty in the analysis of solutions is where the concentration of the analyte element is too low. Such a situation will generally arise in the low or even sub-ppm range and in this instance, some form of preconcentration is necessary. Ion-exchange resins [1] and resin-impregnated filter papers [2], evaporation [3], and coprecipitation [4] have all been successfully employed in that area.

By far the most difficult form of sample to analyze with the X-ray spectrometer is the powder sample. Materials such as cement raw mix, ores, minerals, refractories, and whole rock tend to be heterogeneous over the micrometer range and very careful grinding and pelletizing may be necessary to produce a good specimen. Such preparation may be hindered by the presence of grains of varying hardness and size and in extreme cases it may be necessary to fuse the sample with a flux to form a homogeneous melt which can be cast into a disk or solidified, ground, and pelletized. Lithium and sodium tetraborates have both found great use as fluxing agents and are typically used in the ratio of 1:5 to 1:20 sample to borate.

10.4 TYPES OF COMMERCIAL INSTRUMENTS AND THEIR AREAS OF APPLICATIONS

There are probably in excess of 10,000 X-ray spectrometers in the world representing a wide range of types of instruments from about 6 major and about 15 minor manufacturers. Two major categories exist, namely, wavelength-dispersive systems (see Chap. 1) and energy-dispersive systems (see Chap. 2). Subdivisions are present in each category and Table 10.3 lists the features of each of the major categories. In the wavelength-dispersive systems, these divisions would include sequential systems and multichannel systems. Sequential spectrometers are able to measure one wavelength, i.e., one element

TABLE 10.3

Comparison of X-Ray Emission Methods for Elemental Analysis

	X/EDS[a]	SF/EDS[b]	P/EDS[c]	SC/WDS[d]	MC/WDS[e]
Speed[f]	3 min	5 min	5 min	8 min	1 min
Sensitivity (ppm)	5-500	0.5-20	1-50	0.5-50	5-100
Smallest sample size	100 µg	10 µg	1 µg	10 µg	1 mg
Accuracy	0.5%	0.2%	1%	0.2%	0.2%
Price range (10^3 dollars)	20-60	50-80	?	40-120	80-150

[a]X/EDS: X-ray excited energy-dispersive spectrometry.
[b]SF/EDS: secondary fluorescer/energy-dispersive spectrometry.
[c]P/EDS: proton-excited/energy-dispersive spectrometry.
[d]SC/WDS: single-channel/wavelength-dispersive spectrometry.
[e]MC/WDS: multichannel/wavelength-dispersive spectrometry.
[f]For the analysis of 6 major and 4 trace elements.

at a time, and contain typically two to five analyzing crystals, two collimators, and two detectors, each of which may be selected manually or, in an automatic instrument, under program control. Source conditions such as X-ray tube voltage and current are also variable, thus allowing a good range of excitation and measurement conditions over a wide wavelength range. Sequential systems are relatively slow, however, since the total measurement time is the sum of each of the individual measurements times.

Multichannel wavelength-dispersive systems contain a number of fixed channels, the maximum number being around 25. Each of these channels can, in principle, be tailor-made to satisfy the optimum measurement conditions for each measured wavelength. At first sight, this might suggest improved sensitivity over the sequential systems, but in practice, any such gain in sensitivity is more than lost due to the decrease of the closely coupled optics of the sequential systems because of physical and mechanical limitations. Multichannel systems are extremely fast since all elements are being measured simultaneously. Against this, however, they are less flexible and more costly than sequential spectrometers. More modern multichannel systems may include a sequential (scanning) channel for qualitative identification of unexpected elements in a routine control analysis, or for the analysis of occasional samples for which the spectrometer is not calibrated. Multichannel spectrometers generally employ curved analyzing crystals since sensitivity is generally at a premium. Flat crystals are not unknown in multichannel spectrometers, but such crystals are generally better suited to sequential systems.

Both categories of wavelength-dispersive spectrometers may be available with multiple specimen-handling facilities and/or an integrated digital computer which may serve the dual function of spectrometer control and data manipulation. Wavelength-dispersive spectrometers are mostly X-ray fluorescence spectrometers, i.e., they incorporate a primary X-ray photon source, and have been commercially available since the early 1950s. Wavelength-dispersive spectrometers are more wide-spread than energy-dispersive systems, probably in the ratio of 10:1. A similar ratio exists for the numbers of sequential wavelength-dispersive systems relative to multichannel systems. These figures are, of course, subject to change due to instrumental developments and new applications.

Energy-dispersive spectrometers are much more recent in terms of their use as a routine analytical tool. Instruments started to become available in the late 1960s and the first major analytical application was in the area of electron column instruments (EPMA, SEM; see Chap. 7). Stand-alone energy-dispersive spectrometers came on the scene in the early 1970s and these earlier units mostly contained low-power X-ray tubes as the excitation source. They were fast and flexible in addition to being very cost-competitive with their wavelength-dispersive counterparts. Their ability to do high-speed qualitative as well as quantitative analysis made them very attractive to the general purpose analytical laboratory and their popularity increased rapidly over a very short time period.

Problems occurred, however, particularly in those cases where high-precision or wide-energy range calibration was required. These problems were almost all due to the fact that the earlier EDS systems were very count-rate-limited since the total photon flux at the detector was due to *all* the excited elements in the specimen at any one time. Newer EDS spectrometers incorporate secondary excitation sources which are able to give good selective excitation over a limited energy range and although the high inherent speed of the EDS systems is somewhat reduced in this configuration, greater sensitivity and precision are obtained.

A third type of EDS system is also of increasing importance, particularly in the analysis of airborne particulates. This is the energy-dispersive spectrometer based on the proton source (Chap. 6). Interest in this area started in the United States in the late 1970s when a cutback in research funding terminated several fundamental research projects involving linear accelerators. About 500 of these machines are currently available in the United States. Projects were sought which would have a direct bearing on current rather than long-term problems. All of this coincided with a peaking in public awareness of pollution and ecological problems and this appeared to represent a very potential area for research. Using a proton source and an energy-dispersive spectrometer, very high sensitivities can be obtained on extremely small specimens making this a particularly powerful tool for the analysis of airborne particulates [5].

10.5 TYPICAL APPLICATION AREAS OF THE X-RAY SPECTROMETER

Uses of the X-ray spectrometer in the general service laboratory are best illustrated by a few typical examples. The following have been selected to demonstrate the typical circumstances under which a certain type of spectrometer might be chosen as well as the rationale behind the choice of an actual analytical scheme.

10.5.1 Analysis of Cements

Finished cements and raw mixes contain typically Ca, Si, Al, and O at high concentration levels, plus Fe, K, Mg, and Na in the low percentage levels. Of these elements, all but oxygen are potentially determinable with the X-ray spectrometer. One of the major problems in accurate cement analysis is in obtaining a homogeneous specimen, particularly in the case of raw mixes, where the source of raw material may be variable. Most of the elements to be determined are of low atomic number, hence the penetration of their characteristic lines will be of the order of a few microns only. Careful grinding and pelletizing will suffice in perhaps 80% of the cases, but in the remainder, fusion with lithium tetraborate may be necessary. The consequent dilution will further lower the concentrations of sodium and magnesium, making the determination of these two elements particularly difficult.

Possible matrix effects are minimal and probably the only major interference is that of potassium on calcium. Matrix correction procedures are thus minimal, and a simple intensity correction will be sufficient, i.e., an integrated computer would be an expensive luxury.

One problem in applying the energy-dispersive spectrometer to this type of analysis is that the close proximity and low atomic numbers of most of the matrix elements causes severe line overlaps, and computer deconvolution of the lines is mandatory. This causes some loss in precision with the result that the precision of the EDS data will be about a factor of two worse than the WDS data. Using the wavelength-dispersive technique, the multichannel WDS-type system is ideally suited to this type of analysis and would be the optimum choice where specimen throughput is high. This approach is, however, expensive, thus the final choice of instrumentation would be mainly one of specimen throughput with the following alternatives:

Spectrometer	Data acquisition time (sec)/sample	Accuracy (%)	Cost (10^3 dollars)
Multichannel WDS	40	0.15	110
Sequential WDS	400	0.2	50
Bremsstrahlung source EDS	200	0.3	70

Typical sample preparation time: pellets, 2-4 min/sample and fused beads, 5-10 min/sample.

10.5.2 Analysis of Alloyed Steels

An "alloyed steel" may include a very wide range and variety of different types, so for the purpose of this discussion, we will consider a typical situation of alloy types based on major (>10%) concentrations of Fe, Cr, and Ni, minor (0.5-5%) concentrations of Si, Mn, Nb, Zr, Mo, Ta, and W, and trace (<0.1%) concentrations of Al, P, S, As, and Pb.

Specimen preparation in this particular case is not too critical and a surface finish equivalent to about 600 grit would suffice. Surface contamination by the grinding medium should be avoided with care being especially important in the case of aluminum and silicon (therefore, grind with boron carbide or diamond paste). Matrix effects, on the other hand, are extremely prevalent and a rather sophisticated matrix correction procedure is necessary. Really, only two courses of action are open to the potential analyst, either to go for type standardization, which would require large numbers of standards (tens to hundreds, depending upon the range and number of alloy types), or to use a fundamental-parameter (or perhaps a sophisticated semiempirical) technique, making the use of a computer mandatory.

The energy-dispersive spectrometer would be the cheapest approach, but the range of elements and concentrations would demand the most flexible of the EDS approaches, i.e., the secondary fluorescer approach. The multichannel X-ray spectrometer would be good for this type of work but would be the most expensive. In addition, this type of machine is not too good for the very low concentrations of elements like aluminum (typically 0.005%) and arsenic (typically 0.003% and interfered with by lead). If time is not critical, the best solution would be to use the sequential, computer-controlled WDS system. The final decision would be based on cost, throughput of samples required, turn around time of one sample, and the number of standards available. Again, three alternatives are viable:

Spectrometer	Data acquisition and calculation time (sec)/sample	Accuracy (%)	Cost (10^3 dollars)
Sequential WDS	500	0.15	120
Multichannel WDS	60	0.15	180
Secondary fluorescer (EDS)	300	0.25	90

Typical sample preparation time: 2-4 min/sample.

10.5.3 Analysis of Air Particulates

Air particulates will generally be collected on filter membranes and will be presented to the spectrometer as a rather heterogeneous specimen both in terms of composition as well as particle-size distribution. The total mass of specimen is likely to be small and probably in the milligram range. Although the range of elements to be analyzed may be quite varied, within a given study some rough idea of the number of more important elements will generally be known. On the other hand, the presence of a few nonexpected elements is not only likely, but their presence could be very important. This more than anything else makes analysis of air particulates an ideal application for the energy-dispersive spectrometer. High analytical accuracy is not important and data within the range of ±5% would probably be sufficient. High throughput is also likely to be important since any meaningful study must include large numbers of samples to minimize the sampling error.

Since the specimen is more or less of the form of a thin film, interelement effects are not likely to be too critical, so sophisticated matrix correction procedures are unnecessary. On the other hand, the specimens are bound to exhibit problems related to particle size for which some correction should be made [6].

Probably the best choice of instrument for this type of problem is the energy-dispersive spectrometer using a variable secondary fluorescer source and, preferably, multiple specimen handling facilities. An on-line minicomputer would be invaluable for line-overlap corrections, particle distribution corrections, and data logging. The multichannel WDS system can be used in those cases where the range of elements sought is well defined. The scanning WDS system would be a poor third alternative due to its slow speed.

Spectrometer	Data acquisition and calculation time (sec)/sample	Accuracy (%)	Cost (10^3 dollars)
Secondary fluorescer (EDS)	300	5	90
Multichannel WDS	100	5	130
Sequential WDS	1000	5	60

Typical specimen preparation time: 60 sec, assuming particles are already collected on membranes.

10.5.4 Analysis of Unused Oils

Unused oils are usually analyzed for additive elements including Ba, Zn, Mn, Ca, P, and Cl, plus naturally occurring elements including S and Na. In blended stocks, the concentration level of these elements would typically lie within the range 0.01-2.5%. A special problem is that this analysis will be performed in the liquid phase and a helium atmosphere must be used. Large matrix effects are likely because of the variable concentration levels of relatively heavy elements in a very low average atomic-number matrix.

High accuracy is vital in this work because the cost of the additive is a significant fraction of the total cost of the blended oil. Where a wide range of oil types are to be analyzed, a computer will be almost mandatory for matrix correction.

The requirement in this instance is for high accuracy with a reasonably fast specimen throughput. This is an ideal situation for the computer-controlled sequential WDS system. The multichannel WDS and EDS systems are both poor alternatives. The first because a range of oil types is likely to have a large variation in the combination of elements and the multichannel WDS system does not have this flexibility. The accuracy obtainable with EDS, over the atomic-number range required, is probably marginal in most cases.

Spectrometer	Data acquisition and calculation time (sec)/sample	Accuracy (%)	Cost (10^3 dollars)
Sequential WDS	300	0.1	90
Multichannel WDS	60	0.1	125
Secondary Fluorescer (EDS)	200	0.25	80

10.5.5 Forensic-type Applications

This type of application is likely to be very variable and although it may include much well defined, accurate type standardization, it will more typically include the analysis of small samples for a wide range of elements, usually on a semiquantitative basis. Unexpected elements are of great importance and for this type of work, the multichannel WDS system must be excluded. The scanning WDS system is relatively slow and although a relatively cheap "manual" system will probably suffice, the EDS system with bremsstrahlung source is faster and more versatile. Several "macro"-type EDS systems are available at relatively low cost and these are ideal for this type of work. The choice in this case is really twofold.

Spectrometer	Typical semiquantitative analysis time/sample	Cost (10^3 dollars)
Bremsstrahlung source EDS	300 sec	30
Sequential WDS	1 hr	45

10.6 SETTING OF THE X-RAY ANALYTICAL METHOD

The various steps in setting up an analytical method have already been mentioned and listed in Fig. 10.1. Using as an example one of the typical application areas already discussed, we shall now go through the various steps in detail and explore each of these with the goal of avoiding potential pitfalls and achieving the optimum analytical method. The example chosen is that of the analysis of unused lubricating oils. The pros and cons of the choice of the spectrometer have already been discussed and the optimum choice appeared to be the sequential wavelength-dispersive system. The starting point for this exercise, therefore, will be "given a sequential wavelength-dispersive spectrometer, how do we go about setting this instrument up to perform the analysis of unused lubricating oils for the elements Ba, Zn, Ca, Cl, S, and P?"

10.6.1 Selection of Instrumental Parameters

There are three essential points to be considered in the selection of the optimum instrumental parameters, namely excitation conditions, dispersion conditions, and counting conditions. In each of these three areas, deference must be given to the selected wavelengths for each analyte element. Of the six elements in question, four would be classified as light elements (P, S, Cl, and Ca) and only one of the six, Ba, offers any real choice in the selection of the analytical wavelength. In the case of barium, a choice can be made between one of the K-series lines (for example, Ba Kα = 0.387 Å) or one of the L-series lines (for example, Ba Lα = 2.78 Å). As far as excitation conditions go, it is generally best to choose an X-ray tube which is optimum for long-wavelength (for example, Cr, Ag, or Rh anode) or short-wavelength (for example, W, Au, or Pt anode) excitation. By choosing the Ba Lα line, one can aim for overall optimum long-wavelength excitation by selection of, for example, a chromium-anode X-ray tube. Choice of the Ba Kα would also be unwise for two additional reasons. Firstly, the relatively light matrix could result in a large penetration of the X-ray beam, perhaps up to about 1 cm. This, in turn, might lead to a geometric constraint (shielding) due to the fact that the primary collimator is not able to "see" the whole excited volume of the specimen. Secondly, since the excitation potential V_c of Ba Kα is fairly high (about 37 keV, for high-voltage generators operating at a maximum of 50-60 kV), enhanced instability due to the inherent voltage fluctuation ΔV may result. It can be shown that the fractional change in intensity $\Delta I/I$ is related to the tube potential V by the relationship

$$\frac{\Delta I}{I} = Kx \frac{1.6 \Delta V}{V - V_c} \tag{10.4}$$

where K is a constant. A typical value for ΔV would be in the range 0.02% to 0.1%, and use of a low overvoltage $(V - V_c)$ can triple this value in terms of $\Delta I/I$. One normally aims to achieve a value of V that is 3-5 times V_c.

As far as dispersion is concerned, the initial aim is always to use an analytical line for each element which is as free as possible from spectral interference. In this instance, the only case where this proves difficult is that of phosphorus, the Kα line of which is partially overlapped by a second-order calcium Kβ line. Although, in principle, the influence of calcium on phosphorus should be completely removed by pulse-height selection, it is found, in practice, that the escape peak, which arises when the calcium Kβ radiation excites argon K radiation from the counter gas, partly interferes with the phosphorus Kα pulse-amplitude distribution and this effect is illustrated in Fig. 10.3. Here it will be seen that although the pulse distributions due to P Kα and Ca Kβ are reasonably well resolved, the escape peak associated with the calcium radiation

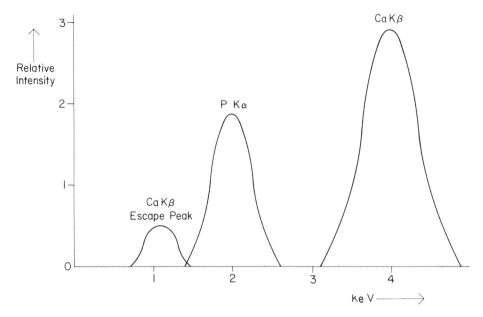

FIG. 10.3 Pulse amplitude distribution of P Kα and Ca Kβ [7].

lies very close to the P Kα distribution. Care must be employed in the setting up of the pulse-height selector particularly since the resolving ability of the pulse-height selector may be greatly reduced by bad detector resolution and/or pulse-amplitude shifts.

Selection of counting conditions is predicated on the sensitivity (i.e., count/sec/%) obtained for each element in question, the concentration range to be covered, and the analytical precision sought. The essential aim is to ensure that the error due to the statistics of counting does not represent the limiting error in the total analytical error.

The data given in Table 10.4 might represent a typical set of operational conditions for the given problem. Note that the spectrometer needs to be operated under an atmosphere of helium since the longer wavelengths are strongly attenuated by air and the volatile nature of the oil, plus potential problems of dissolved air generally precludes the use of vacuum. Table 10.5 lists typical values of counting rates and analysis times predicated by the

TABLE 10.4

Optimum Instrumental Parameters for Lubricating Oil Additives[a,b]

Element	Analyte line	Wavelength (Å)	Crystal	Collimator	Detector[c]
P	Kα	6.155	P. E.	480 µm	Flow
S	Kα	5.373	Graphite	480 µm	Flow
Cl	Kα	4.729	Graphite	480 µm	Flow
Ca	Kα	3.360	LiF(200)	480 µm	Flow
Zn	Kα	1.437	LiF(200)	160 µm	Scint.
Ba	Lα	2.775	LiF(200)	160 µm	Flow + scint.

[a]Cr-anode tube at 50 kV, 60 mA.
[b]Helium flush.
[c]Pulse-height selection applied in all cases.

TABLE 10.5

Concentration Ranges of Standards

Element	Range (%)	Required precision 2σ	Typical values		
			Count/sec/%	Bkg. count/sec	Counting time (sec)
P	0.1-2	1 ± 0.005	3000	60	53
S	1.5-3.5	2.5 ± 0.01	8000	80	31
Cl	0.05-0.5	0.25 ± 0.002	9000	80	28
Ca	0.1-2	1 ± 0.005	16000	120	10
Zn	0.1-2	1 ± 0.005	11000	100	14
Ba	0.05-0.3	0.15 ± 0.002	14000	90	11

required precision at the mid-concentration range. The counting times were calculated using Eq. (10.1). Background can be ignored in all instances since peak-to-background ratios are always in excess of 10:1 except for the very low end of the phosphorus concentration range.

10.6.2 Selection of Calibration Standards

Where some prior knowledge of a potential problem is available, some workers prefer first to select a correction scheme and then go on to select standards for sensitivity, drift correction, matrix evaluation, or whatever is required. In this example, it is assumed that no prior knowledge of the problem is available.

Calibration standards should always be selected such that they reflect the full concentration ranges of all elements to be analyzed. As an example, Table 10.5 indicates the concentration ranges to be covered and at first sight, it might seem that standards could be made by carefully preparing one master standard containing all elements at their maximum concentration ranges, then making perhaps 10 secondary standards by diluting aliquots of the master standard in the ratios 1:1, 1:2, 1:3, ..., 1:10, using an inert diluent (e.g., the base stock of the lube oil in question). Such a procedure would provide eleven standards of steadily decreasing additive element concentration. Unfortunately, these standards would, in fact, be most unsatisfactory for the calibration procedure, because they do not reflect *all* possible matrix variations. For example, both barium and zinc have a very strong absorption effect on phosphorus and each of these elements would tend to suppress the response for P Kα. Since in our 11 standards, all elements increase in concentration together, the calibration curve for P Kα would be of steadily decreasing slope as indicated in Fig. 10.4, curve a. In actual practice, a specimen submitted for analysis might contain low concentrations of zinc and barium, but a high concentration of phosphorus. Use of any calibration scheme based on the use of curve a would then give a high and completely erroneous phosphorus concentration. In this particular instance, the true relationship between measured intensity and concentration for phosphorus is represented by the wedge-shaped area in Fig. 10.4, bounded by the curves a and b. Curve b represents the curve of least absorption and maximum enhancement.

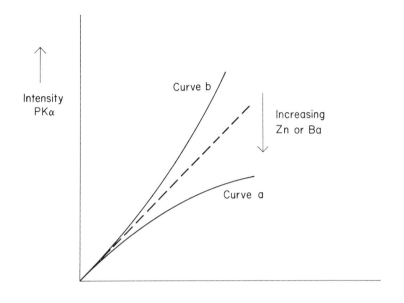

FIG. 10.4 Effect of zinc and barium on the calibration curve for phosphorus.

A far better selection of standards could be made by taking a more or less random selection of each matrix element, e.g., high barium/high zinc/low phosphorus, low barium/medium zinc/high phosphorus. Preparation of standards of the selected concentration ranges is relatively easy in the case of lube oil additive elements since all of these are available as organometallic compounds such as naphthanates and phenates, each of which contain the required element at very high concentration. Thus, preparation of standards is simply a question of blending the appropriate weight of the high-concentration organometallic, with the lube oil-base stock.

Preparation of specimens for analysis is again a very simple procedure; all that is required is the taking of an aliquot of the specimen and placing this in a cell fitted with a thin plastic window. Care must be taken to exceed the critical depth of the characteristic radiation which approximates to $2/(\mu/\rho)\rho$ cm, where ρ is the average specimen density and μ/ρ is the mass absorption coefficient of the *lowest* absorbing specimen to be encountered. In our case, that defined by the lowest concentration levels in Table 10.5 for the shortest wavelength (i.e., Zn Kα) represent this lowest absorbing specimen. The lowest absorption value is then about 10 cm^2 g^{-1}; with a density of about 1 g/cm^3, the critical depth would be about 2 mm.

10.6.3 Compare Replicates and Confirm Precision

A typical midrange standard should be selected and four aliquots placed in different specimen holders. About 25 separate measurements should be taken on all analyte elements in each of the four aliquots, collecting the number of counts indicated in Table 10.4, i.e., count/sec/% × percent concentration × analysis time. In each measurement, it is vital that the complete analytical cycle be executed, i.e., removal, insertion, helium flush, angle selection, counting, etc. Standard deviations should then be calculated for each element and each aliquot, using the appropriate 25 data. Standard deviations from each data set should then be compared and an evaluation made.

As an example, assume that 1.54×10^5 counts were taken on the Zn Kα line in a ratio measurement yielding a theoretical standard deviation of 0.36%. If the four aliquots yielded values of 0.35%, 0.38%, 0.38%, and 0.54%, it would be apparent that something is seriously wrong with the fourth aliquot. A similar bad result on all elements for this particular aliquot would immediately indicate a problem such as a poorly seating sample holder, a sagging cell window, or something similar. Another effect might be consistently poor results on all of the lower atomic number elements, and this might indicate insufficient flushing with helium or problems due to variable cell-window thickness; again, consistently poor results on shorter wavelengths might indicate an insufficient depth of specimen; and so on.

10.6.4 Evaluation of Matrix Effects

Two approaches should be used for the evaluation of potential matrix effects. First, where calibration standards have been selected and run, a plot of raw characteristic-line intensity against analyte concentration will give an immediate indication of potential absorption and enhancement effects, particularly since, by this time, some knowledge of the precision of each datum point is available. Second, study of a list of mass absorption coefficients for each analyte wavelength will quickly reveal strongly interfering elements. Table 10.6 shows such a set of values. It will be remembered that the slope of a calibration curve is roughly inversely proportional to the total absorption of the specimen, of which secondary (characteristic line) absorption is the dominant effect. The total secondary absorption μ_T of a matrix of n elements is given by

$$\mu_T = \sum_{j=1}^{n} \mu_j(\lambda_i) W_j \qquad (10.5)$$

where $\mu_j(\lambda_i)$ is the mass absorption coefficient of matrix element j for the analyte wavelength λ_i, and W_j the weight fraction of element j. Potentially large interferences will be those where $\mu_j(\lambda_i)$ is

TABLE 10.6

Mass Absorption Coefficients

Element	P Kα	S Kα	Cl Kα	Ca Kα	Ba Lα	Zn Kα
P	260	2400	1600	660	390	60
S	355	245	2000	770	460	75
Cl	450	310	225	880	510	81
Ca	790	550	400	160	800	140
Ba	2250	1620	1180	495	300	290
Zn	2220	1600	1220	510	300	50
C	265	175	122	44	26	4
H	10	7	5	2	1	0.2

large, or small, relative to the average matrix absorption, especially in cases where the range of W_j is large. As an example, in the determination of P Kα, the base hydrocarbon matrix C_nH_{2n+1} has a secondary mass-absorption coefficient of about 250 cm^2/g. Both barium and zinc with absorption values in excess of 2,000 are obviously potential interfering elements.

It is also useful to establish potential enhancing elements, even though in practice enhancement effects are invariably less significant than absorption effects. Strong enhancers will be those elements with strong characteristic lines immediately to the short-wavelength side of the absorption edge of the analyte element.

10.6.5 Selection of Correction Scheme

A brief literature search is always useful as a first step in the selection of a matrix correction scheme since it obviously makes little sense to repeat a detailed study of a given problem where this has already been evaluated.

During the last several years, a considerable number of papers have been published describing various methods based on X-ray fluorescence for the analysis of lube oil additives. The more important of these were reviewed a few years ago by Toft [8], and Table 10.7 summarizes the relative merits of each. Briefly, the external standard method [9,10] is rapid but can only be used over relatively small concentration ranges. The internal-standard method [11,12] is potentially very accurate, but since internal standards have to be added for each of the six elements in question, the time involved in sample preparation is considerable. The absorption method [13] is applicable only to fairly simple one- or two-element systems, and the thin-film method [14] is no longer applied in cases where large variations in matrix absorption occur over the calibration range to be covered. Methods based on matrix compensation using coherently and incoherently scattered tube lines [15,16] have been employed with success in light hydrocarbon mixtures, but since these are always established in a rather empirical way to date, no published

TABLE 10.7

Potential X-Ray Methods for the Analysis of Oil Additives

Method	Elements covered	Advantages	Disadvantages
External synthetic [9] standards	Ba, Ca, Zn	Rapid	Only works over narrow concentration range
External synthetic [10] standards	S, Cl	Rapid	Large matrix effects from Ba, Ca, and Zn
Internal standards [11,12]	Ba, Ca, Zn, S, P, Cl	Eliminates matrix effects	Relatively slow methods
Thin film [14]	Ba, Zn, P, S, Cl	Matrix effects are minimal	Difficult to produce films of constant thickness
Absorption [13]	S	Rapid	Large matrix effects from other elements
Absorption correction using tabulated μ data [18]	Ba, Ca, Zn, P, S	Minimum number of standards required	Does not correct for primary absorption
Proposed method	Ba, Ca, Zn, S, P, Cl	Rapid	Initial calibration procedure long

attempt has been made to employ the method for the relatively complex case of lube oil additives. Much the same can be said about the use of the internal-standard disk method [17].

Since the specimens in this case are homogeneous, well defined, and calibration standards readily available, a mathematical correction procedure based on a suitable concentration/intensity algorithm [19] is probably the best approach. Several papers have already been published describing this approach to lube oil-additive elements [20], and a detailed description of the actual method would be redundant in this text.

Briefly, most mathematical correction methods depend upon the assumption that when an interfering element j suppresses the response (i.e., the slope factor of the curve m) of an element i, the

TABLE 10.8

Example of Correction Equations (from Ref. 22)

$$C_{Zn}^x = \frac{\frac{N^x(T^s - N^s t)}{N^s(T^x - N^x t)} - K_{Zn}}{1 - K_{Zn}} C_{Zn}^s [1 + 0.1322(C_{Ba}^x - C_{Ba}^s) + 0.0717(C_{Ca}^x - C_{Ca}^s) + 0.0426(C_{Cl}^x - C_{Cl}^s) + 0.0379(C_S^x - C_S^s) + 0.0349(C_P^x - C_P^s)]$$

$$C_{Ba}^x = \frac{\frac{N^x(T^s - N^s t)}{N^s(T^x - N^x t)} - K_{Ba}}{1 - K_{Ba}} C_{Ba}^s [1 + 0.0278(C_{Zn}^x - C_{Zn}^s) + 0.0949(C_{Ca}^x - C_{Ca}^s) + 0.0632(C_{Cl}^x - C_{Cl}^s) + 0.0492(C_S^x - C_S^s) + 0.0576(C_P^x - C_P^s)]$$

$$C_{Ca}^x = \frac{\frac{N^x(T^s - N^s t)}{N^s(T^x - N^x t)} - K_{Ca}}{1 - K_{Ca}} C_{Ca}^s [1 + 0.0343(C_{Zn}^x - C_{An}^s) + 0.0327(C_{Ba}^x - C_{Ba}^s) + 0.0694(C_{Cl}^x - C_{Cl}^s) + 0.055(C_S^x - C_S^s) + 0.0616(C_P^x - C_P^s)]$$

$$C_{Cl}^x = \frac{\dfrac{N^x(T^s - N^s t)}{N^s(T^x - N^x t)} - K_{Cl}}{1 - K_{Cl}} C_{Cl}^s [1 + 0.0361(C_{Zn}^x - C_{Zn}^s) + 0.0129(C_{Ba}^x - C_{Ba}^s) + 0.0046(C_{Ca}^x - C_{Ca}^s)$$
$$+ 0.0688(C_S^x - C_S^s) + 0.0637(C_P^x - C_P^s)]$$

$$C_S^x = \frac{\dfrac{N^x(T^s - N^s t)}{N^s(T^x - N^x t)} - K_S}{1 - K_S} C_S^s [1 + 0.0557(C_{Zn}^x - C_{Zn}^s) + 0.0769(C_{Ba}^x - C_{Ba}^s) + 0.0174(C_{Ca}^x - C_{Ca}^s)$$
$$+ 0.0138(C_{Cl}^x - C_{Cl}^s) + 0.0641(C_P^x - C_P^s)]$$

$$C_P^x = \frac{\dfrac{N^x(T^s - N^s t)}{N^s(T^x - N^x t)} - K_P}{1 - K_P} C_P^s [1 + 0.0423(C_{Zn}^x - C_{Zn}^s) + 0.0307(C_{Ba}^x - C_{Ba}^s) + 0.0091(C_{Ca}^x - C_{Ca}^s)$$
$$+ 0.0002(C_{Cl}^x - C_{Cl}^s) + 0.006(C_S^x - C_S^s)]$$

change in the slope of the calibration curve Δm is proportional to the concentration of C_j of the interfering element. Enhancement is simply treated as negative absorption. A general expression can then be written relating the concentration C_i of an element i with its peak counting rate R_i (ignoring for the moment the background) in the presence of interfering elements. Several such algorithms are available, but for simplicity, we will assume a concentration correction procedure based on the Lachance-Traill approach [21]. A general form of this is as follows:

$$C_i = \frac{R_i}{m_i} (1 + \sum_j K_{ij} W_j) \qquad (10.6)$$

where m_i is the slope factor for i in the absence of element j; W_i is the weight fraction of the element j; and K_{ij} is the respective influence factor of element j on i. K_{ij} will be positive in the case of positive absorption and negative in the case of negative absorption or enhancement. In practice, one never tries to correct an elemental response to zero matrix effect since this is a purely hypothetical concept. The standard practice is to compare the counting rate from an unknown R_i^x with that from a ratio standard R_i^s and known concentration C_i^s. Substitution in Eq. (10.6) gives

$$C_i^x = \frac{R_i^x}{R_i^s} C_i^s (1 + \sum_j K_{ij} \Delta W_j) \qquad (10.7)$$

where ΔW_j is the concentration difference of element j between the ratio standard and the analyzed specimen. If the same ratio standard is always used, Eq. (10.7) can be rewritten in the form

$$C_i^x = R'[K_i + \sum_j (K_{ij} \Delta W_j)] \qquad (10.8)$$

where R' is the ratio of counts between sample and standard for element i. Thus in the case of barium, zinc, calcium, chlorine, sulphur, and phosphorus, six equations can be written:

$$Ba = R'_{Ba}(K_{1a} + K_{2a}Zn + K_{3a}Ca + K_{4a}Cl + K_{5a}S + K_{6a}P)$$
$$Zn = R'_{Zn}(K_{1b}Ba + K_{2b} + K_{3b}Ca + K_{4b}Cl + K_{5b}S + K_{6b}P)$$
$$Ca = R'_{Ca}(K_{1c}Ba + K_{2c}Zn + K_{3c} + K_{4c}Cl + K_{5c}S + K_{6c}P)$$
$$Cl = R'_{Cl}(K_{1d}Ba + K_{2d}Zn + K_{3d}Ca + K_{4d} + K_{5d}S + K_{6d}P)$$
$$S = R'_{S}(K_{1e}Ba + K_{2e}Zn + K_{3e}Ca + K_{4e}Cl + K_{5e} + K_{6e}P)$$
$$P = R'_{P}(K_{1f}Ba + K_{2f}Zn + K_{3f}Ca + K_{4f}Cl + K_{5f}S + K_{6f})$$

where Ba, Zn, ..., represent the concentration of these elements and R'_{Ba}, R'_{Zn}, ..., their respective count ratios. All of the 36 K_{ij} terms are constant and, once determined, should only change when the ratio standard is changed. Table 10.8 lists such a set of equations, which were actually derived in a series of real experiments [22].

10.6.6 Long-Term Evaluation of the Method

Now that the analytical method has been set up and the spectrometer calibrated, it remains only to perform a long-term evaluation. In the routine control laboratory, this is best done by submitting standards "blind" to the laboratory at irregular intervals and intermixed with routine unknown samples. Only by such a means can long-term precision and accuracy be established, particularly in those cases where several operators are using the method. In this type of instrumental procedure, it is not uncommon to find that long-term precision involving data from several different users is worse by perhaps a factor of two than short-term tests might indicate.

REFERENCES

1. W. T. Grubb and P. O. Zemany Nature, 176, 221 (1955).
2. W. T. Campbell, T. E. Green, and S. L. Law, Amer. Lab., 2, 28 (1970).
3. D. E. Leyden, Advan. X-Ray Anal., 17, 293 (1973).
4. C. L. Luke, Anal. Chim. Acta, 45, 377 (1969).

5. T. B. Johansson, R. Akselsson, and S. A. E. Johansson, *Advan. X-Ray Anal.*, *15*, 373 (1971).
6. J. W. Criss, *Anal. Chem.*, *48*, 179 (1976).
7. R. Jenkins and P. W. Hurley, *Can. Spectrosc.*, *13*, 35 (1968).
8. R. W. Toft, *Proc. 4th Conf. on X-Ray Anal. Methods*, Philips: Eindhoven (Sheffield, 1964).
9. E. N. Davis and R. A. van Nordstrand, *Anal. Chem.*, *26*, 973 (1954).
10. T. C. Yao and F. W. Porsche, *Anal. Chem.*, *31*, 2010 (1959).
11. R. Louis, *Z. Anal. Chemie*, *201*, 336 (1964).
12. R. F. Haycock, *J. Inst. Pet.*, *50*, 123 (1964).
13. R. W. Cranston, F. W. H. Matthews, and N. Evans, *J. Inst. Pet.*, *40*, 55 (1954).
14. T. P. Schreiber, A. C. Ottoline, and J. L. Johnson, *Appl. Spectrosc.*, *17*, 17 (1963).
15. C. W. Dwiggins, *Anal. Chem.*, *33*, 67 (1961).
16. R. Jenkins, *J. Inst. Pet.*, *48*, 246 (1962).
17. E. L. Gunn, *Appl. Spectrosc.*, *19*, 99 (1965).
18. H. V. Carter, *Norelco Reporter*, *13*, 45 (1966).
19. R. Jenkins, *Advan. X-Ray Anal.*, *19* (1975).
20. R. J. Bird and R. W. Toft, *J. Inst. Pet.*, *56*, 169 (1970).
21. G. R. Lachance and R. J. Traill *Can. Spectrosc.*, *11*, 43 (1966).
22. R. Jenkins and G. N. Thorne, unpublished results.

Chapter 11

METALS AND ALLOYS

R. W. Gould

Department of Materials Science and Engineering
University of Florida
Gainesville, Florida

11.1	Introduction	278
11.2	Sample Preparation: Metal and Alloy Surface Treatment	278
11.3	Microstructure of Metals	280
11.4	Surface Preparation of Metals	282
11.5	Other Sample Preparation Methods: Pressed Fragments	284
11.6	Liquid-Solution Methods	285
11.7	Borate-Glass Fusion	285
11.8	Thin Films	285
11.9	Use of Comparison Standards in Metallurgical X-Ray Spectroscopy	287
11.10	Homemade Standards	288
11.11	Other Standards for the Analysis of Metals	288
11.12	Mathematical Correction Procedures	289
11.13	Application of X-Ray Spectrometric Analysis to Metals	289
References		291

11.1 INTRODUCTION

Chemical analysis of metallic samples by X-ray spectrometry was discussed by von Hevesy [1] over 40 years ago. Rapid growth of this technique began in the mid-1950s and has continued at a steady pace since that time. Process control in today's highly automated production facilities is strongly dependent upon fast, precise, and accurate chemical analysis, and X-ray spectrometry has been found to be widely applicable in the metals industries. Metallic samples do pose several unique problems for the X-ray spectroscopist, especially in the areas of sample preparation and interelement effects. In general, metallic samples must be used without severe sample alteration such as comunition, fusion, or other dissolution methods. When metals are analyzed as a bulk solid, X-ray physics predicts that the X-ray information will be derived from a thin surface layer. In some cases--silicon in iron, for example--99.9% of the measured Si Kα X-ray intensity comes from a layer approximately 16 µm (0.0006 in.) thick. Thus surface preparation requirements and microstructure must be critically examined and understood.

11.2 SAMPLE PREPARATION: METAL AND ALLOY SURFACE TREATMENT

Figure 11.1 is a schematic flow chart for possible sample preparation alternatives in metallic systems. While Fig. 11.1 indicates that many procedures are available, time considerations generally dictate the use of the most rapid and reproducible techniques, i.e., the massive solid form with minimal surface treatment. We will therefore discuss this procedure first and consider alternative metallic sample preparation methods in a subsequent section of this chapter.

Table 11.1 shows the surface depth from which 99.9% of the emerging radiation originates. For ease of calculation the excitation wavelength (primary absorption) was set equal to two-thirds of the absorption-edge wavelength of the analyte. For additional convenience the incident primary radiation angle (ϕ) as well as the

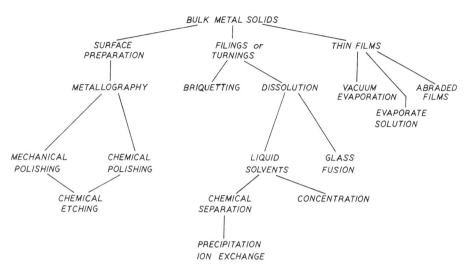

FIG. 11.1 Flow chart of metallic sample preparation methods applicable to X-ray fluorescent analysis.

take-off angle (ψ) were chosen to be 45°. Under these assumptions the following relationship holds [2].

$$t(99.9\%) = \frac{6.91 \sin 45}{\mu(\lambda_{eff}) + \mu(\lambda_{analyte})}$$

The mass absorption coefficients were taken from McMasters [3].

Table 11.1 illustrates an inherent limitation in the use of

TABLE 11.1

Surface Depth from Which 99.9% of the Measured Analyte Radiation Originates

Analyte and matrix	Mass absorption Coefficients		Effective thickness		
	λ_{eff}	$\lambda_{analyte}$	cm	μm	in.
Si in Fe	4.50	7.13	0.00158	15.7	0.00062
Ni in Fe	0.993	1.66	0.01069	106.9	0.0042
Si in Al	4.50	7.13	0.0014	11.4	0.00045
Cu in Al	0.920	1.54	0.07816		0.0307

bulk metal specimens, that is, the volume of material analyzed can be very small owing to the shallow effective penetration depth. Thus the sample preparation procedures must expose or produce a surface which truly represents the bulk metal sample.

11.3 MICROSTRUCTURE OF METALS

The space allotted for this chapter is not sufficient to fully discuss metallic microstructures. Nevertheless, the X-ray spectroscopists who wish to analyze metals must be able to answer the basic question, "Where is the analyte element located within the specimen?" Metals are notoriously heterogeneous with the exception of fully homogenized metallic solid solutions and ultrafine grained samples.

Figure 11.2 is a schematic equilibrium phase diagram of a simple binary system consisting of a mixture of metals A and B.

This figure is drawn to illustrate some basic phenomena which can occur when metals are mixed. Let us examine several compositions and note the effect of various thermal treatments on the resulting microstructures.* For more details the reader is referred to Guy [4], Rhines [5], and Hansen [6].

An alloy consisting of 10% B in A is raised to a temperature T_1 where it will be a liquid solution. If this alloy is rapidly cooled to room temperature as in a casting operation it will contain heterogeneities known as cored dendrites. The "tree-like" grain structure (dendrites) will contain compositional variations from the center to the edge of each dendrite arm. Such composition variations are generally not harmful to X-ray analysis since the B atoms are in a solid solution. If this alloy is heated for several hours at some temperature such as T_2 these composition fluctuations will become homogenized. If the homogenized 10% alloy at a temperature T_2 is rapidly cooled to room temperature it will be thermodynamically

---

*The term *microstructure* refers to the microscopically observed image of the polished metal surface and includes grain size and shape, grain boundaries, second phase inclusions, defects, and the like.

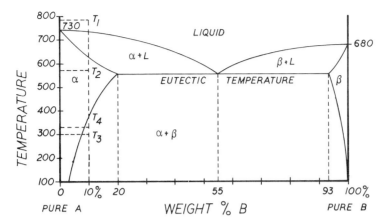

FIG. 11.2 Schematic binary phase diagram for a mixture of elements A and B.

unstable. At room temperature this alloy would prefer to consist of two phases α and β. α is a solid solution of B in A while β is a solid solution of A in B. Using the so-called lever rule and tie line principles [5], the alloy should consist predominantly of α (~95%) with some β as a second phase. Immediately after the quench, however, the alloy, if polished, will appear to consist of only a single phase. If this alloy is subsequently heat treated at some temperature T_3 for a period of time the second phase will appear at the grain boundaries and within the grains. The second-phase particles will grow in size and volume fraction with time until equilibrium is achieved. The quantity will approach that predicted by the lever rule while the grain size is controlled by surface energy factors. The important point to be made, however, is that particle size and quantity of β phase change with time. At some higher temperature T_4 the equilibrium quantity of β phase would be different and the rate of change would be faster. Of course it is true that the total quantity of B atoms in the alloy is unchanged (10%). However, the local environment of the B atoms has changed from a solid solution (being surrounded by A atoms) to a β phase precipitate (being surrounded predominantly by other B atoms). This change when combined with a sufficiently large β phase particle size can cause marked primary and secondary X-ray absorption anomalies [7]. Changes such as this can occur at room temperature in low melting-point alloys such as Pb-Sn solders. This effect can cause standards in these alloy systems to change microstructure and X-ray fluorescent intensity with time [8].

An alloy containing 55% B has an unusual, so-called "eutectic" microstructure. In this lowest melting point alloy (typical of many solders), the α and β phases solidify simultaneously from the liquid solution producing a microstructure of alternating α and β platelets. Heating for long periods below the eutectic temperature will cause the platelets to round off at the ends and spherodize to produce a minimum surface to volume ratio.

11.4 SURFACE PREPARATION OF METALS

Metallographic quality preparation is not generally required for X-ray spectrometric analysis. What is required is a reproducible surface, free of any unwanted defects. Reproducibility alone is not sufficient since improper preparation may produce and reproduce a totally erroneous result as in the case of soft-phase over-smearing. When a new metallic system is being examined a good policy is to become familiar with the available literature and monitor your sample preparation procedures metallographically while they are being developed. The concept of effective penetration depth is most useful (Table 11.1) in the early stages of sample preparation. Specimens should not contain grooves or scratches whose depth exceeds the effective penetration depth. Table 11.2 lists some critical factors to consider when preparing metallic samples for X-ray spectrometric analysis.

Bertin [14] has discussed the preparation of metallic specimens in his excellent text on X-ray spectrometry. For those unfamiliar with metallographic surface preparation procedures, Samuels [15], Buehler [16], and Metals Handbook Vol. 8 [13] are recommended reading. Chemical polishing and etching procedures are also covered by

TABLE 11.2

Factors to Consider When Preparing the Surface
of a Metal for X-Ray Spectrometric Analysis

Factor	Comments
Effective penetration depth surface roughness	Very critical for low-energy (long wavelength) radiation.
Microstructural changes	Consult phase diagrams and perform metallographic study, see Kemper [9] Al-Si alloys.
Imbedded polishing abrasives	Mechanical polishing can imbed SiC, Al_2O_3; select abrasive not to interfere with analyte.
Etching	May selectively remove phase or selectively corrode phase of interest.
Storage of samples	Oxidation and corrosion of clean metal surfaces occur very rapidly in ambient air store in vacuum desiccator. Changes in standards [8].
Grain size	Large grain samples generally unsuitable for X-ray analysis--beware of castings.
Cutting specimens	Rough cutting can produce severe surface damage which polishing will not remove. Spark cutting recommended [10].
Plucking out inclusions	Polishing can remove inclusions containing analytes; special precautions and visual observation is required.
Electropolishing	Same as chemical polishing.
Smearing of soft phases	May require gentle polishing with fine diamond abrasive. Good technique exists; see Manners [11] and Kilday [12] and metallurgical literature for polishing solder and similar alloys [13].

Kehl [17], while Jenkins and Hurley [10] discuss spark cutting. As stated previously mechanical polishing is generally adequate for most X-ray spectrometric analysis. Table 11.3 lists the common abrasive designations and their sizes in μm. Thus the analysis of silicon in iron (effective depth of 16 μm) would require mechanical polishing down to at least 4.0 emery paper.

TABLE 11.3
Polishing Abrasives and Their Sizes[a]

Grit size	Approximate equivalent size (μm)
80	200
100	150
200	60
300	35
400	25
600	15
Emery Paper Grade	
2	60
1	40
1/0 (0)	22
2/0 (00)	20
3/0 (000)	18
4/0 (0000)	16
Al_2O_3 Powder	
AB Levigate	15
AB No. 1	5
AB No. 2	0.3
AB No. 3	0.05

[a] Courtesy Adolph Buehler [16].

11.5 OTHER SAMPLE PREPARATION METHODS: PRESSED FRAGMENTS

Lathe turnings, drillings, and filings may be briquetted under high pressure to form a disk suitable for the analysis of elements whose analytic lines are of low energy. It will have its best success on alloys where the analyte is in solid solution rather than distributed in a second phase whose quantity and distribution can change with thermal treatment. Soft-phase smearing can also present problems in this type of analytical technique.

11.6 LIQUID-SOLUTION METHODS

The method of chemical dissolution is widely used in spectrometric analysis of metals and alloys. Dissolution is often followed by some method of chemical separation; precipitation, chelation, ion exchange, etc., to achieve high sensitivity and selectivity of analysis. Dissolution chemistry is beyond the scope of this chapter and the reader should consult the analytical chemistry literature for assistance in this area. Liquid samples present unique problems, some of which are discussed in Sec. 10.3. Bertin [14] has also discussed liquid sample preparation in great detail.

11.7 BORATE-GLASS FUSION

The borate-glass fusion method is widely used for the direct analysis of oxides [18,20] but can be applied to the analysis of metal and alloy powders if prefusion oxidation methods are utilized. These procedures have been reviewed recently by Jenkins et al. [19]. Solid-glass solutions like liquid solutions eliminate heterogeneity effects and most interelement effects but at the expense of time and sensitivity. Metallic powders may be roasted in air or excess oxygen to induce oxidation. Alternatively [18,19] the metal may be heated in the presence of excess sulfur, dissolved in fused potassium pyrosulphate or acid prior to dissolution in borate glass.

11.8 THIN FILMS

Thin-film specimens offer several distinct advantages to the analyst (Giaugue [21]). Chief among these is the reduction or elimination of matrix effects if the film is sufficiently thin. Chung [22] has quantitatively discussed the necessary criteria for deciding whether a specimen meets the minimum thickness requirements. Thin films can be prepared from metallic samples in a variety of ways; chemical solution and deposition, vacuum evaporation, flash evaporation, and ion bombardment (sputtering) [23]. Unfortunately, the most

TABLE 11.4

Rubbing Film Analysis; Steels Florestan [24]

Chemical Cr	X-Ray Cr
16.77	16.79
17.00	17.41
17.02	16.80
16.44	16.41
17.69	17.50
16.72	16.71
17.39	17.08
16.85	16.63

rapid and direct evaporation techniques produce a thin film whose chemical composition may not be characteristic of the bulk metal from which it was obtained. At present the sputtering method offers the most promise for producing a thin film whose composition is chemically identical to the bulk solid from which it was obtained.

A novel thin-film technique is described by Florestan [24] which involves lightly rubbing the metal specimen onto a pure Al_2O_3 substrate disk to produce a thin metallic smear. Table 11.4, taken from this paper, shows the nature of his results (Cr in steel).

This technique was used by this author to analyze Cu_3Au which had been lightly rubbed on 600 grit SiC paper. The X-ray intensities from this film were measured by the fundamental parameters method of Sparks [25] and the results are shown in Table 11.5.

TABLE 11.5

Analysis of Cu_3Au by Rubbing Film Methods

Chemical		X-Ray	
Cu	Au	Cu	Au
49.15	50.85	48.81	51.18

11.9 USE OF COMPARISON STANDARDS IN METALLURGICAL X-RAY SPECTROSCOPY

Direct comparison standards are widely used in metallurgical analysis. Speed, precision, and accuracy are optimized when a standard having nearly identical physical and chemical properties to those of the unknown can be "compared" in the X-ray spectrometer. The need for strict similarity in chemistry and physical properties (microstructure, surface preparation) poses some difficulties for the analyst unless he is working with specific commercial metallic alloys for which such standards do exist. The National Bureau of Standards (NBS) has been active for many years preparing standards specifically for use by the X-ray fluorescence community. NBS special publication 260 [26] lists all of these available standards, their physical form, nominal chemical composition, and cost. Up-to-date information on the availability of NBS standards can be obtained by writing directly to the NBS. Yolken [27] has recently discussed the use of standard reference materials for X-ray measurements. Table 11.6 is

TABLE 11.6

Availability of NBS Disk Standards for X-Ray Spectrometric Analysis [26]

Alloy type	Number of disk standards available
Low alloy steels and ingot irons	16
Stainless steels	6
Tool steels	5
Maraging steel and high nickel	2
High-temperature Cr-Ni-Mo-Co steels	10
Cast steels and irons	10
Copper base (brasses and bronzes)	22
Lead base (solder and bearings)	2
Nickel base	2
Titanium base	7
Zirconium Base	1

an abbreviated listing of the standards available in disk form
(1-1/4 in. diameter by 3/4 in. thick) suitable for X-ray spectrometry. For more details consult the most recent copy of the NBS catalog of standard reference materials [26].

Michaelis [28] has reviewed the available sources of standard reference materials for X-ray spectrochemical analysis. Metallurgical reference standards are available from agencies similar to NBS in foreign countries such as the Bureau of Analyzed Samples Ltd. [29]. Alcoa produces aluminum alloy standards [30] and other standards are available from specific metals industries and technical societies.

11.10 HOMEMADE STANDARDS

There is obvious danger in preparing one's own standards unless a reference method is available for certification. Metallic samples can be melted and cast but the nominal (as prepared) composition seldom equals the analyzed composition due to volatilization and slag formation involving specific constituents. Powder metallurgy methods are in use [31] in which pure metal powders are mixed, sintered and hot pressed to form a fully dense metal. Though this is better than melting, the nominal powder weights cannot be relied upon to provide high precision, and a referee analysis is required.

11.11 OTHER STANDARDS FOR THE ANALYSIS OF METALS

Pure metal standards can be used for the analysis of metals and alloys if the fundamental-parameters method is employed. These standards are available from chemical supply houses but seldom in large disk form suitable for X-ray spectrometric analysis. Most of the common metals can be obtained in pure bulk form from manufacturers of specific metals. Information of the availability of pure metals may be obtained from the Research Materials Information Center at Oak Ridge [32].

Liquid and fusion reference standards can obviously be prepared when direct bulk analysis is not required. If the glass fusion method is used, care must be taken to characterize the volatilization losses which can occur during the preparation of fusion standards. Even when using low dilution ratios interelement effects are not entirely eliminated, thus standards containing the diluent and only the one element of interest may not be satisfactory to compare with the similarly dissolved alloy.

11.12 MATHEMATICAL CORRECTION PROCEDURES

This voluminous subject has been briefly covered in previous chapters (3 and 4) and extensively reviewed at the 1975 Denver X-Ray Conference [33]. In the metallurgical industries the empirical coefficient methods have received the widest use of any of the mathematical correction procedures. Inherent inaccuracy and basic mistrust cause these methods to be not readily accepted in industries where well-characterized reference standards can be used effectively. The increasing availability and lower cost of dedicated on-line minicomputers coupled with theoretical advances has served to stimulate broader use of mathematical methods. The next decade should see widespread adoption of many of these methods in the metals industries.

11.13 APPLICATION OF X-RAY SPECTROMETRIC ANALYSIS TO METALS

This section of the chapter will serve as a literature guide for the application of X-ray spectrometry to the analysis of metallic systems. For brevity we will restrict the coverage to specific alloys and refer the reader to several excellent reviews which provide comprehensive coverage of metals analysis by X-ray spectrometry (see Table 11.7). Table 11.8 lists groups of references covering commercially important alloy systems. These lists are not exhaustive, hopefully they will illustrate the types of difficulties which have been solved by previous researchers.

TABLE 11.7

Selected Review Sources for Metals Analysis by X-Ray Spectrometry

Jenkins and DeVries Metallurgical Review [29]	Excellent review of application of X-rays to metallurgical industry, 309 references to 1969.
Muller [34]	Eighty-two selected abstracts of application to various metals systems and ores (to 1964).
XRFS Abstracts [35]	Specific section devoted to industrial metallurgical analysis. Abstracts in English, 1959-present.
GAMS [36]	Abstracts mostly in French; excellent coverage of European research, not specific to metals and alloys.
Analytical Chemistry Annual Reviews [37]	Topical review coverage of X-ray applications. In earlier years (pre-1972) references to specific analyte metals are given.
Daniels [38] (in French)	X-Ray fluorescent analysis in metallurgy, 1965.

TABLE 11.8

References Covering X-Ray Spectrometric Analysis of Specific Metals and Alloy Systems

Sodium and magnesium alloys [39-43]

Aluminum alloys [9,44-56]

Titanium alloys [51,57-61]

Ferrous alloys [24,41,49,59,62-73]

Nickel alloys [41,42,50,54,64,67,74-77]

Copper alloys [59,67,78-90]

Zirconium alloys [51,91-98]

Tungsten alloys [99,100]

Gold alloys [81,101]

REFERENCES

1. G. von Hevesy, *Chemical Analysis by X-Rays and Its Applications*, McGraw-Hill, New York, 1932.
2. B. D. Cullity, *Elements of X-Ray Diffraction*, Addison Wesley, Reading, 1956.
3. W. H. McMaster, N. K. Del Grande, J. H. Mallett, and J. H. Hubbel, U. S. Atomic Energy Comm. Report UCRL-50174, 1969.
4. A. G. Guy, *Physical Metallurgy for Engineers*, Addison Wesley, Reading, 1962.
5. F. N. Rhines, *Phase Diagrams in Metallurgy*, McGraw-Hill, New York, 1956.
6. M. Hansen, *Constitution of Binary Alloys*, 2nd ed., McGraw-Hill, New York, 1958.
7. F. Claisse and Samson, *Advan. X-Ray Anal.*, 5, 335 (1962).
8. G. H. Glade and H. R. Post, *Appl. Spectrosc.*, 22 (2), 123 (1968).
9. M. A. Kemper, *X-Ray Spectrometry*, 3 (3), 111 (1974).
10. R. Jenkins and P. W. Hurley, *Proc. 5th Congress X-Ray Analytical Methods*, Eindohoven Philips, 1966, p. 88.
11. V. J. Manners, J. V. Craig, and F. H. Scott, *J. Inst. Metals*, 95, 173 (1967).
12. B. E. Kilday and R. E. Michaelis, *Appl. Spectrosc.*, 16 (4), 136 (1962).
13. American Society of Metals, *Metals Handbook*, Vol. 8, Metals Park, 1973.
14. E. P. Bertin, *Principles and Practice of X-Ray Spectrometric Analysis*, 2nd ed., Plenum Press, New York, 1974.
15. L. E. Samuels, *Metallurgical Polishing by Mechanical Methods*, Elsevier, 1971.
16. The AB Metals Digest, Adolph Buehler Ltd., Evanston, Illinois.
17. G. L. Kehl, *The Principles of Metallographic Laboratory Practice*, McGraw-Hill, New York, 1943.
18. F. Claisse, *Norelco Reporter*, 4, 3 (1957).
19. R. Jenkins, R. W. Gould, and D. Gedcke, *Advanced X-ray Spectrometry*, Marcel Dekker, New York (to be published), 1978.
20. R. Tertian, *Spectrochemica Acta*, 24 B, 447 (1969).
21. R. D. Giaugue and J. M. Jaklevic, *Advan. X-Ray Anal.*, 15, 164 (1971).
22. F. H. Chung, A. J. Lentz, and R. W. Scott, *X-Ray Spectrometry*, 3 (4), 172.

23. K. L. Chopra, *Thin Film Phenomena*, McGraw-Hill, New York, 1969.
24. J. Florestan, *Methodes Physiques d' Analyses* (GAMS), 1967, April, 1965, p. 118.
25. C. J. Sparks, Quantitative X-Ray Fluorescence Analysis Using Fundamental Parameters, *Advan. X-Ray Anal.*, *19* (1976), p. 19.
26. Catalog of NBS Standard Reference Material, NBS Special Publication 260, 1975-1976 Edition, U. S. Department of Commerce, National Bureau of Standards.
27. H. T. Yolken, *Advan. X-Ray Anal.*, *17*, 1 (1974).
28. R. E. Michaelis, *Amer. Soc. Testing and Materials Data Serv.*, *DS-2*, 156 (1964).
29. R. Jenkins and J. L. DeVries, *Metals Rev.*, *16*, 125 (1971).
30. Alcoa Spectrochemical Standards for Analysis of Aluminum and Its Alloys, Alcoa Research Laboratories, New Kensington, 1969.
31. G. E. Hicho, H. Yakowitz, S. D. Rasberry, and R. E. Michaelis, *Advan. X-Ray Anal.*, *14*, 78 (1971).
32. Research Materials Information Center, Solid State Division, Oak Ridge National Laboratory, Oak Ridge.
33. Proc. 1975 Denver X-Ray Conf. *Advan. X-Ray Anal.*, *19*, Kendall/Hunt Publishing Company, Dubuque (1976).
34. R. O. Muller, *Spectrochemical Analysis by X-Ray Fluorescence*, Plenum Press, New York, 1972.
35. X-Ray Fluorescence Spectrometry Abstracts, Science and Technology Agency, 3 Dyers Bldg., London, E. C. 1, 1970.
36. Methodes Physique d'Analyse Groupement pour l' Avancement de Methodes Physique d'Analyse (GAMS) 10 Rue de Delta, 75 Paris 9e, France.
37. Analytical Chemistry (Annual and Bienniel Reviews), American Chemical Society, 1155 16 St. N. W., Washington, D. C.
38. A. Daniels, *Chemie. Anal.*, *47* (11), 571 (1965).
39. H. Schneider, D. Schonwals, and H. Schumann, Determination of Calcium in Sodium Metal in the PPM Range with the Aid of X-Ray Fluorescent Analysis, *Z. Anal. Chemie.*, *247*, 175-176 (1970).
40. G. A. Stoner, Rapid Automatic Analysis of Magnesium Alloys, *Anal. Chem.*, *34*, 123 (1962).
41. C. A. Kienberger and A. R. Flynn, Report K 1638, Union Carbide Corp., Oak Ridge, January, 1966.
42. E. F. Spano and T. F. Green, *Anal. Chem.*, *38*, 1341 (1966).
43. H. Kessler and H. Z. Rammernsee, *Metallkunde*, *51*, 548 (1960).
44. J. V. Gilfrich and D. C. Sullivan, *Norelco Reporter*, *10*, 127 (1963).

45. F. Wagner, Z. Anal. Chem., 198, 98 (1963).
46. L. J. Christiansen, J. M. Khan, and W. F. Brunner, Rev. Sci. Inst., 38, 20 (1967).
47. J. E. Cline and S. Schwarts, J. Electrochem Soc., 114, 605 (1967).
48. L. Fergason, Rev. Sci. Inst., 37, 964 (1966).
49. W. L. Baun and D. W. Fischer, J. Appl. Phys., 38, 2092 (1967).
50. D. W. Fischer and W. L. Baun, Phys. Rev., 145, 555 (1966).
51. H. De Laffolie, Arch. Evenhuttenw, 7, 535 (1967).
52. K. Hirokawa and A. Saito, Z. Anal. Chem., 237, 419 (1968).
53. J. G. Dick and A. R. Fraser, Can. J. Spectrosc., 17, No. 5, 135 (1972).
54. D. Duzevic and T. Gacesa, X-Ray Spectrometry, 3, No. 4, 143 (1974).
55. H. Goto and A. A. Saito, Sci. Res. Inst., Tohoku University, Ser. A, 20, 59 (1968).
56. E. Davidson, J. Metals, 22, 48 (1969).
57. F. Creton and B. Maschin, Met. Ital., 8, 425 (1963).
58. N. M. Sine and C. L. Lewis, Talanta, 12, 389 (1965).
59. C. L. Lewis, W. L. Ott, and N. M. Sine, The Analysis of Nickel, Pergamon Press, New York, 1966, p. 85.
60. R. F. Stoops and K. H. McKee, Anal. Chem., 33, 589 (1961).
61. G. L. Vassilaros and J. P. McKaveney, Talanta, 16, No. Z, 195 (1969).
62. F. Wagner, Z. Anal. Chem., 198, 98 (1963).
63. G. Bonissoni and M. Paganelli, Met. Ital., 8, 268 (1966).
64. E. Chow and E. P. Cocozza, Appl. Spectrosc., 21, 290 (1967).
65. E. P. Cocozza and A. Ferguson, Appl. Spectrosc., 21, 286 (1967).
66. E. A. Hakkila and G. R. Waterbury, Anal. Chem., 37, 1773 (1965).
67. H. Goto and A. A. Saito, Sci. Res. Inst., Tokoku University, Ser. A, 20, 59 (1968).
68. A. Tsukamoto, I. Shrimizu, and M. Ohata, Nippon Kinzoku Gakkaishii, 32, 473 (1968).
69. W. Becker, W. Dobner, and G. Wronka, Siemens Rev., XXXIX (1972).
70. R. Berger and P. Deceuleneer, Rev. Univ. Mines, French, 17, 207 (1961).
71. R. W. Jones and R. W. Asley, Anal. Chem., 31 (10), 1629 (1959).
72. H. Goto, K. Hirokawa, A. Saito, and F. Haeda, Z. Anal. Chemie., 247, 306 (1969).

73. E. Gilliam and H. T. Heal, *British J. Appl. Phys.*, *3*, 353 (1952).
74. F. Creton and B. Moschen, *Met. Ital.*, *8*, 425 (1963).
75. E. C. R. Hunt, *Lab Methods*, 77, 135 (1968).
76. J. M. Griffiths and H. R. Whitehead, *X-Ray Spectrometry*, *4*, No. 4 (1975).
77. M. Pluchery, *Spectrochem. Acta*, *19*, 533 (1963).
78. R. Alvarez and R. Flitsch, Nat. Bur. Stand., U. S. Misc. Publication 260-5 (1965).
79. B. W. Mulligan, H. J. Caul, S. D. Rasberry, and B. F. Scribner, *J. Res. Nat. Bur. Stand.*, *68A*, 5 (1964).
80. L. Backerud, *Appl. Spec.*, *21*, 315 (1967).
81. J. D. Eick, H. J. Cave, D. L. Smith, and S. E. Rasberry, *Appl. Spectrosc.*, *21*, 324 (1967).
82. D. W. Fischer, *J. Appl. Phys.*, *36*, 2048 (1965).
83. J. Kinnunen, P. Rautavalta, and M. Keponen, *Metallurgia*, *75*, 189 (1967).
84. R. H. Myers, D. Womeldoph, and B. J. Alley, *Anal. Chem.*, *39*, 1031 (1967).
85. A. Carnevale and A. J. Lincoln, *Dev. Appl. Spectrosc.*, *5*, 45 (1966).
86. S. D. Rasberry, H. J. Carl, and A. Yezer, *Spectrochem. Acta*, *23B*, 340 (1968).
87. S. J. Zanin and G. E. Hooser, *Appl. Spectrosc.*, *22*, 105 (1968).
88. V. J. Manners, J. V. Craig, and F. H. Scott, *J. Inst. Metals.*, *95*, 173 (1967).
89. L. Backerud, *X-Ray Spectroscopy*, *1*, No. 1, 3 (1972).
90. Ortec Tefa, *Analysis of High Purity Copper*, 1974.
91. T. J. Cullen, *Dev. Appl. Spectrosc.*, *3*, 97 (1964).
92. E. A. Hakkila, H. Hurley, and G. R. Waterbury, *Anal. Chem.*, 2094 (1964).
93. J. S. Rudolph, O. H. Kriege, and R. J. Nadalin, *Dev. Appl. Spectrosc.*, *4*, 57 (1965).
94. O. R. Gates and E. J. Brooks, NRL Report 6427 (1966).
95. B. L. Taylor, *Proc. SAC Conf.*, Nottingham, W. Heffer and Sons, Ltd., Cambridge, England, 1965, p. 81.
96. M. Brill, *Z. Anal. Chem.*, *244*, 36 (1969).
97. C. L. Luke, *Anal. Chem. Acta*, *41*, 453 (1968).
98. G. L. Vassilaros and J. P. McKaveney, *Talanta*, *16*, 195 (1969).

99. F. Maeda and T. Hayasaka, *Sci. Rep. Res. Inst.*, Tohoku University. A., Volume 23, No. 3-4, 1972.
100. C. L. Luke, *Anal. Chem.*, *35* (1), 56 (1963).
101. B. W. Mulligan, H. J. Caul, S. D. Rasberry, and B. F. Scribner, *J. Res. Nat. Bur. Stand.*, *68A* (1), 5 (1964).

Chapter 12

GEOLOGY

Brent P. Fabbi

Branch of Analytical Laboratories
U. S. Geological Survey
Reston, Virginia

12.1	Introduction	298
	12.1.1 Historical Development	298
	12.1.2 General Capabilities, Advantages, Limitations, and Range of Concentration	300
12.2	Sample Preparation	303
	12.2.1 Solids	303
	12.2.2 Powders	306
12.3	Qualitative and Quantitative Analysis	318
	12.3.1 Calibration Standards	318
	12.3.2 Mathematical Correction for Matrix Effects	323
12.4	Examples	328
	12.4.1 Silicate Rocks	328
	12.4.2 Carbonate-Phosphate Rocks	333
	12.4.3 Minerals	339
	12.4.4 Lunar Samples	340
Acknowledgments		342
References		342

12.1 INTRODUCTION

The application of X-ray spectrometry to the analysis of geologic samples has become increasingly competitive and sophisticated. Because rapid multielement analysis is possible, large quantities of analytical data can be supplied to support geological investigations. Dedicated and time-share computers have increased manifold the capabilities of the analyst to develop, examine, and apply complex mathematical corrections to the data. Data are acquired faster by interfacing the computer to X-ray spectrometers for instrument control.

Qualitative, semiquantitative, and quantitative analyses of geological samples are made with high precision, accuracy, and at costs competitive with other analytical techniques. Because of the wide variety of geological samples submitted for analysis, the broad range of elemental concentrations encountered, and the increasing need for greater sensitivity, the X-ray spectroscopist is constantly challenged to develop new X-ray spectrochemical techniques.

12.1.1 Historical Development

As with any technique applied to geochemical samples, X-ray spectrometry has developed from a limited capability to one of versatility. Hadding first applied X-ray spectrometry to the analysis of minerals in 1922 [1]. Important contributions have been made in the development of instrumentation, mathematical corrections, and sample preparation. (See Table 12.1.).

The attainment of the present state of the art has been dependent on several factors. Instrumental improvements have enabled the analyst to determine elements in the hard, soft, and ultra-soft X-ray region with good sensitivity through the use of helium and vacuum chambers, high reflectivity and high-resolution analyzing crystals, high-intensity X-ray target tubes, demountable anodes, secondary targets, etc. The analyst's understanding of the X-ray spectra has improved the accuracy of geochemical analysis through the development of diverse mathematical approaches to correct for absorption-

TABLE 12.1

Historical Development

Instrumental

1958	H. Bizouard and C. Roering examined sphalerite with a microprobe [2].
1961	B. L. Henke developed a demountable X-ray tube for light element rock and mineral analysis [3].
1974	A. J. Hebert and K. Street, Jr., developed a rotating anode energy dispersive spectrometer for quantitative major element analysis [4].

Mathematical Corrections

1954	J. H. Beattie and R. M. Brissey developed a binary influence coefficient method [5].
1966	G. R. Lachance and R. J. Traill developed a multielement influence coefficient method [160].
1968	J. W. Criss and L. S. Birks introduced the fundamental parameter method [6].
1974	S. O. Rasberry and K. F. J. Heinrich incorporated a correction for secondary fluorescence effects to influence coefficient methods [7].

Sample Preparation

1952	J. Despujols analyzed trace metals in mineral powders [8].
1955	I. Adler and J. M. Axlerod applied the internal standard method to silicate analyses [9].
1956	F. Claisse studied the effects of particle size and developed a fusion method for analyzing geological samples [10].
1963	H. J. Rose et al. applied the Lanthanum heavy absorber method to geological samples [11].
1964	K. Norrish and J. T. Hutton [12] and Welday et al. [13] developed similar briquetting techniques for rock powders.
1973-1975	Automatic fusion devices for preparing homogeneous glass disks were developed [105,106].

enhancement, line-interference, dead-time, and X-ray scattering effects. Ingenuity in sample preparation by a host of investigators has made possible the analysis of every variety of geological sample.

12.1.2 General Capabilities, Advantages, Limitations, and Range of Concentrations

12.1.2.1 General Capabilities

With careful sample preparation, proper standard calibration, astute selection of instrumental parameters, and a knowledge of matrix effects, it is possible to quantitatively analyze geological samples with an accuracy of 3-10% relative for trace elements and 1% or better for the major rock-forming elements regardless of sample type, size, or concentration range. Elements from Na to U are routinely determined with conventional X-ray spectrometers. The very light elements C, N, O, and F have been determined in geological samples with a specially designed X-ray spectrograph [14]. Sensitivity is dependent on the method of sample preparation, the sample matrix, and the instrumental conditions of analysis [15]. Sensitivities on the order of 20-1000 ppm for elements $Z = 9$ to $Z = 21$, 5-10 ppm for elements $Z = 22$ to $Z = 42$, and 1-20 ppm for elements $Z > 42$ have been reported by many investigators.

12.1.2.2 Advantages

The primary advantage of X-ray spectrometry to the geochemist is that once the sample is prepared, the method is nondestructive. Samples prepared by a variety of methods (see Sec. 12.2) can be analyzed for several elements, placed in storage, and retrieved for further analyses. The shape, size, or quantity of the geologic samples is not a limiting factor for analysis. A few milligrams to several grams can be placed on a suitable substrate or backing material for support prior to analysis.

Owing to the versatility in selection of operating parameters and sample-handling devices, X-ray spectrometry offers a rapid, accurate, low cost per determination method of analysis. With the advent of fully automated X-ray spectrometers, sequential and multichannel wavelength-dispersive instruments, and energy-dispersive instruments, a severalfold increase in productivity is possible. Five hundred or more quantitative determinations per week are not unusual [16], and as many as 1500 determinations per week have been achieved.

Because of the diverse nature of geochemical investigations, X-ray spectrometry is a most useful method for determining many elements sequentially or simultaneously. The X-ray spectra are generally less complex than other optical spectra. Line interferences can be eliminated through the use of higher resolution crystals (LiF 200 to separate Rb and Sr Kα from U Lα), and with crystals which do not reflect second-order lines (Ge, analysis of P with Ca Kα, 2 unreflected). Adjacent energies (KeV) are resolved with pulse-height analysis or through the proper selection of detector gases (70% Ne and 30% methane) to reduce the effect of Ca on the determination of F [17]. Interferences are negated by reducing operating kV below the excitation potential of the interfering element and through the use of absorbing filters. Mathematical equations are used to correct the analyte intensity for line interference.

Absorption-enhancement effects are predictable in X-ray analysis. These effects can be dealt with through mathematical corrections (Sec. 12.3.2) or through several methods of sample preparation including internal standard, dilution, standard addition, and thin film.

12.1.2.3 Limitations

The high initial cost of instrumentation is the principal economic disadvantage. However, as pointed out, the speed and multiplicity of elements determinable can result in a lower unit cost ($1 to $5 per determination).

Specimen surface, homogeneity, particle-size effects, and sample thickness can adversely affect the precision and accuracy of analysis. Elements of $Z < 22$ are subject to variations in X-ray intensity if the surface is not completely smooth, flat, and of the same density. Sample inhomogeneity will produce errors as high as 20% relative. Particle size affects the specimen surface and homogeneity. Bernstein examined particle-size effects and related variations in X-ray intensity directly to variations in particle size [18]. As line intensities of an element increase with the depth of penetration of the primary X rays, samples must be prepared of a uniform thickness which exceeds critical thickness (see Sec. 12.2.2).

The oxidation state of an element is generally thought to affect only slight nonmeasurable changes in 2θ. However, the several oxidation states of the sulfur species have been shown to vary in X-ray intensity by as much as 36% [19] (see also Chap. 8).

Matrix effects of absorption enhancement and line interference have been shown to result in orders of magnitude errors in trace element analysis and 10-20% relative errors for major element analysis within suites of rock types (see Secs. 12.2 and 12.3).

Until recently the determination of the light elements in geologic samples was difficult because of the low natural abundance or low X-ray intensities of elements such as F and Na. With higher reflective analyzing crystals such as Thallium Acid Phthalate (TAP) and a rhodium-target X-ray tube, 0.054% F and 0.004% Na_2O have been detected in undiluted geologic samples [20], and 0.01% Na_2O has been detected with a chromium-target tube [20]. Techniques for the determination of Na_2O have been reported [21,22].

12.1.2.4 Range of Concentration

The analysis of geochemical samples often involves the analysis of a host of sample types having concentrations of elements of interest ranging from 0.0001% to 80%. Within the limits of sensitivity for a given element, X-ray spectrometry has the capability of meeting these requirements on a routine basis provided proper selection of instrumental parameters is made and sample preparation methods are optimized.

GEOLOGY

12.2 SAMPLE PREPARATION

Without proper and careful sample preparation, all analyses, interpretations of data, calculations of matrix effects, and experimentation are meaningless. At least 80% of all analytical error can be attributed to sample preparation.

12.2.1 Solids

12.2.1.1 Field Collection

Preparation of geochemical samples begins in the field. No attempt is made here to discuss the sample size or methods of selection necessary to collect a representative field sample. Geochemical field samples are usually 0.5 kg or larger (up to 10 kg), but can also be chips from channel cuts, cuttings or cores from drill holes, sediments, etc.

Contamination of the field samples during collection, storage, and transit is a source of analytical error. The analyst and field geologist should be aware of possible contamination of samples during collection caused by the breaking or cutting tool which may contain Co, Ni, Fe, Mn, Cr, or W. The heavy mud used to cool drill bits contains Ba and S.

Contamination of samples in transit by vehicle tires (S, Sb), vehicle exhausts (Pb, S, Pt), and cross-contamination of exposed samples is sometimes overlooked [79]. All samples should be placed in a bag or container which itself will not contaminate the samples. Durable plastic bags are preferable to cloth. Paper bags can contaminate the samples with Hg, S, P, Ca, and Ti. Inks used in marking samples should not contain inorganic material [23].

12.2.1.2 Preparation of Solid Samples

Solid specimens of geological samples are cut to have a flat surface and a size suitable to fit into the X-ray spectrometer. The flat surface to be analyzed is abraded by an abrasive grinder to remove Cr or Cu smeared onto the surface by the cutoff saw and to produce a relatively smooth surface. Analyses made with these specimens are generally qualitative owing to the heterogeneity of rock samples.

12.2.1.3 Primary Grinding of Solid Samples

Preparation of mineral separates from rock and ore specimens have been discussed in detail elsewhere [24,25]. Any grinding component which comes in contact with the sample is a potential source of contamination or dilution (see Sec. 12.2.2.4). A variety of rock-crushing methods have been used to break the sample down to a grain size suitable for chemical or spectrochemical analysis. It is essential that the sample analyzed in the laboratory is homogeneous and representative of the field sample. Each time a sample is reduced in mesh size, it should be well mixed either by coning and quartering, rolling on mats, or by V-blenders before the sample is split for further comminution.

Samples which are too large to be broken by a jaw crusher can be reduced to a workable size with a minimum of contamination in a hydraulic mechanism having large tool-steel chisel-shaped platens. The sample, thus sufficiently reduced in size, is passed through a jaw crusher with the gap set for 3/8 in. The face plates of the jaw crusher should be knurled and not flat [25]. Flat plates require more pressure to break the rock and abrade contaminating metals into the sample.

Two methods are in common use for grinding the samples less than 3/8 in. In the first method, all of the 3/8-in. sample is passed through a set of crushing rolls with the gap set at 20 mesh. The sample is then coned and quartered to assure homogeneity before splitting out a portion to be ground to 100 mesh in alumina (mullite) or tool-steel rotary-plate grinders [26]. After the first pass through the rotary-plate grinders, the sample is sieved through a 100-mesh silk screen to avoid overgrinding, and the plus 100-mesh material is reground. Grinding beyond 100 mesh can cause absorption or loss of H_2O, CO_2, and S. Excessive grinding may oxidize ferrous iron to ferric iron [23,27]. The sample is then split into 50 to 100-g portions using riffles or by scooping 2 to 3-g portions of material from all parts of the sample in order to obtain a representative sample. It is then bottled in clean bottles and distributed for analysis.

The previously described procedure to obtain 100 mesh material is lengthy and tedious. A faster approach for batch grinding is the use of rotary disk mills provided the following conditions are satisfied:

1. The sample size is 3/8 in. or less.
2. The total sample can be ground all at once, or ground in batches, homogenized, and split.
3. Contamination of the grinding media will not affect the final analysis.
4. The ground sample will be analyzed immediately (see the preceding paragraph).

The 3/8 in. or smaller material can be ground rapidly (< 1 min) to 100 or 325 mesh in the rotary disk mills. Variations in sample hardness may affect the time required for grinding. The sample is split and bottled as described. At 325 mesh, the sample can be briquetted or inserted into Mylar cups for immediate analysis. However, materials containing high concentrations of SiO_2 (> 65%) do not briquet well. This method is effective where particle size does not effect the X-ray analysis, standards can be matched to the samples, matrix effects are not severe, and speed in grinding and analysis are of primary importance.

If the analytical application permits, materials (Borax, Dreft, etc.) which bind the sample together during briquetting and which also act as grinder-cleansing agents can be ground with the sample. It is sufficient in these applications to scoop the well-mixed 20-mesh sample and binder to avoid weighing procedures. The ratio of sample to binder volume should be about 4:1. The resulting powder is usually homogeneous enough for semiquantitative analysis (see Sec. 12.2.2.4).

12.2.2 Powders

Samples submitted for analysis in the powdered form can range from ground samples at 100 mesh to silica sands at 20-30 mesh, or clay samples at 1200 mesh. It is difficult to quantitatively analyze most geochemical samples of powdered material without grinding the samples to 200-400 mesh, fusing the samples, digesting, or dissolving the samples in order to overcome the effects of heterogeneity, particle size, mineralogy, oxidation state, and absorption enhancement.

12.2.2.1 Heterogeneities

Floor vibrations and handling of powdered rock samples between the grinding laboratory and the analytical laboratory cause the heavier minerals to settle toward the bottom of the sample bottle, resulting in unmixing. Vibration problems are of special concern when ground samples are stored for long periods. Mixing of the 100-mesh or finer powdered sample by tumbling in V-blenders, hand mixing, "rolling" on noncontaminating paper, or mixing in plastic vials in mechanical mixers prior to splitting of the sample to be taken for analysis reduces sampling errors significantly. Errors as high as 15% relative occur if horizontal mechanical mixers are used. The heavy minerals at the bottom of the bottle are not mixed with the lighter minerals unless some type of tumbling action occurs [19].

Splitting of the sample taken for analysis should be done through the use of micro splitters, by taking portions from all parts of the sample in several increments or by coning and quartering the sample down to the approximate amount for analysis. Discussion of the effects of laboratory sampling error is given by several investigators [28-30].

12.2.2.2 Grinding Size

Calculations of matrix effects are doubtful unless particle size of the sample (loose powder, briquette, or thin film) is consistent within the sample and from sample to sample [31]. Differences in the particle size of the sample cause:

1. A decrease in X-ray intensity as particle size increases, especially for low-atomic number elements [18,31]
2. An increase in X-ray intensity with increased briquetting pressure for a given particle size
3. A decrease of X-ray intensity caused by mutual shielding of small particles by larger particles
4. Heterogeneity of the sample because the finer particles generally are the softer, lighter minerals and the larger particles are the harder, heavier minerals or the micas
5. Unmixing of coarse and fine particles when pouring the sample into sample cells or briquetting dies
6. Absorption of X-ray intensity of the low-atomic number elements due to sample surface roughness

Accurate analyses of powdered geological samples in the form of rock powders, fusion powders, ion-exchange beads, etc., are therefore best achieved if the irradiated sample is finely ground, homogeneous, completely flat, and of a constant packing density.

Grinding. Particle size effects are not serious in the short-wavelength X-ray region provided the sample is less than 200 mesh. For this reason, loose powders or briquettes of homogeneous silica sand are readily analyzed for elements of $Z > 25$ in the glass industry, often without grinding. However, elements of $Z < 25$ cannot be determined without grinding geological samples to less than 400 mesh. Theoretical calculations and experimentation of particle-size effects have been made by several authors [10,18,31-38].

Grinding of the samples to less than 400 mesh is most efficiently done with mechanical grinders since both a mixing and a grinding action occurs. Ball mills have less capacity (1-10 g) than the rotary disk mills already described, but a wider variety of inexpensive grinding media is available (alumina ceramic, tungsten carbide, tool steel, stainless steel, agate, and plastic). Since the control of particle size is critical, greater control of both sample weight (1-10 g) and grinding time (5-10 min) is necessary for quantitative analysis [26,39].

Sieve tests should be made on several samples to determine if the finely ground material passes 400 mesh. Nylon or silk screen do not contaminate the samples with the metals found in brass and stainless steel screens (Pb, Cu, Zn, Fe, Ni, Cr, Mn). After grinding, the sample is again homogenized.

Preparation of the Finely Ground Powder for X-Ray Spectrochemical Analysis. The ground homogeneous sample may be irradiated as a loose powder, briquette, or thin film. Selection of form depends on the analytical requirements for accuracy, speed, amount of material available for analysis, and the effects of particle size and packing density on the X-ray analyte intensity.

Where speed and minimal accuracy are required, loose powder preparation is most convenient. Disposable plastic cups and aluminum, steel, or glass planchets are used [40,41]. In inverted optic spectrometers, the sample cup is covered with Mylar. The disadvantages to this method of preparation include variability of packing density from sample to sample, difficulty of analysis of light elements in vacuum, and X-radiation from elements lighter than Si is severely attenuated by the Mylar film. It is most difficult to analyze loose powders in spectrographs having nonhorizontal sample changers. The powders settle with the rotation of the inclined samples in the changer which causes voids to appear at the top and surface of the sample cell. A uniform packing density can be approached with loose powders by carefully filling sample cells with a constant weight of sample which overfills the sample cell and by packing the powder without loss of sample. Back-filling the sample cell forces the larger particles toward the surface to be analyzed, thereby displacing the smaller particles which filtered down through the larger particles during filling of the sample cell.

The most consistent analyses of trace elements are obtained by briquetting powdered geochemical samples [9-13,15,16,18-22,26,39, 42-50]. Several investigators have determined the major rock-forming elements using powdered rock samples with some success [9,13, 16,26,46,51]. To reduce preparation errors, the briquette should have a uniform particle size of less than 400 mesh, be homogeneous, be packed to a uniform density and thickness, and have a smooth, flat surface. Uniform particle size and homogeneity is obtained by use of grinding techniques already described. Homogeneity can be maintained by using the back-loading technique previously described to minimize segregation of particles on the analytical surface. The briquette itself can be supported variously as by a backing of starch, cellulose, etc. [11,42,44], by a metal ring which forms a jacket [4, 44] or a combination backing and jacket of these materials [42-46], and by a metal jacket and backing also of these materials. Planchetting offers a rapid method of briquetting, but is less desirable as a support because the larger particles are forced to the bottom of the disk by compression, leaving an overabundance of smaller particles on the surface of the briquette.

The use of binders is suggested for preparing briquettes of samples which are not cohesive under pressure and to produce durable briquettes. Binders also reduce matrix effects, improve homogeneity and uniform packing density, and enable one to obtain a more uniform particle size in grinding [42,44,52]. Binders may contain contaminants, and must be weighed. Binders reduce X-ray analyte intensity by dilution and absorb low-Z X-ray intensities. Binders also increase background due to an increase in the scattered radiation. Many materials such as methyl or ethyl cellulose, starch, boric acid, detergents, and commercially available inert powders have been used for binders [11,42,44,52]. Boric acid is undesirable as a binder, as it forms interstices around the powdered geological samples causing heterogeneity [53]. This is due to absorption and loss of water under repeated vacuum conditions of analysis. Detergents may contain contaminants (P, Na, Ca).

Ratios of sample to binder of 1:1 to 85:15 have been used. The 1:1 preparation reduces matrix effects and produces a solid briquette. However, 5-20% of binder is recommended [15]. Indeed, the use of 1:1 preparation methods induces errors when the X-radiation of new high-intensity X-ray targets "burn" off the organic binders during repeated exposure to X rays. A backing material of 70% methyl cellulose and 30% spectrographic pure wax adheres well to the ground powder (85 parts sample and 15 parts chromatographic cellulose) and forms stable briquettes [54].

X-Ray intensity is affected by briquetting pressure. A pressure of 21 kg/mm^2 (15 ton/in.^{2}) is usually sufficient to achieve uniform packing density [15,55].* Vent holes in the briquetting die allow entrapped air to escape as the pressure is raised from 0 to 21 kg/mm^2 (15 ton/in.^{2}). Solid cohesive briquettes are formed by maintaining pressure for 1 min. Briquetting also maintains uniform briquette thickness required for exceeding the critical thickness of the sample, as sample thickness affects analyte intensity.

When sample size is less than that needed to meet critical thickness needs, corrections for intensity loss can be made if the sample weight is known. Figure 12.1 illustrates the percent intensity loss related to counts/sec when Sr is measured at sample weights of 0.2-0.5 g to which an equal amount of cellulose binder was added. The maximum intensity loss of 35% was observed for the 0.2 g sample. Similar measurements were made for CaO and the maximum intensity loss observed was 1.7%. As only a few micrometers depth of the sample are effectively analyzed for CaO, this was expected.

Attenuation of the light elements occurs if the analyzed surface is not completely smooth and flat. Highly polished tool-steel platens or clean glass lenses produce smooth, flat briquette surfaces [40,42-46]. Tool steel platens are repolished as the surface becomes pitted with use. Care should be taken to prevent oxidation on the platen surface. Glass lenses are cleaned to remove small glass droplets extruded onto the surface during manufacturing [42]. The lenses have a tendency to shatter if the pressure is not raised slowly from 0 to 21 kg/mm^2 (15 ton/in.^{2}).

Thin-film techniques are useful when a limited amount of sample is available for analysis, when absorption-enhancement effects are to be minimized, and when recovery of the sample on a thin film is the only method of collecting the sample [56-62]. The loose powders may be supported by Scotch tape, by distributing the powder between two layers of thin Mylar film supported by a ring or cylinder, or on filter paper. An even distribution of the powder on the support medium

---

*In the now internationally recommended system SI, 1 tan/in.^{2} corresponds to 1.52 × 10^7 Pa (Pascal = Newton/m^2).

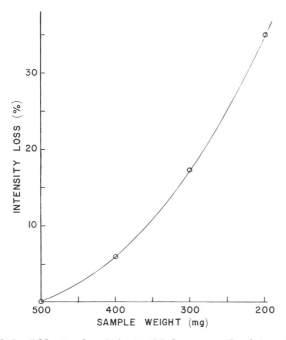

FIG. 12.1 Effect of critical thickness on Sr determinations.

is necessary to obtain quantitative data. Several thin-film methods of determining trace and major elements in geological samples have employed ion-exchange methods [62-69]. Other workers have used coprecipitation techniques or extraction techniques [56,70-73]. The work of Giauque et al. represents a unique method of filtering rock powders onto filter paper for trace element analysis [61]. The effects of particle size on analyte intensities prepared as thin films [74,75] and as ion-exchange resin thin films [76] can be significant.

12.2.2.3 Fusion of Geological Samples

The effects of mineral composition, oxidation state, heterogeneity, and absorption enhancement on the quantitative major element analysis of geochemical samples are overcome by fusion techniques when combined with mathematical corrections. Many techniques have been devised, adopted, or modified to meet particular analytical needs.

The analyst selects that combination of fluxes, crucibles, melting methods, and casting-briquetting methods which meet his analytical requirements. Selection is based on the following considerations:

1. All of the sample should be dissolved by the flux.
2. Elements in the flux do not negate or interfere in the analysis, e.g., Na, K, La, Ce, Br.
3. The method reduces mineralogical, oxidation state, and absorption-enhancement effects.
4. The method does not seriously affect the detectability of the analytes.
5. The briquette, pressed powder or glass disk, is homogeneous, flat, of constant density, and has no surface effects.
6. The glass disk is strain-free, free of bubbles, and requires little or no polishing.
7. The cost of reagents, crucibles, furnaces, etc., are not excessive.
8. The method is easily performed.

There is no panacea fusion method. Salient characteristics of the principal fusion methods are described here. Application of fusion methods to geochemical analysis and applied mathematics are discussed in Secs. 12.3 and 12.4.

Samples have been fused with fluxes at various ratios from 1:1 to 1:100. Minimum dilution fusion techniques, 1 part sample to 1-5 parts flux, yield high X-ray intensity for the low-Z elements, and minor-trace element analysis is possible [12,13,22,49.77,78,80-84, 136]. Matrix effects are only minimally reduced [85]. Norrish and Hutton [12] combined 38 g $Li_2B_4O_7$, 29.6 g $LiCO_3$, and 13.2 g La_2O_3 to prepare a flux. 1.5 g of flux, 0.02 g of $NaNO_3$, and 0.28 g of sample were weighed, mixed, placed in a Pt-Au crucible, and fused at 980°C. The melt was stirred during fusion with a Pt wire, poured onto a graphite disk, and an Al plunger was used to compress the melt inside a brass ring which sits on the graphite.

Moderate dilution fusion techniques, 1 part sample to 6-10 parts flux, with or without a heavy absorber, are useful when a compromise is made between reduction of matrix effects and analyte intensity [11,86-94]. Rose et al. [11] adapted the lanthanum-heavy absorber technique to a wide range of samples. They combined 1.0 g of $Li_2B_4O_7$ with 0.125 g of La_2O_3 as a flux to which 0.125 g of sample was added for fusion in a graphite crucible at 1100°C for 10 min. The bead was allowed to cool after removal from the furnace, and weighed. H_3BO_3 was added to the bead to a total weight of 1.3 g and the components ground in a mixing mill. The finely ground powder was pressed into a briquette having a backing of H_3BO_3 at 50,000 pounds per square inch (psi). These authors have since substituted cellulose powder for H_3BO_3. Kodoma et al. [94] combined 4 g of $Li_2B_4O_7$ and 1 g of $LiCO_3$ with 0.5 g of sample. The mixture was fused at 1000°C over a Meker burner for 10 min with continuous agitation to remove bubbles. The melt was quenched in water, the glass crushed, and the powder transferred to a disk-shaped graphite mold. The mold was placed in a furnace at 1000°C for 1 hr, removed, covered with a hot graphite cap, and allowed to cool slowly. They also modified the method of Norrish and Hutton [12] by pouring the melt onto the heated (300°C) graphite mold, which was placed in a cool furnace, heated to 450°C for 1 hr, and allowed to cool.

High-dilution fusion techniques, 1 part sample to 15-100 parts flux, are useful in significantly reducing matrix effects. However, longer counting times are required and sensitivity is reduced [10, 19,21,53,93-96]. Claisse [10] diluted 0.1 g of sample with 10 g of $Na_2B_4O_7$. The mixture was fused in Pt crucibles at 800-1000°C in a furnace or over a Meker burner, and agitated to remove bubbles. The melt was poured onto an Al plate at 450°C for preliminary annealing. The glass disk was placed on a heated Transite plate (450°C) and allowed to cool. 1.5-2.0 g of BaO_2 was added to the flux to dissolve sulfide ores. 1 g of potassium pyrosulfate and 0.1 g sample were fused at 1000°C before adding the $Na_2B_4O_7$ when complex sulfide ores were fused. Fabbi [53] fused 0.1 g of sample with 1.4 g of $LiBO_2$ in a graphite crucible at 950°C. The bead was allowed to cool, weighed, placed in a grinding vial, and chromatographic cellulose was added (total weight, 1.8 g) before grinding in a mixer mill.

The ground powder was pressed at 30,000 psi into briquettes having a backing and edge of methyl cellulose.

Tertian [98] has applied a double-dilution fusion method which allows interelement correction [99] by double measurement of the two preparations. 9.2 g of flux (90% $Li_2B_4O_7$ and 10% LiF) was mixed with 0.8 g of sample in the first dilution, and 8.0 g of flux is mixed with 2.0 g of sample in the second dilution. The samples are fused in Pt crucibles in a furnace at 1100°C for 50 min, and agitated occasionally. The melts are poured onto a heated (370°C) plate to form glass disks and allowed to cool.

Modifications to the above methods include the use of other reagents, crucibles, and manual or automatic fusion devices. The primary fluxes in use are $LiBO_2$, $Li_2B_4O_7$, $K_2S_2O_7$, $Na_2B_4O_7$, and Na_2CO_3. The alkaline fluxes more readily attack and dissolve the refractory rock-forming minerals [100] $LiBO_2$ [4,53,100], $Li_2B_4O_7$ + Li_2CO_3 [81, 90,94], $Li_2B_4O_7$ + B_2O_2 [78], and $Na_2B_4O_7$ + $Li_2B_4O_7$ + LiF. $K_2S_2O_7$ has special application to the dissolution of ores [10,97]. $LiNO_3$ [101], $Ce(NH_4)_2(NO_3)_6$ [19], $NaNO_3$ [12,22,83], and NH_4NO_3 are added to the flux to oxidize sulfides, carbides, and metals. La_2O_3 [10-12,86-89] as well as Ce_2O_3 [87,90] and $Ce(NH_4)_2(NO_3)_6$ [19] are also used as heavy absorbers to moderate matrix effects. Strain-free glass disks are prepared by adding Na_2CO_3 to the flux or by procedures outlined by Drummond [102] or Kodoma et al. [94].

Graphite and platinum crucibles were the forerunners of modern fusion vessels. More recently palau [94] and Pt-Au-Rh [93] and Pt-Au [22,82,91,103-105] crucibles have gained preeminence because of their nonwetting characteristics. Fusion melts are either poured easily into casting molds, or the melt is allowed to solidify in the crucible. After cooling, the solid disk disengages readily from the crucible. Surfaces of the Pt-Au crucibles may be attacked by high concentrations of transition metals during fusion. The surface is restored after cleaning by heating the crucible, containing a 10% $AuCl_2$ solution, until red hot. The Au amalgamates with the Pt as the solution is heated and swirled.

The time-consuming aspect of fusion methods has been a disadvantage. Analysts have recently developed semiautomatic [22] and automatic fusion devices [92,93,105] which markedly increase productivity over manual methods and also reduce the tedium of manual methods. Harvey et al. [22] and Hebert [4] have incorporated a multiple sample fusion apparatus combined with the method developed by Norrish and Hutton [12] of casting glass disks in rings. LeHouillier and Turmel [92] have applied the six Meker-burner fusion device developed by Claisse [106] to mineral investigations. Whittman et al. [93] have developed a high-frequency furnace which melts the sample, transfers the disk for annealing, and slides the cooled finished disk down a chute where it is ready for analysis. Matocha [105] developed an automatic fusion device which controls the time and temperature of a Meker burner required for low heat (550°C), high heat (1100°C), mixing, and cooling. Two additional fusion units can be controlled by a master control unit. Costs of these devices vary from $1500 to $5000, but a significant saving in time (4-10 times) is achieved over manual methods.

Certain precautions should be noted with regard to fusion methods. Beads ground to fine powders and briquetted should be stored in desiccators as most are hygroscopic. Three to five percent relative errors may result if the briquettes absorb moisture prior to analysis because the surface becomes convex and the sample to X-ray target distance is lessened. Preliminary pumping of the briquettes in a vacuum bell jar outgasses the briquettes sufficiently to hasten spectrograph pump down time. Glass disks poured into rings are more durable than disks which are not because strain fracturing is reduced. Furthermore, they are not easily contaminated by handling. Heterogeneity may occur in glass disks during fusion, if mixing or other agitation is insufficient [84,92,107]. The glass disk may have surface irregularities. Polishing with grinding disks [90] imbedded with 600-grit diamonds gives a relatively smooth surface. Il'in and Loseva [108] recommend polishing with diamond dust having a particle-size of 0.25 μm. Hattori [91] has periodically shaped the bottom of crucibles with a plunger to keep the crucible surface mirror flat and to maintain the shape of the crucible. Temperatures in excess of 1100°C may cause volatilization of alkali elements from the sample.

Preparation methods for microprobe analysis of rocks [227,228] and minerals are discussed in Chap. 7. Worth noting, however, is a novel method of preparing silicate rock glass for microprobe analysis. Nicholls [109] fused 10-20 mg of rock powders on an iridium-strip heater at 1600-1800°C. The temperatures are obtained by passing a 10 to 40-A current at 3-5 V ac through the iridium strip. The melt is stirred with a Pt rod during fusion requiring 20-30 sec.

12.2.2.4 Cleanliness and Contamination

Cleanliness of the laboratory and grinding equipment is essential to avoid cross contamination of the samples. All grinding surfaces, mixing apparatus, and splitting devices should be carefully cleaned. Recent work by Agus and Hesp [110] describe excellent cleaning procedures.

The choice of the grinding medium (Al ceramic, tungsten carbide, tool steel, agate, etc.) is dictated by the maximum allowable contamination for a given geochemical investigation. Al and Si contamination of samples by ceramic rotary plate grinders and mortars and pestles seriously affect the analysis of mafic* rock samples [26,50,110]. Contamination of samples with Fe, Ti, Cr, Cu, and Ni by hardened-steel rotary plate grinders is also significant [50,110]. Agus and Hesp have documented the contaminants, Fe, Mn, Cr, Co, and W, which are introduced during grinding by agate mortar and pestles, agate and steel rotary ball mills, Fe rotary plate grinders, and by rotary disk mills having chrome-steel or tungsten carbide liners. The minimum contamination of these elements is 25 ppm Fe with the agate mortar and pestle. Maximum contamination of 1.35% Fe was observed with the rotary ball mill. The rotary disk mill introduced insignificant amounts of Fe, Mn, and Cr (90, 2, and 2 ppm, respectively). It did introduce 200 ppm Co and 1200 ppm of W, however. Grinding time of

---

*Mg and Fe are major constituents.

less than 1 min is needed to obtain a minus-200 mesh sample, whereas the other devices required from 5 to 45 min. The rotary disk mills can be considered advantageous in speed and minimal contamination.

Samples received in the X-ray spectroscopy laboratory require grinding to less than 400 mesh for reasons given earlier. The largest surface area of the sample is exposed to the grinding medium when grinding the sample to minus-400 mesh or smaller. Hence, the potential for contamination is greater. Hardened tool steel grinding balls and tungsten carbide grinding balls can add many contaminants to samples. Table 12.2 indicates the levels of contamination introduced by both types of grinding balls when 1 g of spectroscopically pure quartz was ground in polystyrene vials in a mixer mill.

The tool-steel ball contaminates the quartz with high levels of Fe, Cr, and W and low levels of Mn, Mo, Ni, Ca, and Cu. An increase of 30% contamination of these elements is observed when three 3/8-in. steel balls are used. Of total consternation to the analyst, however,

TABLE 12.2

Contamination of Spectroscopically Pure Quartz by Grinding Media (ppm)[a]

Element	Unground quartz	Tool steel 1 ball 1/4 in.	Tool steel 3 balls 1/4 in.	WC-1 1 ball 3/8 in.	WC-2 1 ball 3/8 in.
Fe	<10	1500	2000	<10	<10
Ti	<2	<2	<2	300	<2
Mn	<1	7	10	<1	<1
Cr	<1	200	300	<1	<1
Mo	<2	10	15	7	<2
Ni	<1	7	10	15	3
Cu	<1	2	2	<1	<1
Co	<2	20	30	500	100
W	<30	700	1000	850	700
Nb	<7	<7	<7	70	<7

[a]Emission spectroscopic analysis, R. E. Mays (analyst), *U. S. Geol. Survey*, Menlo Park, California.

is the difference and high levels of contamination introduced by the WC-1 ball vs. the WC-2 ball. The presence of Ti, Mo, Nb, and the high levels of Ni, Co, and W can be ruinous in geochemical investigations. The balls supplied by different vendors were of supposedly high purity WC. WC-1 was found to be made from scrap material. Although not shown in this investigation, Ni is also used as a binder instead of Co in manufacturing WC balls. The analyst can not assume that grinding media or reagents will be the same purity from lot to lot.

12.3 QUALITATIVE AND QUANTITATIVE ANALYSIS

X-Ray spectrometry is very adaptable to the qualitative and quantitative analysis of geological samples. The technique has versatility through selection of optimum spectrometer parameters, flexibility in sample preparation, selectivity in calibration methods, and a variety of mathematical methods to correct for matrix effects.

Energy-dispersive spectrometers (EDS), Chap. 2, are effective in providing rapid simultaneous qualitative and semiquantitative analysis. Wavelength-dispersive spectrometers (WDS), Chap. 1, provide qualitative analyses at a slower rate because of sequential scanning. WDS provide more accurate semiquantitative analyses and quantitative analyses than EDS. This is because the resolution of WDS (1-3 eV) is so much better than EDS (140-180 eV). Quantitative analysis with EDS is limited to about 2-10% relative [111].

12.3.1 Calibration Standards

Selection of calibration standards and of the method of calibration is dependent on the analytical requirements of the geochemical investigation, the availability of analyzed or artificial standards, and the type or amount of material to be analyzed.

12.3.1.1 Qualitative Analysis Standards

Spiking a carrier (cellulose, boric acid, water, starch, etc.) with the analyte of interest provides a calibration standard which is free of interfering lines. Single element briquettes are useful in peaking spectrometer analyzing crystals or scanning energy ranges with the pulse-height analyzer. By adding about 10 elements of interest to the carrier, the standard has more uses and the spectra is not too complex. Standards are available which contain multielements at several ranges of concentration. Line interferences from multielement standards may affect qualitative interpretation of the spectra.

Multielement artificial standards useful in trace and major element analysis, the number of elements in the standards and the concentration range are given in Table 12.3. Standards used at the U. S. Geological Survey (U.S.G.S.) are powdered material to which the trace elements have been added to a silicate matrix containing seven other major elements. The National Association of Technical Research (ANRT) standard is a powdered material containing the trace elements in a silicate matrix having eight other major elements. The National Bureau of Standards (NBS) and Standard Reference Materials (SRM) are glass disks. The Spex standards are available as powders in a graphite matrix.

TABLE 12.3

Multielement Artificial Standards

Source	Standard number	Number of trace elements	Concentration range
U.S.G.S. [112][a]	GSA-E	50	0.0-500 ppm
A.N.R.T. [113]	VS-N	28	500-1000 ppm
N.B.S. [114]	SRM 610, 612, 614, 616	61	0.02-500 ppm
Spex Ind. [115]	Rare earth	16	5.28%
Spex Ind. [115]	Multielement	49	0.0001-1.29%

[a]These artificial glass standards were prepared for internal use.

12.3.1.2 Quantitative Analysis

Geochemical samples are extremely variable in elemental composition and mineral species. Standardization directly affects accuracy because X-ray spectrometry is a secondary or comparative method of analysis. Several authors [15,49,116-118] have discussed in detail the theory and methods of standardization and calibration. Applications of these methods to geological samples are discussed here.

Artificial Standards. Artificial standards are prepared by combining the major and trace elements of interest, usually as the oxides, at concentration levels which approximate those of the samples to be analyzed. Because the composition of these standards and of the samples to be analyzed are matched closely, matrix effects are slight. However, because the reagents used are powders recovered by precipitation, the particle size may be less than that of the samples. Hence, particle-size effects may introduce some error. Additionally, oxidation state differences of an element between the standard analytes and the sample analytes may yield nonproportional X-ray intensities. The analysis of sulfide samples for trace elements is especially susceptible to such variations [119]. Wood and Bingham [120] overcame this difficulty by roasting the standards and samples at 500°C.

Rare earth standards [121] have been prepared by adding aliquots of standard solutions at desired concentrations to 10-ml beakers. The solutions are evaporated to dryness, 1 ml of 5 + 95 HNO_3 added, the solution absorbed onto chromatographic cellulose, and mixed. After drying overnight, the dried cellulose is mixed in a boron carbide mortar and pressed into a briquette at 52.7 kg/mm^2 (37.5 $ton/in.^2$).

The addition of spec pure reagents to a SiO_2 base for trace element analysis was suggested by Hower [122] and adopted by others [47,50,51,123-125]. Webber and Newbury [47] found that a multielement standard containing 49 elements mixed with SiO_2 had more interferences than naturally occur in geochemical standards.

Major element calibration with artificial standards has been applied to the analysis of silicate rocks [82,90,94], metallurgical slags [82], and cements [126].

Internal Standards. When only a few analytes are to be determined and the range of analyte concentration does not vary excessively, the internal-standard method is applicable to the determination of elements heavier than Ca. The wavelength of the analyte and that of the internal standard must be close together and on the short-wavelength side of the absorption edge of matrix elements. The ratio of the X-ray intensity of the analyte to that of the internal standard is multiplied by the concentration of the internal standard to calculate the analyte concentration. The internal standard method has been used to determine Th with Tl as the internal standard [127]; Ba, Ti, and Zn in sediments with La and As as internal standards [128]; Rb and Cs with Sr and I as internal standards [129]; Cu with Pb as the internal standard in ores [120]; and Zr in rocks and minerals with Mo as the internal standard [130]. Several sets of international silicate standards were analyzed for 14 trace elements using two sets of internal standards [131,132]. One set contained Te, Ce, Mo, and Co as internal standards, and the other contained Y and Ge.

Standard Addition. Known amounts of the analyte are added to a portion of the sample. The ratio of the measured intensities from the spiked and unspiked samples is related to the amount added to the spiked sample to calculate the concentration in the untreated sample. Application of the technique to silicate rock standards trace analysis has been used with some limited success [133-135]. Analytical errors of about 10% of the amount present occur with this method of calibration.

Scattered X Rays. The use of scattered X rays to correct for matrix effects was suggested in 1958 [136]. The ratio of coherent to incoherent radiation scatter was also found useful [137]. The method has been applied to the determination of Zr in rocks [138], Mn in rocks and marine sediments [139], and to the accurate determination of Rb and Sr in silicate rock standards [140].

Analyzed Samples and Standard Reference Materials (SRM). The predominant calibration method employed by X-ray spectroscopists for geochemical analysis is the method of calibration standardization. The X-ray intensity of the analyte from the samples is referenced to that of one or more standards.

Geochemical samples are often analyzed by classical chemical or spectrophotometric methods. These samples may provide the X-ray spectroscopist with useful secondary standards. Caution in their use is suggested. The secondary standard obtained by the X-ray analyst may not be the same split as that which was analyzed by other methods. It may be of a different mesh size or may not be sampleable for the analyte of interest [28,29]. The precision and accuracy of the other methods may be less than that needed for good X-ray spectroscopic calibration. Accurate secondary standards are mandatory if they are to be used for calculating correction coefficients.

Less than ten years ago there was only a handful of well-analyzed geochemical SRM. Recognizing this dearth, several institutions and agencies initiated programs which have since made available a wide range of geochemical standards. These sources of geochemical standards have been summarized by Flanagan [141,142]. The preparation and use of SRM is discussed by Gillieson [143]. Methods of determining the usefulness and quality of SRM should be considered by the analyst [28,29,144]. Chemical data for SRM reported in the literature may be recommended or average values [145]. The average values may not, in fact, be the "true" values. Abbey has made critical evaluations of the several sets of international standards. On the basis of empirical calculations, he rejected 15% of the data which were farthest removed from the mean values [146-150]. When values reported for SRM are questionable, the analyst must carefully assess those techniques used in arriving at average values. Primary methods of analysis such as isotope dilution provide excellent data for trace element calibration [140,151,152,192]. Data obtained by spark-source mass spectrography [153] and neutron activation are also useful [154,155]. The analyst may have no choice in selecting values for major elements except to utilize classical data obtained by recognized classical chemists.

Highly accurate X-ray spectrometric analyses of geochemical samples have been reported by many investigators using the calibration standardization method. Examples of this method are given in Sec. 12.4 since there is such a profusion of papers.

12.3.2 Mathematical Correction for Matrix Effects

Mathematical corrections for matrix effects are basically of two types: coefficient methods [5-7,15,49,82,99,117-118,124,125,156-179] and the fundamental-parameter method [6]. Tertian combined the two methods into a generalized iteration method [178]. Boniforti et al. [179] has presented a critique of all of these methods. Derivations of these corrections are found in the references given and are also in Chaps. 2 and 3. Application of the methods to spectral-interference and absorption-enhancement effects as related to geological samples are discussed here.

12.3.2.1 Spectral Interference Background

Spectral line overlap of an adjacent element on the emission line of the analyte can often be overcome through proper selection of instruments, through proper choice of crystal, kV, collimators, filters, pulse-height analyzer settings, counter gas, etc. Whenever instrumental manipulation is ineffective, mathematical corrections are used to generate accurate analyses. When one element interferes, the correction is made as follows:

$$I_a = I_m \times k \qquad (12.1)$$

where I_a is the intensity in count/sec of the analyte corrected for spectral interference, I_m is the measured intensity of the analyte, and k is negative and equal to $I_{i,m}/I_i$, where $I_{i,m}$ is the intensity of the interfering element measured at the analyte 2θ angle, and I_i is the intensity of the interfering element measured at its 2θ angle.

For this method to be successful, the geologic sample must contain low levels of the analyte. This method compensates for errors due to matrix effects, whereas the alternative method of measuring a spiked blank does not [15]. This method or variations of it have been reported in the determination of many trace elements in SRM [19,47,125,180-186]. It has also been applied to the correction of multiple interferences in determining the rare earth elements [121]. A method of correcting background by solving a series of simultaneous equations has been used to correct for the mutual interferences of Ba-Ce, Cr-V, Ti-V, and Sr-Zr [187]. Multiple regression methods have also been reported [16,180].

The analysis of trace elements is influenced by the background radiation which varies from sample to sample. Correction for background is necessary since the integration of the analyte signal includes background.

Background measurements are often made by subtracting the average of the background measured on both sides of the analyte peak.

$$TB = \frac{B_1 + B_2}{2} \tag{12.2}$$

where TB = true background, and B_1 and B_2 are the two background intensities. When the spectrum is curved, the expression given below is more accurate [47,186].

$$TB = B_1 + f(B_2 - B_1) \tag{12.3}$$

where f is the correction factor for spectrum curvature. The foregoing equations are given to point out to the reader that background measurements which are used to calculate correction coefficients for spectral interference and absorption enhancement may be in error unless spectrum curvature is corrected. A method for correcting background using the Reynold's method [137] has also been described [212].

12.3.2.2 Absorption Enhancement

Preparation techniques are not sufficient or are too cumbersome to entirely correct for matrix effects because of the diverse mineralogy and chemical composition of geological samples. Spectroscopists have utilized existing mathematical methods or devised additional expressions to quantify analyses of geological samples.

Absorption Correction Method. Hower [122] developed curves of relative mass-absorption coefficients by relating various minerals to Al_2O_3. Trace element analyses were obtained by determining the absorption of the mineral sample relative to Al_2O_3 at the wavelength of the analyte. The method was also adapted to traces by relating absorption coefficients to SiO_2 [123]. More recently, the technique has been applied to major [125] as well as trace elements [124,125, 140,188]. A method of directly measuring the absorption of a thin film of sample was proposed by Salmon [189] and later refined by Norrish and coworkers [12,49,83,190,191]. By measuring the intensity of the analyte wavelength, with the absorber in and out of position, the mass absorption coefficient was then calculated. The acceptance of this method was evidenced by the large number of analysts who are using the method [182,186,192]. The method is highly accurate and useful for the determination of analytes having short wavelengths. It does require the measurement of the total absorption of the sample prepared as a thin specimen. And, it also requires the measurement of analyte intensities in a separate split of the sample prepared as a pressed powder briquette.

Fundamental-Parameters Method. This method requires few standards and has application to varying matrices [6]. It does require the use of known factors for calculation of percent concentration (the spectral distribution of the primary beam, estimated or measured, the absorption tables, and the fluorescent yield). The disadvantages are the necessity of measuring the primary beam spectrum and the need for a large computer memory as many iterations are required. This method has been applied to the analysis of alloys and phosphate rocks [177]. Extensive use of the method was made by Toulmin et al. in developing an energy-dispersive spectrometer to characterize the elemental composition of the surface materials of Mars [197].

Effective Wavelength Method. A method proposed to correct for matrix effects is independent of the spectral distribution of the X-ray source, and requires one standard per analyte. Stephenson applied the method to a variety of sample types through the use of a FORTRAN program CORSET [168]. Both primary and secondary fluorescence corrections are made. The weight fraction of each analyte is iteratively adjusted until observed and calculated intensities agree in order to obtain the best estimate of the weight fraction. There is some controversy as to the equivalence of the terms effective wavelength and equivalent wavelengths [169-172,193,194]. Tertian has applied an "analytically equivalent wavelength" with Lachance-Traill equations to the analysis of geological samples by means of the double-dilution preparation method [98,99]. This method has been applied to solutions of iron ores [195] and may have application to solvent extraction methods [196].

Influence Coefficient Methods. Influence coefficient methods relate measured X-ray intensity to concentration in multicomponent systems. Correction of matrix effects is made by calculating the correction constants through the use of mass-absorption coefficient data and by polynomial multiple regression.

A modification of expressions derived by Beattie and Brissey [5] was applied to multicomponent major element analysis through the use of dual preparations with satisfactory results [78], and the method has been applied to silicate trace analysis [47,50,51]. Because the influence coefficients are obtained by first approximations, they are not entirely accurate.

By summing the effect of all coefficients on weight fractions, which involves all interfering elements, Lachance and Traill [159] effectively determined all α influence coefficients simultaneously. The method has been applied to limestones [198], solutions [199], and silicates [77] with good results. By increasing the number of standards and sets of simultaneous equations, improved influence coefficients are calculated. This may require more computer memory than is available in dedicated minicomputers.

Lucas-Tooth and Pyne [163] simplified the equation through summations of interfering element intensities. Their method is only useful over limited concentration ranges, requires large numbers of standards of known composition, and requires the solution of many simultaneous equations, hence a large computer is needed. Analysis of silicate rocks is reported using this method [94]. Other analysts [198] determined that the method was not applicable to limestone analysis using unfused samples.

Norrish and Hutton developed a formula having the same basic form as that of Lucas-Tooth and Pyne, but using iterations to improve the calculations. They derived correction coefficients by experimental measurements of intensities and by calculation, utilizing published mass-absorption coefficients. The agreement between the measured and calculated coefficients was good. As their coefficients do not rely on direct count rates, the coefficients have been used by several analysts who also prepare samples in the same manner [12]. The method has been successfully employed for major element analysis using wavelength-dispersive [22] and energy-dispersive [4] spectrometers.

Rasberry and Heinrich [7] have proposed a method which is applicable to samples having wide ranges of concentrations and which requires few standards. Their method includes an influence coefficient β which accounts for secondary fluorescent effects.

Multiple regression methods have been used to calculate influence coefficients [166,167]. Analyte X-ray intensity is correlated to concentration with linear, multiple-linear, and higher order polynomial equations. The influence coefficients are generally evaluated on the basis of the best fit of data to the regression line. Large numbers of well-analyzed standards are required ($2n^2 + 1$, where n is the number of elements affecting the analyte). The method has found use in silicate analysis [19,52,53,180,200], but was inadequate for the analysis of phosphate rocks [201].

Albee and Ray [202] have extended the binary correction influence coefficient to multicomponent systems by using the concentration-weighted average method for use in analyzing naturally occurring silicates, oxides, carbonates, phosphates, and sulfates with the electron probe. Large computer programs for rigorous data reduction of quantitative electron probe analysis are in current use [229]. A smaller version for on-line correction of X-ray data is made with a minicomputer [230].

12.4 EXAMPLES

Accurate X-ray spectrometric analyses of geological or other samples are dependent on sampling, sample preparation, instrumental parameter selection, calibration, correction of spectral line interferences, and matrix effects, contamination, and counting precision. Recent advances have vastly improved the ability of analysts to quantify analyses, increase sensitivity, and expand the range of elements determined in geochemical samples.

12.4.1 Silicate Rocks

12.4.1.1 Major Elements

Initially, major element analyses of silicate rocks were made by calibration standardization methods on loose powders [203] and briquettes [26,45,46,51,77,91,204]. Because of its platy structure, biotite mica does not grind easily, resulting in both positive and negative deviations of major element intensities [205]. The deviations may also depend on the composition of the biotite species [206]. Relative errors of 5%, 15%, and 25% for SiO_2, Al_2O_3, and total Fe_2O_3, respectively, are common. The use of multiple regression mathematics improved the analyses somewhat [200], as did the use of calculated influence coefficients based on the Lachance-Traill method [77] (2%, 4%, 6% for SiO_2, Al_2O_3, and total Fe_2O_3, respectively). More recent work [125,180] indicates that improved calculation methods have also improved determinations of SiO_2, Al_2O_3, and Fe_2O_3 to 1%, 2%, and 3%, respectively. These three oxides are discussed since they are most affected by matrix effects in silicate rocks.

Fusion techniques which overcome mineralogical and matrix effects improve accuracy significantly. Low and moderate dilution fusion methods [13,85,91] are improved with the use of mathematical corrections [22,49,77,78,80,82,83,94,98], high dilution ratios [53], heavy absorbers [11,86-88], and larger numbers of well-analyzed calibration standards. These exciting advances in methodology have carried X-ray spectroscopy to the forefront of rapid analyses which are competitive with conventional chemical methods. Table 12.4 indicates the improvement obtained in analytical accuracy, especially for SiO_2, Al_2O_3, and Fe_2O_3, as the progression is made from the method of unfused powdered briquettes to fused briquettes on which the analyte concentration is corrected for matrix effects.

Even though energy-dispersive spectrometers are routinely applied to semiquantitative analysis of silicate rocks [111,207], the work of Hebert and Street [4] is an excellent example of optimizing all analytical conditions to effect quantitative silicate analyses. Similarly good results were obtained on fused samples with an energy-dispersive spectrometer on an electron microprobe [109].

With the advent of the demountable X-ray target tube developed by Henke [3], analysis of the light elements Na-C in geological samples became a reality. He and other investigators determined Na_2O, MgO [208], and oxygen [209,210] in silicate rocks with this instrument. Bernstein and Mattson [211] used a curved-crystal high vacuum spectrometer to determine Mg-F in geological samples. A light field portable soft X-ray spectrometer has been used to determine Na_2O, F, and C [14]. Conventional X-ray spectrometers equipped with KAP [22,83,98], RAP [21,82,88], or TAP analyzing crystals are now used with good precision, accuracy, and sensitivity to determine Na_2O in geological samples.

12.4.1.2 Trace Elements

Semiquantitative analysis of 38 elements in silicate rocks has been reported [213]. Utilizing wavelength-dispersive spectrometers and the methods described previously, analysts have reported trace

TABLE 12.4

Silicate Rocks; Comparative Results of Preparation/Mathematical Methods of Analysis (%)[f]

Oxide	BR Basalt					G-2 Granite				
	A[a]	B[b]	C[c]	D[d]	E[e]	A[a]	B[b]	C[c]	D[d]	E[e]
SiO_2	38.20	38.83	38.64	38.89	38.29	69.11	69.85	69.56	69.36	69.01
Al_2O_3	10.20	9.16	9.68	10.28	10.10	15.40	16.07	15.64	14.95	15.23
$Fe_2O_3(T)$	12.88	12.20	12.91	12.71	12.85	2.65	2.85	2.81	2.76	2.70
MgO	13.28	13.29	13.16	13.29	13.30	0.76	0.85	0.84	0.81	0.81
CaO	13.80	13.63	13.74	13.84	13.71	1.94	2.07	2.02	1.97	1.89
Na_2O	3.05	3.16	3.05	3.20	3.00	4.07	4.38	4.23	4.16	4.00
K_2O	1.40	1.44	1.40	1.41	1.39	4.51	4.55	4.54	4.38	4.48
TiO_2	2.60	2.56	2.60	2.62	2.58	0.50	0.52	0.51	0.50	0.52
P_2O_5	1.04	1.08	1.04	1.11	1.04	0.14	0.16	0.14	0.12	0.15
MnO	0.20	0.18	0.19	0.19	0.19	0.03	0.04	0.04	0.03	0.04

[a]F. J. Flanagan, *Geochim. Cosmochim. Acta*, 37, 1189 (1973); average values.
[b]Briquetted rock powder, least squares calculation.
[c]Briquetted rock powder, multiple regression calculation.
[d]Briquetted fusion powder, least squares calculation.
[e]Briquetted fusion powder, multiple regression calculation.
[f]*X-Ray Spectroscopy*, B. P. Fabbi (analyst), U. S. Geological Survey, Menlo Park, California.

determinations of many elements in silicate rocks, primarily in the SRM. Elements determined by many analysts include Cu, Ni, Rb, Sr, Zn, and Zr [123]; Ti, V, Cr, Mn, Co, Ni, Cu, Zn, Ga, Rb, Sr, Y, Zr, Nb, Sn, Ba, La, and Pb [133]; Sr, Rb, Th, Pb, Zn, and Ni [134]; As, Ba, Cd, Ce, Cs, Cr, Co, Gd, Ga, Ge, Ni, Nd, Nb, Rb, Sm, Sc, Sr, Th, V, Yb, Y, and Zn [51]; Zr [138]; Ba, Ce, Co, Cr, Cu, Ga, La, Mn, Nd, Ni, Pb, Pr, Rb, Sm, Sr, Th, Ti, V, Y, Zn, and Zr [182]; Co, Cr, Ga, Mn, Nb, Ni, Rb, Sr, V, Y, and Zr [186]; Cr, Cu, Mn, Ni, Pb, Rb, Sr, Zn, and Zr [47]; Ba and Sr [151,184]; As, Sb, Ni, Rb, Sc, V, and Zn [185]; As, Cr, Cu, Mo, Nb, Ni, Pb, Rb, Sn, Sr, U, Y, Zn, and Zr [131]; Ba, Cr, Mn, Ni, Rb, Sr, Zn, and Zr [52]; K, S, Rb, Sr, Y, and Zr [19]; Ba, Ce, Cr, Cu, Ga, La, Pb, Sr, V, Zn, and Zr [124, 125]; and Ba, Ce, Cr, Cu, La, Nd, Ni, Rb, Sc, Sr, Ti, V, and Y [180]. The understanding of the variables which affect the quantification of X-ray spectrometric trace determinations have improved the quality of the data. Table 12.5 indicates the accuracies obtainable when the effects of background, spectral line interference and absorption enhancement are corrected on selected analytes in powder briquette standards. Multiple regression expressions and Lachance-Traill influence coefficients were used to calculate the corrected values.

Energy-dispersive spectrometers offer a rapid means of multi-element analysis. Low levels (0.005-0.5%) of Mo have been determined in ore processing [214]. Total Fe, Rb, Sr, Y, Zr, Nb, and Mo were reported for standard rocks [215], as were the elements K, Ca, Ti, Fe, As, Cr, Cu, Ga, Mn, Ni, Pb, Rb, Sr, V, and Zn [61].

Extraction and separation methods are useful techniques for preconcentration of elements which are difficult to detect or which are masked by other elements. Co, Cu, and Ni [73] and Bi, Co, and Ni [69] have been determined by ion-exchange X-ray spectrometric methods. Application of coprecipitation methods has been made to the determination of the precious metals Pt, Pd, Rb, and Au [216]. Separation of Re from molybdenite by precipitation followed by X-ray analysis is preferred to colorimetric methods [217].

TABLE 12.5

Silicate Rocks; Comparison of Multiple Linear Regression[a]/Lachance-Traill[b] Influence Coefficients Applied to Trace Element Determinations (ppm)[c]

Element		W-1 Diabase	GSP-1 Grano-diorite	AGV-1 Andesite	BCR-1 Basalt	G-2 Granite	GA Granite	GH Granite	DR-N Diorite
Rb	Chem.[d]	22	241	68	48	174	175	390	80
	MLR[a]	24	241	71	46	175	177	389	74
	L-T[b]	24	242	71	47	171	172	391	76
Sr	Chem.[d]	198	247	674	342	477	301	8	400
	MLR[a]	198	233	677	331	476	300	21	410
	L-T[b]	194	234	675	335	480	304	19	415
Zr	Chem.[e]	100	544	227	185	316	164	147	135
	MLR[a]	97	526	228	177	307	158	165	136
	L-T[b]	93	542	237	185	314	157	154	136
Ba	Chem.[e]	169	1320	1240	690	1860	840	20	360
	MLR[a]	154	1317	1219	699	1948	850	17	391
	L-T[b]	122	1288	1250	673	1907	844	33	384

[a] Multiple linear regression.
[b] Lachance-Traill influence coefficients.
[c] X-Ray Spectroscopy, B. P. Fabbi (analyst), U. S. Geological Survey, Menlo Park, California.
[d] Isotope dilution values.
[e] Recommended values.

GEOLOGY

The importance of quantitative X-ray spectrometric trace element analysis is best illustrated through application of the method to geochronological age dating. Evaluation of several calibration methods regarding the accuracy of Rb and Sr [140,191,192,218-220] X-ray spectrometric determinations has been thorough.

Rare earth determinations of milligram quantities chemically separated from rare earth minerals have demonstrated high accuracy [121]. Fusion methods of determining microgram quantities in siliceous residues [221] and ion-exchange preconcentration techniques applied to silicate rocks [63] have broadened the capability of the analyst.

12.4.2 Carbonate-Phosphate Rocks

The pattern of quantitatively analyzing carbonate rocks (limestones) is akin to that of silicate rocks, and with the same results. Initial analyses of briquetted powders were poor (10%, 10%, 25%, and 20% relative errors) for CaO, SiO_2, Al_2O_3, and Fe_2O_3, respectively [222]. Fusion of the sample [104,223] improved the data, and use of the exponential absorption law also helped [222]. Accuracies of 1% relative were reported for CaO and SiO_2, and 2% for Al_2O_3 and Fe_2O_3. The analysis of Na_2O has become routine [224]. High-order polynomials [225] and utilization of Lachance-Traill influence coefficients [199] enabled analysts to determine the major constituents with accuracies of 1% relative or better [225,226]. Major and trace element associations in limestones and dolomites were determined after acetic acid separation (CaO, MgO, MnO, Fe_2O_3, Sr, Cr, Cu, Ni, Pb, V, and Zn [126]. The analysis of phosphate rocks [223] is markedly improved if mathematical corrections for matrix effects are fully utilized [201]. Again relative errors of less than 1% are observed. Analyses of several briquetted powder limestone standards are given in Table 12.6.

Fusion methods and mathematical algorithms are established primarily for a particular suite of sample types, for example, silicate rocks. When it is determined that the concentrations of analytes

TABLE 12.6

Carbonate Rocks; Major and Minor Oxide Analysis (%)[e]

		400[a] Dolomite	401[a] Limestone	402[a] Limestone	403[a] Dolomite + limestone	1a[b] Limestone	1b[b] Limestone	88a[b] Dolomite
SiO_2	Chem.	0.075	2.09	2.63	1.81	14.11	4.92	1.20
	XRF[c]	0.029	1.87	2.99	1.80	14.18	5.38	0.95
	XRF[d]	0.13	1.76	2.83	1.85	14.11	4.91	1.07
Al_2O_3	Chem.	0.029	0.22	0.53	0.43	4.16	1.12	0.19
	XRF[c]	0.054	0.44	0.46	0.27	4.16	1.25	0.14
	XRF[d]	0.073	0.29	0.51	0.32	4.16	1.14	0.21
$Fe_2O_3(T)$	Chem.	0.053	0.199	0.370	0.308	1.63	0.75	0.28
	XRF[c]	0.055	0.169	0.351	0.321	1.64	0.725	0.369
	XRF[d]	0.063	0.195	0.358	0.293	1.63	0.758	0.289
MgO	Chem.	21.50	3.60	5.74	13.78	2.19	0.36	21.3
	XRF[c]	21.71	3.10	6.14	13.70	2.13	0.49	21.09
	XRF[d]	21.54	3.69	5.92	13.79	2.20	0.30	21.24
CaO	Chem.	30.51	50.07	46.65	37.99	41.32	50.90	30.1
	XRF[c]	30.97	49.94	46.86	39.05	38.60	49.72	30.10
	XRF[d]	30.51	49.63	46.64	38.48	41.41	50.60	29.78

K_2O	Chem.	0.016	0.062	0.164	0.138	0.71	0.25	0.12
	XRF[c]	0.012	0.065	0.168	0.133	0.709	0.259	0.102
	XRF[d]	0.010	0.060	0.168	0.143	0.710	0.247	0.116
TiO_2	Chem.	0.004	0.020	0.032	0.026	0.16	0.046	0.02
	XRF[c]	0.003	0.021	0.031	0.029	0.160	0.047	0.015
	XRF[d]	0.003	0.020	0.031	0.030	0.160	0.044	0.016
P_2O_5	Chem.	0.00	0.035	0.048	0.044	0.15	0.08	0.01
	XRF[c]	0.014	0.051	0.045	0.029	0.150	0.051	0.012
	XRF[d]	0.010	0.042	0.052	0.025	0.150	0.077	0.012
MnO	Chem.	0.0052	0.011	0.019	0.018	0.038	0.20	0.03
	XRF[c]	0.012	0.014	0.016	0.014	0.037	0.199	0.025
	XRF[d]	0.006	0.013	0.020	0.013	0.039	0.200	0.031

[a] G. F. Smith Co., Columbus, Ohio, standards; C. O. Ingamells and N. H. Suhr, *Geochim. Cosmochim. Acta*, 31, 1347 (1967), analysts.
[b] National Bureau of Standards, SRM; certificate of analysis.
[c] Least squares calculation.
[d] Multiple linear regression calculation.
[e] X-Ray Spectroscopy, B. P. Fabbi (analyst), U. S. Geological Survey, Menlo Park, California.

exceed linear calibration, the analyst must completely recalibrate for the new rock type. An alternative method, which has produced satisfactory results, is to analyze a small number of secondary standards having similar composition to the unknown sample using the original fusion method and algorithms. The X-ray-calculated concentrations are plotted graphically vs. chemical concentrations. Least squares regression of line segments of the new extrapolation are calculated from the X-ray vs. chemical concentrations to derive the slope and intercept which cover the concentration range of interest. The method has been applied to sedimentary clay analysis when determining SiO_2, MnO, and CaO. The fused clay samples contained lower levels of SiO_2, <35%, and higher levels of MnO, >1%, and CaO, >20%, than for which the multiple regression algorithm was calibrated. Figures 12.2, 12.3, and 12.4 illustrate the extrapolation method. The line segments in Figs. 12.2 and 12.3 indicate the magnitude of the change in slope. Table 12.7 gives comparative results for the three oxides obtained by the multiple regression and extrapolation calibration methods. Unacceptable errors occur when analyte concentration exceeds linear calibration.

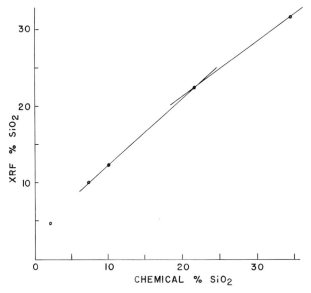

FIG. 12.2 SiO_2 extrapolation-calibration line segments.

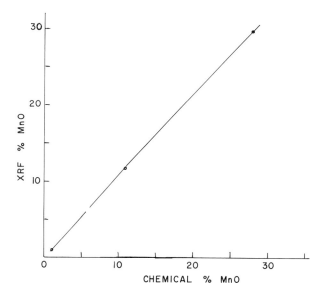

FIG. 12.3 MnO extrapolation-calibration line segments.

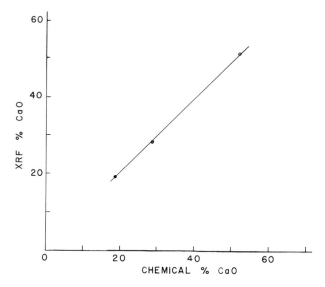

FIG. 12.4 CaO extrapolation-calibration line segments.

TABLE 12.7

Extrapolation Calibration Results for
SiO_2, MnO, and CaO in Sedimentary Clays (%)[a]

Standards					
SiO_2					
Chem.	2.29	7.40	10.10	34.40	21.52
XRF[b]	4.75	10.04	12.27	31.92	22.63
XRF[c]	2.29	7.37	10.13	34.39	21.40
MnO					
Chem.	1.04	1.21	10.82	27.90	
XRF[b]	1.03	1.18	11.73	29.57	
XRF[c]	1.06	1.19	10.82	27.82	
CaO					
Chem.	19.89	25.70	51.44		
XRF[b]	18.87	23.90	47.12		
XRF[c]	19.85	25.48	51.45		
Sedimentary clays					
SiO_2					
XRF[b]	12.52	25.08	34.59		
XRF[c]	10.36	25.95	37.70		
MnO					
XRF[b]	2.28	4.81	6.65	15.17	
XRF[c]	2.16	4.54	6.27	14.28	
CaO					
XRF[b]	43.15	27.13	18.27		
XRF[c]	47.01	29.09	19.18		

[a] *X-Ray Spectroscopy*, B. King (analyst), U. S. Geological Survey, Menlo Park, California.

[b] Oxide value calculated by multiple regression methods.

[c] Oxide value calculated by least squares regression of line segments.

12.4.3 Minerals

The analysis of minerals is more complex than that of rocks owing to greater compositional ranges and abundances of only a few elements. Matrix and oxidation effects can be severe, and an extremely large library of well-analyzed standards is needed. Unfortunately, few standards are available as SRM. Development of mineral SRM is complicated by the heterogeneity of many mineral species owing to incomplete separation of a given mineral. Separation of minerals is hampered by intergrowths, inclusions, and the formation of two or more species of the same mineral [206], each having slightly different compositions, during different periods of crystallization. These difficulties are overcome by fusion or dissolution methods and calibration methods previously described, and point by point counting several grains of sample with the electron microprobe [231].

Major element analysis of 5-8 mg of biotite and garnet has been reported utilizing finely ground pressed powders [232]. Control of iron contamination has resulted in accurate analysis of iron in garnets [233]. Iron has been determined in sphalerite [234]. Selenium was determined in sulfide minerals of powdered samples [235]. Trace levels of Rb have been determined in biotites [236]. Improved accuracies were achieved with fusion methods [11] and have been applied to the analysis of ilmenites [237], biotites, feldspars, muscovites, and hornblendes [238]. Solution methods [121] have been applied to the analysis of feldspar-pyroxene minerals, chromium minerals [239], rare earth minerals [121,240], and sulfide minerals [240,241].

The electron probe has preempted the X-ray spectrometer in the field of mineral analysis. This is because "bulk" material is not often recovered in separation procedures. Microprobe analysis is a broad subject in itself. Hence, application of the method to selected specimens is given here. The subject is discussed in Chap. 7, and Keil has reviewed the subject [242].

The electron probe has aided investigators in discovering and characterizing over 120 new minerals [242,243]. Analyses have been reported for clinopyroxenes, orthopyroxenes, garnets [244,246], chromites, magnetite, chlorite [245], pyroxene, amphibole, and biotite [247]. This method of mineral analysis has been extremely useful in characterizing minerals associated with ore deposits. Ag-Bi-bearing galena ores have been examined [248]. Investigations of sphalerite, chalcopyrite, pyrite, and bornite equilibria have been made [249]. Quantitative studies of inhomogeneities in native gold grains have revealed profiles of natural diffusion plus original chemical heterogeneity [250]. And, ruthenarsenite, iridarsenite, laurite, and cubic iron-bearing platinum have been analyzed [251]. Energy-dispersive analyzers have added a new dimension to electron probe analysis. Clinopyroxenes [252], hornblende, and other silicates [253] have been determined with energy-dispersive X-ray analysis and electron microbeam excitation. Scanning electron microscopes have been utilized to investigate pyrite, chalcopyrite, barite, chromite, and microcline [254]. The ion microprobe mass analyzer has been applied to the analysis of trace elements in amphibole minerals in situ in standard polished thin sections [255]. Without question, the electron probe will continue to dominate the realm of mineral analysis.

12.4.4 Lunar Samples

Every type and combination of X-ray spectroscopic instrumentation, method of calibration, method of sample preparation, and mathematical correction has been useful in analyzing lunar samples.

12.4.4.1 Meteorites

Before lunar exploration began, X-ray spectroscopists were analyzing extraterrestrial samples, primarily meteorites. Reports of these investigations are extensive and continue up to the present time. Major element analysis of tektites [256] have been reported.

GEOLOGY

Determinations of Fe, Ni, Co, Ca, Cr, and Mn in meteorites [257], Fe, Si, Mg, Ca, Al, K, Ti, Mn, P, Zr, Sr, and Rb in meteorites [258], and Ni, Ga, and Ge in meteorites [259,260] have been reported using X-ray fluorescence methods. The microprobe was found useful in determining Ni, Fe, and P in meteorites [261]. The FeS contents of sphalerites, a minor phase in some meteorites, were useful in calculating the pressures of formation of meteorites [262].

12.4.4.2 Samples Obtained from the Moon

X-Ray spectroscopy played an important role, as did other methods of analysis, in identifying the elemental abundances of geological material recovered from the surface of the moon during the Apollo mission program. Analyses of these materials were performed under almost clinical conditions to prevent contamination from any source.

X-Ray fluorescence analyses of lunar material collected by the Apollo 11 mission include major element analysis of rock fragments, pulverized rocks, and soils [263], major and trace element determinations (Ba, Cu, Nb, Ni, Pb, Sr, Y, Zn, and Zr) [264,267], and trace element determinations (Cr, Nb, Sr, Y, and Zr [265] and Ni [266]). Geochronological age dating, as well as major element analyses, were reported [219]. Fe, Ni, and Co were determined in metal grains with an electron microprobe [268] as was FeS in a troilite crystal [269]. Lunar samples collected by the Apollo 12 and 14 missions were found to be more mafic* in character than were those collected by the Apollo 11 mission when analyzed by combined X-ray spectroscopic, emission spectroscopic, and chemical methods [270,271]. Over 1,000 microprobe analyses of glasses from the Apollo 14, 15, and 16 missions were found useful in characterizing the composition of the rocks from which the glasses were derived based upon Si, Al, Fe, Mg, Ca, Na, K, Ti, and Cr determinations [272]. Analyses of pyroxene with the microprobe [273], studies of vapor-phase crystallization in plagioclase, pyroxene, ilmenite, apatite, whitlockite, iron, and troilite with a scanning electron microscope and an energy-dispersive

---

*See Sec. 12.2.2.4.

X-ray spectrometer [274], and analysis of iron and troilite with an energy-dispersive spectrometer [275] have aided scientists in the deduction of the high reducing capacity of lunar materials and in defining the highly feldspathic nature of the lunar highlands.

Numerous experiments involving X-ray spectroscopic methods of analysis and instrumentation in unmanned space flights were designed and carried out. Adler discusses in detail the use of wavelength-dispersive spectrometers, alpha-emitter radionuclide-excitation pulse-height analyzer spectrometers and orbital X-ray fluorescence spectrometers [276,277]. An energy-dispersive spectrometer used in the Mars Lander has been discussed previously [197].

ACKNOWLEDGMENTS

The author wishes to thank H. J. Rose, J. R. Lindsay, and I. May for critical reviews of this chapter. Special thanks are due to F. Pickthorn for typing and collating all material.

REFERENCES

1. A. Hadding, Z. Anorg. Allg. Chem., 122, 195 (1922).
2. H. Bizouard and C. Roering, Geol. För. Stokh. Förh., 80, 309 (1958).
3. B. L. Henke, Advan. X-Ray Anal., 4, 244 (1961).
4. A. J. Hebert and K. Street, Jr., Anal. Chem., 46, 203 (1974).
5. H. J. Beattie and R. M. Brissey, Anal. Chem., 26, 980 (1954).
6. J. W. Criss and L. S. Birks, Anal. Chem., 40, 1080 (1968).
7. S. D. Rasberry and K. F. J. Heinrich, Anal. Chem., 46, 81 (1974).
8. J. Despujols, J. Phys. Radium, 13, 31A (1952).
9. I. Adler and J. M. Axlerod, Spectrochim. Acta, 7, 91 (1955).
10. F. Claisse, Province of Quebec, Canada, P. R. 327 (1956).
11. H. J. Rose, I. Adler, and F. J. Flanagan, Appl. Spectrosc., 17, 81 (1963).
12. K. Norrish and J. T. Hutton, Divl. Rep. Div. Soils, CSIRO, 3 (1964).

13. E. E. Welday, A. K. Baird, D. B. McIntyre, and K. W. Madlem, *Amer. Mineral.*, *49*, 889 (1964).
14. J. J. Sahores, E. P. Larribau, and J. Mihura, *Advan. X-Ray Anal.*, *16*, 27 (1973).
15. E. P. Bertin, in *Principles and Practice of X-Ray Spectrometric Analysis* (E. Bertin, ed.), 2nd ed., Plenum Press, New York, 1974.
16. B. P. Fabbi, H. N. Elsheimer, and L. F. Espos, *Advan. X-Ray Anal.*, *19*, 273 (1976).
17. G. Loch, *X-Ray Spectrom.*, *2*, 125 (1973).
18. F. Bernstein, *Advan. X-Ray Anal.*, *5*, 486 (1962).
19. H. N. Elsheimer and B. P. Fabbi, *Advan. X-Ray Anal.*, *17*, 236 (1974).
20. B. P. Fabbi (unpublished data).
21. B. P. Fabbi, *X-Ray Spectrom.*, *2*, 15 (1973).
22. P. K. Harvey, D. M. Taylor, R. D. Hendry, and F. Bancroft, *X-Ray Spectrom.*, *2*, 33 (1973).
23. L. C. Peck, U. S. Geol. Survey Bull. 1170 (1964).
24. C. O. Hutton, *Bull. Geol. Soc. Amer.*, *61*, 635 (1950).
25. L. D. Muller, in *Physical Methods of Determinative Mineralogy* (J. Zussman, ed.), Academic Press, London, 1967.
26. A. V. Volborth, *Appl. Spectrosc.*, *19*, 1 (1965).
27. I. J. Lin, S. Nadiv, and D. S. M. Grodjian, *Minerals Sci. Engr.*, *7*, 314 (1975).
28. C. O. Ingamells and P. Switzer, *Talanta*, *20*, 547 (1973).
29. C. O. Ingamells, J. C. Engels, and P. Switzer *24th Int. Geochem. Conf.*, *10*, 405 (1972).
30. P. F. Berry, *Appl. Low-Energy X-Gamma Rays*, 429 (1971).
31. K. W. Madlem, *Advan. X-Ray Anal.*, *9*, 441 (1966).
32. F. Claisse and C. Samson, *Advan. X-Ray Anal.*, *5*, 335 (1962).
33. P. F. Berry, T. Furata, and J. R. Rhodes, *Advan. X-Ray Anal.*, *12*, 612 (1969).
34. F. Claisse, *Spectrochim. Acta*, *25B*, 209 (1970).
35. R. H. Myers, D. Womeldorph, and B. J. Alley, *Anal. Chem.*, *39*, 1031 (1967).
36. F. Bernstein, *Advan. X-Ray Anal.*, *6*, 436 (1963).
37. E. L. Gunn, *Advan. X-Ray Anal.*, *4*, 382 (1961).
38. F. Bernstein, *Advan. X-Ray Anal.*, *7*, 555 (1964).
39. B. P. Fabbi and W. J. Moore *Appl. Spectrosc.*, *24*, 426 (1970).

40. C. K. Matocha, *Appl. Spectrosc.*, *20*, 252 (1966).
41. H. D. Spitz, *Appl. Spectrosc.*, *22*, 206 (1968).
42. B. P. Fabbi, in *Geological Survey Research 1970*, U. S. Geol. Survey Prof. Paper *700B*, B187 (1970); *X-Ray Spectrom.*, *1*, 39 (1972).
43. L. Bean, *Appl. Spectrosc.*, *20*, 191 (1966).
44. E. P. Bertin and R. J. Longobucco, *Norelco Reporter*, *2*, 31 (1962).
45. A. K. Baird, *Norelco Reporter*, *8*, 108 (1961).
46. A. Volborth, Nevada Bureau of Mines, Reno, Report *6*, A-1 (1963).
47. G. R. Webber and M. L. Newbury, *Can. Spectrosc.*, *16*, 90 (1971).
48. B. M. Gunn, *Can. Spectrosc.*, *14*, 2 (1969).
49. K. Norrish and B. W. Chappell, in *Physical Methods of Determinative Mineralogy* (J. Zussman, ed.), Academic Press, London, 1967, p. 161.
50. B. P. Fabbi and A. V. Volborth, Nevada Bureau of Mines, Reno, Report *6C*, C31 (1970).
51. A. V. Volborth, B. P. Fabbi, and H. A. Vincent, *Advan. X-Ray Anal.*, *11*, 158 (1968).
52. I. B. Brenner, L. Argov and H. Eldad, *Appl. Spectrosc.*, *29*, 423 (1975).
53. B. P. Fabbi, *Amer. Mineral.*, *57*, 237 (1972).
54. L. F. Espos and B. P. Fabbi (private communication).
55. K. Togel, in *Zerstörungsfreie Materialprüfung* (E. A. W. Muller, ed.), Munich, Sec. U152, 1961.
56. W. J. Campbell, E. F. Spano, and T. E. Green, *Anal. Chem.*, *38*, 987 (1966).
57. V. R. Darashkevich, B. A. Malyukov, and Yu. M. Ukrainskii, *Zhurnal Analiticheskoi Khimii*, *27*, 1578 (1972).
58. N. B. Price and G. R. Angell, *Anal. Chem.*, *40*, 660 (1968).
59. H. A. Liebhafsky and P. D. Zemany, *Anal. Chem.*, *28*, 455 (1956).
60. E. L. Gunn, *Anal. Chem.*, *33*, 921 (1961).
61. R. D. Giauque, F. S. Goulding, J. M. Jaklevic, and R. H. Pehl, *Anal. Chem.*, *45*, 671 (1973).
62. K. Govindaraju, *X-Ray Spectrom.*, *2*, 57 (1973).
63. G. N. Eby, *Anal. Chem.*, *44*, 2137 (1972).
64. V. A. Ryabukhin, N. S. Stroganova, N. G. Gatinskaya, and A. N. Ermakov, *Zhurnal Analiticheskoi Khimii*, *28*, 2166 (1973).
65. M. Murata and M. Noguchi, *Anal. Chim. Acta*, *71*, 295 (1974).

66. D. E. Leyden and R. E. Channell, *Anal. Chem.*, *44*, 607 (1972).
67. C. W. Blount, R. E. Channell, and D. E. Leyden, *Anal. Chim. Acta*, *56*, 456 (1971).
68. D. E. Leyden, *Advan. X-Ray Anal.*, *17* 293 (1974).
69. C. W. Blount, D. E. Leyden, T. L. Thomas, and S. M. Guill, *Anal. Chem.*, *45*, 1045 (1973).
70. C. L. Luke, *Anal. Chim. Acta*, *41*, 237 (1968).
71. T. M. Reymont and R. J. Dubois, *Anal. Chem. Acta*, *56*, 1 (1971).
72. J. M. Mathiesen, *Advan. X-Ray Anal.*, *17*, 318 (1974).
73. O. F. Gulacar, *Anal. Chim. Acta*, *73*, 255 (1974).
74. C. B. Hunter and J. R. Rhodes, *X-Ray Spectrom*, *1*, 107 (1972).
75. J. R. Rhodes and C. B. Hunter, *X-Ray Spectrom.*, *1*, 113 (1972).
76. A. L. Allen and C. V. Rose, *Advan. X-Ray Anal.*, *15*, 534 (1972).
77. O. H. J. Christie and S. Bergstol, *Acta Chem. Scand.*, *22*, 421 (1968).
78. A. Strasheim and M. P. Brandt, *Spectrochim. Acta*, *23B*, 183 (1967).
79. A. V. Volborth, in *Elemental Analysis Geochemistry,* Elsevier, Amsterdam, London, and New York, 1969.
80. P. R. Hooper and L. Atkins, *Mineral. Magazine*, *37*, 409 (1969).
81. T. K. Smith, *Earth Sci.*, *81*, B156 (1972).
82. C. E. Austen and B. G. Russell, *Nat. Inst. Metallurgy Report* No. 1599 (1974).
83. K. Norrish and J. T. Hutton, *Geochim. Cosmochim. Acta*, *33*, 431 (1969).
84. A. Parker, *Chim. Acta*, *40*, 513 (1968).
85. G. K. Czamanske, J. Hower, and R. C. Millard, *Geochim. Cosmochim. Acta*, *30*, 745 (1966).
86. P. Richter, *N. Jb. Miner. Mh.*, *7*, 209 (1968).
87. M. Bojic, J. M. Bourdieu, G. Jecko, and A. Whitman, *Chim. Anal.*, *52*, 855 (1970).
88. W. B. Stern, *Schweiz. Min. Petr. Mitt.*, *52* (1972).
89. A. Whitman and J. Chmeleff, *Analysis*, *2*, 271 (1973).
90. D. A. Stephenson, *Anal. Chem.*, *41*, 967 (1969).
91. H. Hattori, *Bull. Geol. Survey of Japan*, *22*, 103 (1971).
92. R. LeHouillier and S. Turmel, *Anal. Chem.*, *46*, 734 (1974).
93. A. Whitman, J. Chmeleff, and H. Herrmann, *X-Ray Spectrom.*, *3*, 137 (1974).

94. H. Kodoma, J. E. Brydon, and B. C. Stone, *Geochim. Cosmochim. Acta*, *31*, 649 (1967).
95. R. J. Longobucco, *Anal. Chem.*, *34*, 1263 (1962).
96. J. E. Townsend, *Appl. Spectrosc.*, *17*, 37 (1963).
97. M. J. O'Neill and F. J. Fitzsimons, *Appl. Earth Sci.*, *81*, B209 (1972).
98. R. Tertian and R. Geninasca, *X-Ray Spectrom.*, *1*, 83 (1972).
99. R. Tertian, *Advan. X-Ray Anal.*, *12*, 546 (1969).
100. C. O. Ingamells, *Anal. Chim. Acta*, *52*, 323 (1970).
101. C. B. Belcher (private communication).
102. C. H. Drummond, *Appl. Spectrosc.*, *20*, 252 (1966).
103. U. Kraeft, *G. I. T.*, *16*, 679 (1972).
104. U. Kraeft, *Zement-Kalk-Gips*, *9*, 449 (1972).
105. C. Matocha, *Proc. 6th Int. Conf. on Light Metals*, Leoben, Vienna, Austria, 123 (1975).
106. F. Claisse, *The Spex Speaker*, *18*, 2 (1973).
107. H. J. Rose, *Appl. Spectrosc.*, *17*, 81 (1963).
108. N. P. Il'in and L. E. Loseva, *Zhurnal Analit. Khimii*, *27*, 2314 (1972).
109. I. A. Nichols, *Chem. Geol.*, *14*, 151 (1974).
110. F. Agus and W. R. Hesp, *Div. Miner.*, CSIRO IR, *100* (1974).
111. R. Dewolfs, R. deNeve, and F. Adams, *Anal. Chim. Acta*, *75*, 47 (1975).
112. A. T. Myers, R. G. Havens, and W. W. Niles, *Dev. Appl. Spectrosc.*, *8*, 132 (1970).
113. H. de la Roche, Centre de Recherches Petrographiques et Geochimiques, 15, Rue N. D. des Pauvres, Case officielle n°1, 54500 Vandoeuvre les Nancy, France.
114. Office of Standard Reference Materials, National Bureau of Standards, Washington, D. C.
115. Spex Industries, Inc., P. O. Box 798, Metuchen, New Jersey.
116. I. Adler, *X-Ray Emission Spectrography in Geology*, Elsevier, Amsterdam, London, and New York, 1966.
117. E. L. Gunn, *Handbook of X Rays* (E. F. Kaelble, ed.), McGraw Hill, New York, 1967.
118. H. A. Liebhafsky, H. G. Pfeiffer, E. H. Winslow, and P. D. Zemany, *X-Ray Absorption and Emission in Analytical Chemistry*, John Wiley, New York, London, 1960.
119. H. J. Rose (private communication).

120. R. E. Wood and E. R. Bingham, Anal. Chem., 33, 1344 (1961).
121. H. J. Rose and F. Cuttitta, Appl. Spectrosc., 22, 426 (1968).
122. J. Hower, Amer. Mineral., 44, 19 (1959).
123. M. J. Kaye, Geochim. Cosmochim. Acta, 29, 139 (1965).
124. M. Franzini, L. Leoni, and M. Saitta, X-Ray Spectrom., 1, 151 (1972).
125. L. Leoni and M. Saitta, X-Ray Spectrom., 3, 74 (1974).
126. C. Barber, Chem. Geol., 14, 273 (1974).
127. I. Adler and J. M. Axelrod, Anal. Chem., 27, 1002 (1955).
128. G. J. Lewis and E. D. Goldberg, Anal. Chem., 28, 1282 (1956).
129. J. M. Axelrod and I. Adler, Anal. Chem., 29, 1280 (1957).
130. C. K. Brooks, Geochim. Cosmochim. Acta, 34, 411 (1970).
131. V. Machaček, Časopis Mineral. Geol., 17, 171 (1972).
132. N. A. Al-Saudi and N. W. Al-Derzi, Chem. Geol., 15, 229 (1975).
133. A. Parker, Chem. Geol., 4, 445 (1969).
134. B. M. Gunn, Can. Spectrosc. 12, 3 (1967).
135. E. Donderer, N. Jb. Miner. Mh., 4, 167 (1972).
136. G. Andermann and J. W. Kemp, Anal. Chem., 30, 1306 (1958).
137. R. C. Reynolds, Amer. Mineral., 38, 1133 (1963).
138. G. R. Webber and J. D. Volrath, Can. Spectrosc., 12, 3 (1967).
139. I. A. Wolfe and H. Zeitlin, Anal. Chem. Acta, 51, 349 (1970).
140. H. P. Fairbairn and P. M. Hurley, Geochim. Cosmochim. Acta, 35, 149 (1971).
141. F. J. Flanagan, Geochim. Cosmochim. Acta, 34, 121 (1970).
142. F. J. Flanagan, Geochim. Cosmochim. Acta, 38, 1731 (1974).
143. A. H. Gillieson, Advan. X-Ray Anal., 17, 16 (1974).
144. R. Sutarno and G. H. Faye, Talanta, 22, 675 (1975).
145. F. J. Flanagan, Geochim. Cosmochim. Acta, 37, 1189 (1973).
146. S. Abbey, Can. Spectrosc., 15, 3 (1970).
147. S. Abbey, Geol. Survey Can., Paper 72-30 (1972).
148. S. Abbey, Geol. Survey Can., Paper 73-36 (1973).
149. S. Abbey, Geochim. Cosmochim. Acta, 39, 535 (1975).
150. S. Abbey, Can. Spectrosc., 20, 113 (1975).
151. C. O. Ingamells, N. H. Suhr, F. C. Tan, and D. H. Anderson, Anal. Chim. Acta, 53, 345 (1971).
152. L. B. Owen and S. Faure, Anal. Chem., 46, 1323 (1974).

153. P. F. S. Jackson and F. W. E. Strelow, *Chem. Geol.*, *15*, 303 (1975).
154. O. Johansen and E. Steinnes, *Geochim. Cosmochim. Acta*, *31*, 1107 (1969).
155. K. Randle, *Chem. Geol.*, *13*, 237 (1974).
156. J. Sherman, *Spectrochim. Acta*, *7*, 283 (1955).
157. J. Sherman, *Spectrochim. Acta*, *15*, 466 (1959).
158. R. J. Traill and G. R. Lachance, *Geol. Survey Can.*, Paper 64-57 (1964).
159. G. R. Lachance and R. J. Traill, *Can. Spectrosc.*, *11*, 43 (1966).
160. R. J. Traill and G. R. Lachance, *Can. Spectrosc.*, *11*, 63 (1966).
161. F. Claisse and M. Quintin, *Can. Spectrosc.*, *12*, 129 (1967).
162. R. Rouseau and F. Claisse, *X-Ray Spectrom.*, *3*, 31 (1974).
163. J. Lucas-Tooth and C. Pyne, *Advan. X-Ray Anal.*, *7*, 523 (1964).
164. B. J. Alley and R. H. Myers, *Anal. Chem.*, *37*, 1685 (1965).
165. B. J. Alley and R. H. Myers, *Norelco Reporter*, *15*, 87 (1968).
166. B. J. Mitchell, *Anal. Chem.*, *33*, 917 (1961).
167. B. J. Mitchell and F. N. Hopper, *Appl. Spectrosc.*, *20*, 172 (1966).
168. D. A. Stephenson, *Anal. Chem.*, *43*, 1761 (1971).
169. R. Tertian, *Spectrochim. Acta*, *23B*, 305 (1968).
170. R. Tertian, *Spectrochim. Acta*, *24B*, 447 (1969).
171. R. Tertian, *Spectrochim. Acta*, *26B*, 71 (1971).
172. R. Tertian, *Spectrochim. Acta*, *27B*, 159 (1972).
173. R. Gwozdz, *X-Ray Spectrom.*, *3*, 2 (1974).
174. G. R. Holden, Eighth Australian Spectroscopy Conference, Melbourne, Australia, August 16-19, 1971.
175. V. P. Afonin, T. N. Gunicheva, A. M. Kharchenko, and L. F. Piskunova, *Zavod. Lab.*, *40*, 809 (1974).
176. B. J. Mitchell, *Advan. X-Ray Anal.*, *11*, 129 (1968).
177. R. W. Gould and S. R. Bates, *X-Ray Spectrom.*, *1*, 29 (1972).
178. R. Tertian, *X-Ray Spectrom.*, *2*, 95 (1973).
179. R. Boniforti, G. Buffoni, C. Colella, and R. Riccardi, *X-Ray Spectrom.*, *3*, 115 (1974).
180. G. C. Brown, D. J. Hughes, and J. Esson, *Chem. Geol.*, *11*, 223 (1973).

181. B. E. Leake, G. L. Hendry, A. Kemp, A. G. Plant, P. K. Harvey, J. R. Wilson, J. S. Coats, J. W. Aucott, T. Lunel, and R. J. Howarth, *Chem. Geol.*, 5, 7 (1969).

182. I. S. E. Carmichael, J. Humpel, and R. N. Jack, *Chem. Geol.*, 3, 59 (1968).

183. B. P. Fabbi, *Appl. Spectrosc.*, 25, 41 (1971).

184. B. P. Fabbi. *Appl. Spectrosc.*, 25, 316 (1971).

185. B. P. Fabbi and L. F. Espos, in *Geol. Survey Research*, U. S. Geol. Survey Prof. Paper 80-B, B147 (1972).

186. R. J. Goodman, *Can. Spectrosc.*, 16, 97 (1971).

187. J. L. Brandle and M. I. Cerquiera, *X-Ray Spectrom.*, 3, 130 (1974).

188. L. Leoni and M. Saitta, *X-Ray Spectrom.*, 5, 29 (1976).

189. M. L. Salmon, *Advan. X-Ray Anal.*, 2, 305 (1958).

190. K. Norrish and R. M. Taylor, *Clay Minerals Bull.*, 5, 98 (1962).

191. B. W. Chappell, W. Compston, P. A. Arriens, and M. J. Vernon, *Geochim. Cosmochim. Acta*, 33, 1002 (1969).

192. R. J. Pankhurst and R. K. O'Nions, *Chem. Geol.*, 12, 127 (1973).

193. D. A. Stephenson, *Spectrochim. Acta*, 27B, 153 (1972).

194. R. Tertian, *Spectrochim. Acta*, 27B, 155 (1972).

195. I. V. Nelson and C. E. Urdy, *Anal. Chem. Acta*, 62, 449 (1972).

196. J. B. Zimmerman, *Can. Spectrosc.*, 18, 147 (1973).

197. P. Toulmin III, A. K. Baird, B. C. Clark, K. Keil, and H. J. Rose, *Icarus*, 20, 153 (1973).

198. C. H. Anderson, J. E. Mander, and J. W. Leitner, *Advan. X-Ray Anal.*, 17, 214 (1974).

199. J. G. Dick and A. D. Nguyen, *Can. Spectrosc.*, 19, 110 (1974).

200. A. B. Poole and S. M. Holloway, *Advan. X-Ray Anal.*, 12, 534 (1969).

201. C. N. McKinney and A. S. Rosenberg, *Advan. X-Ray Anal.*, 13, 125 (1970).

202. A. L. Albee and L. Ray, *Anal. Chem.*, 42, 1408 (1970).

203. J. Nicolas and M. Quintin, *Methodes Physiques d'Analyses*, 3, 94 (1967).

204. D. F. Ball, *Analyst*, 90, 258 (1965).

205. A. V. Volborth, *Amer. Mineral.*, 49, 634 (1964).

206. P. K. Hormann and G. Morteani, *Amer. Mineral.*, 54, 1491 (1969).

207. P. R. Gregor and J. A. Ereiser, *Can. Spectrosc.*, 19, 52 (1974).

208. B. L. Henke, *Advan. X-Ray Anal.*, *6*, 361 (1964).

209. A. K. Baird and B. L. Henke, *Anal. Chem.*, *37*, 727 (1965).

210. B. P. Fabbi and A. V. Volborth, *Norelco Reporter*, *15*, 93 (1968).

211. F. Bernstein and R. A. Mattson, *Advan. X-Ray Anal.*, *10*, 494 (1966).

212. J. J. Wilband, *Amer. Mineral.*, *60*, 320 (1975).

213. A. V. Volborth and B. P. Fabbi, Nevada Bureau of Mines, Reno, Report *6C*, C1 (1970).

214. A. P. Langheinrich and J. W. Forster, *Advan. X-Ray Anal.*, *11*, 275 (1967).

215. H. Kunzendorf, *Radioanal. Chem.*, *9*, 311 (1971).

216. P. W. Gerrard and W. Westwood, *J. South African Chem. Inst.*, *25*, 285 (1972).

217. M. W. Solt, J. S. Wahlberg, and A. T. Myers, *Talanta*, *16*, 37 (1969).

218. Z. E. Peterman, I. S. E. Carmichael, and A. L. Smith, *Earth Planet. Sci. Lett.*, *7*, 381 (1970).

219. W. Compston, P. Arriens, M. Vernon, and B. W. Chappell, *Science*, *167*, 474 (1970).

220. J. L. Powell and K. Bell, *Contr. Mineral. Petrol.*, *27*, 1 (1970).

221. S. S. Berman, P. Semeniuk, and D. S. Russell, *Can. Spectrosc.*, *14*, 3 (1969).

222. G. Andermann and J. D. Allen, *Advan. X-Ray Anal.*, *4*, 414 (1961).

223. H. T. Dryer, *Advan. X-Ray Anal.*, *6*, 447 (1963).

224. H. Uchikaw and M. Numata, *Japan Analyst*, *19*, 812 (1970).

225. R. Plesch, *Z. Anal. Chem.*, *262*, 84 (1972).

226. J. C. Russ, *X-Ray Spectrom.*, *1*, 119 (1972).

227. B. L. Gulson and J. F. Lovering, *Geochim. Cosmochim. Acta*, *32*, 119 (1968).

228. T. Mori, P. Jakes, and M. Nagaoka, *Sci. Rep.*, Kanazawa University, Japan, *16*, 113 (1971).

229. J. Hénoc, K. F. J. Heinrich, and R. L. Myklebust, Nat. Bur. Stand. Pub. TN-769 (1973).

230. H. Yakowitz, R. L. Myklebust, and K. F. J. Heinrich, Nat. Bur. Stand. Pub. TN-796 (1973).

231. I. Adler, *A.S.T.M. Spcl. Tech. Pub. 269*, 47 (1959).

232. M. H. Naggar and M. P. Atherton, *Chem. Geol.*, *5*, 107 (1969).

233. Y. S. Kim, *Anal. Chem.*, *39*, 664 (1967).

234. B. R. Doe, A. A. Chodos, A. W. Rose, and E. Godijn, *Amer. Mineral.*, *46*, 1056 (1961).

235. G. D. Loftus-Hills, D. I. Groves, and M. Solomon, *Aust. Inst. Min. Metal. Proc.*, *232*, 55 (1969).

236. D. R. Hunter, *Trans. Geol. Soc.*, South Africa, *75*, 293 (1972).

237. K. Hisano and K. Oyama, *Japan Analyst*, *18*, 1508 (1969).

238. F. C. W. Dodge, B. P. Fabbi, and D. C. Ross, in *Geol. Survey Research 1970*, U. S. Geol. Survey Prof. Paper 700-D, D108 (1970).

239. F. Cuttitta and H. J. Rose, *Appl. Spectrosc.*, *22*, 423 (1968).

240. H. J. Rose and F. Cuttitta, *Advan. X-Ray Anal.*, *11*, 23 (1968).

241. G. K. Czamanske and R. Larson, *Amer. Mineral.*, *54*, 1198 (1969).

242. K. Keil, in *Microprobe Analysis* (C. A. Andersen, ed.), J. Wiley, New York, 1973.

243. K. Keil, *Proc. 7th National Conf. Electron Probe Analysis*, San Francisco, 1972.

244. M. H. Beeson and E. D. Jackson, Pub. Mineralogical Society of America, Special Pub. No. 3, 95 (1970).

245. M. H. Beeson and E. D. Jackson, *Amer. Mineral.*, *54*, 1084 (1969).

246. W. A. Duffield and M. H. Beeson, in *J. Research U. S. Geol. Survey*, *1*, 665 (1973).

247. G. K. Czamanske and D. R. Wones, *J. Petrol.*, *14*, 349 (1973).

248. G. K. Czamanske and W. E. Hall, *Econ. Geol.*, *70*, 1092 (1975).

249. G. K. Czamanske, *Econ. Geol.*, *69*, 1328 (1974).

250. G. K. Czamanske, G. A. Desborough, and F. E. Goff, *Econ. Geol.*, *68*, 1275 (1973).

251. D. C. Harris, *Can. Mineral.*, *12*, 280 (1974).

252. S. J. B. Reed and N. G. Ware, *X-Ray Spectrom.*, *2*, 69 (1973).

253. R. J. Gehrke and R. C. Davies, *Anal. Chem.*, *47*, 1537 (1975).

254. F. Blum and M. P. Brandt, *X-Ray Spectrom.*, *2*, 121 (1973).

255. J. R. Hinthorne and C. A. Andersen, *Proc. 8th National Conf. Electron Probe Analysis*, New Orleans, 1973.

256. H. J. Rose, F. Cuttitta, M. K. Carron, and R. Brown, in *U. S. Geol. Survey Prof.*, Paper 475D, D171 (1964).

257. W. Nichiporuk, A. Chodos, E. Helin, and H. Brown, *Geochim. Cosmochim. Acta*, *31*, 1911 (1967).

258. J. P. Willis, A. J. Erlank, L. H. Ahrens, *Earth Planetary Sci. Lett.*, *5*, 383 (1969).

259. S. J. B. Reed, *Meteoritics*, 7, 257 (1972).
260. W. W. Thomas and J. R. DeLaeter, *X-Ray Spectrom.*, 1, 143 (1972).
261. G. J. Taylor and D. Heymann, *Geochim. Cosmochim. Acta*, 34, 677 (1970).
262. H. P. Schwartz, S. D. Scott, and S. A. Kissin, *Geochim. Cosmochim. Acta*, 39, 1457 (1975).
263. H. J. Rose, F. Cuttitta, E. J. Dwornik, M. K. Carron, R. P. Christian, J. R. Lindsay, D. T. Lignon, and R. R. Larson, *Science*, 167, 520 (1970).
264. C. M. Brown, C. H. Emeleus, J. G. Holland, and R. Phillips, *Science*, 167, 599 (1970).
265. J. C. Bailey, P. E. Champness, A. C. Dunham, J. Esson, W. S. Fyffe, W. S. MacHenzie, E. F. Stumpel, and J. Zussman, *Science*, 167, 592 (1970).
266. P. W. Gast and N. J. Hubbard, *Science*, 167, 485 (1970).
267. A. A. Smales, D. Mapper, M. S. W. Webb, R. K. Webster, and J. D. Wilson, *Science*, 167, 509 (1970).
268. A. M. Reid, C. Meyer, R. S. Harmon, and R. Brett, *Earth Planetary Sci. Lett.*, 9, 1 (1970).
269. K. Keil, T. E. Bunch, and M. Prinz, in *Proc. Apollo 11 Lunar Sci. Conf.* (A. A. Levinson, ed.), Pergamon, New York, Vol. 1, 1970, p. 399.
270. F. Cuttitta, H. J. Rose, C. S. Annell, M. K. Carron, R. P. Christian, E. J. Dwornik, L. P. Greenland, A. W. Helz, and D. T. Lignon, in *Proc. 2nd Lunar Sci. Conf., Geochim. Cosmochim. Acta, Suppl. 2*, 2, 1217 (1971).
271. H. J. Rose, F. Cuttitta, C. S. Annell, M. K. Carron, R. P. Christian, E. J. Dwornik, L. P. Greenland, and D. T. Lignon, in *Proc. 3rd Lunar Sci. Conf., Geochim. Cosmochim. Acta, Suppl. 3*, 2, 1215 (1972).
272. R. W. Brown, W. I. Ridley, J. L. Warner, and A. M. Reid, in *Proc. 7th Nat. Conf. Electron Probe Analysis*, San Francisco, 1972.
273. A. E. Bence, *Proc. 7th Nat. Conf. Electron Probe Analysis*, San Francisco, 1972.
274. D. S. McKay, U. S. Clanton, and G. H. Ladle, *Proc. 7th Nat. Conf. Electron Probe Analysis*, San Francisco, 1972.
275. U. S. Clanton, D. S. McKay, R. B. Laughon, and G. H. Ladle, *Proc. 8th Nat. Conf. Electron Probe Analysis*, New Orleans, 1973.

276. I. Adler and J. I. Trombka, in *Physics and Chemistry in Space* (J. G. Roederer, and J. Zahringer, eds.), Springer-Verlag, New York, Vol. 3, 1970.

277. I. Adler, J. Trombka, J. Gerald, P. Lowman, R. Schamadabek, H. Blodgett, E. Eller, L. Yin, and R. Lamothe, *Science, 175*, 436 (1972).

Chapter 13

MINING AND ORE PROCESSING

Armin P. Langheinrich
W. M. Tuddenham

Metal Mining Division-Research Center
Kennecott Copper Corporation
Salt Lake City, Utah

13.1	Introduction	356
	13.1.1 Historical Development	356
	13.1.2 General Capabilities	358
13.2	In-Plant Analysis	361
13.3	Specimen Preparation	368
	13.3.1 Grinding	368
	13.3.2 Heterogeneities	369
13.4	Quantitative Analysis	370
	13.4.1 Calibration Standards	370
	13.4.2 Internal Standards	371
	13.4.3 Scatter Background	371
	13.4.4 Mathematical Correction	372
13.5	Specific Examples	373
	13.5.1 Exploration	373
	13.5.2 Mining	377
	13.5.3 Concentrating	378
	13.5.4 Smelting	380
	13.5.5 Refining	382
	13.5.6 Environmental	384

References 384
Additional References 388

13.1 INTRODUCTION

The common term in use throughout the mineral industries for the analysis of materials is *assaying*. This term implies the acquisition of important data in a straightforward manner over wide ranges of concentration, from levels of interest to the prospector to levels of importance in the final product. X-Ray instruments more than many others might be called assay devices.

In mining and ore processing, X-ray spectrometry is used for quality control, process control, and occasionally accounting. Quality control and accounting are generally performed in established service laboratories of the type discussed in Chap. 10; process control with X-ray instruments is done in service laboratories or on site. Types of X-ray instruments and instrument assemblies used in the mineral industries are tabulated in Table 13.1.

13.1.1 Historical Development

Elemental X-ray analysis in industry had its beginnings in the late 1940s, with a strong developmental surge in the early 1950s. On-stream analysis with sample transport to a centralized location was first used in 1957. With the advent of compact X-ray instruments, centralization is no longer a requirement. Since early this decade, equipment has been developed for the on-stream, on-site analysis of by-pass samples and for in-stream analysis by direct insertion of X-ray sensors into process streams.

In a recent paper, H. R. Cooper [1] provided a summary of on-stream and in-stream X-ray installations. His up-to-date chronology lists 4 installations, worldwide, for the pioneering period prior to 1960, 41 for the development period from 1960 to 1969, and 35 since

TABLE 13.1

Examples of X-Ray Instrumentation
Used in the Mineral Industries

Laboratory X-ray spectrometer: water-cooled X-ray tube, wavelength dispersion, proportional and/or scintillation counter(s), goniometer for sequential determinations or fixed spectrometer(s) for single-element or simultaneous multielement determinations, air and/or vacuum or helium path, nonportable.

Laboratory X-ray spectrometer: air-cooled small X-ray tube, energy dispersion with Si(Li) detector, single-channel analyzer(s) or multichannel analyzer for single-element or simultaneous multielement determinations, air and/or vacuum path, nonportable.

On-stream X-ray spectrometer: centralized in-plant location, fixed spectrometer(s) for single-element or simultaneous multielement analysis, single- or multistream use, slurry flow cell(s), stationary or moving, spectrometer head, stationary or moving, water-cooled X-ray tube, wavelength dispersion, air path, nonportable.

On-stream X-ray spectrometer: in-plant, on-site location, energy dispersion with Si(Li) detector or scintillation counter(s), flow cell for one stream, single-element or multielement determinations through use of single-channel analyzer(s) or multichannel analyzer, radioisotope excitation, semiportable.

In-stream X-ray probe: X-ray sensors in slurry stream, energy dispersion, radioisotope excitation, scintillation counter for single-element determination or Si(Li) detector for single- or simultaneous multielement determinations, semiportable.

Field X-ray instrument: battery operated, radioisotope excitation, balanced filters for energy discrimination, scintillation or proportional counters, portable--or Si(Li) detectors with generator power supply, semiportable.

1970. The geographic distribution of these 80 installations is summarized in Table 13.2. Not all of them are in use presently.

For the ferrous industry areas of main analytical concern are mining and steel making. In the nonferrous industries, additional process steps require analytical control. A typical process sequence is as follows:

 mining → crushing and grinding → concentration → smelting → refining

TABLE 13.2

Geographic Distribution of 80
On-Stream and In-Stream Analyzers

Australia	9	South Africa	1
Canada	18	Sweden	2
Chile	1	Tasmania	1
Finland	10	United States	24
Iran	1	USSR	3
Ireland	2	Yugoslavia	1
Japan	3	Zambia	3
Poland	1		

X-Ray applications in mining and concentrating predate applications in smelting. Uses in refineries are less common. The potential of X-ray spectrometry as an analytical tool in the mineral industries was greatly enhanced through the development of the following: radioisotope excitation (1960s), silicon detectors (1966), air-cooled tubes (early 1970s), small data processors (early 1970s).

13.1.2 General Capabilities

X-Ray spectrometry in mining and ore processing has been accepted mainly because of its relative simplicity and speed. Speed is more important than the nondestructive nature of the technique. Usually, sufficient sample material is available for several analytical determinations; only in special cases are precautions necessary to preserve a sample. Qualitative, semiquantitative, and even quantitative determinations of many elements are faster and cheaper than by most other methods. In the laboratory, X-ray spectrometry complements other high work-volume techniques, such as emission spectroscopy and atomic absorption; in the plant it is without a rival. It lends itself to automation or semiautomation. Spectral specificity and operating simplicity are additional advantages.

Disadvantages include the following: the necessity to deal with wide matrix variations; the existence of interelement effects; radiation from X-ray tubes and radioisotope sources; geographic distances between plant and vendor or service organization; plumbing (sample transport) problems for centralized on-stream systems.

The usefulness of X-ray instrumentation in assaying is well established, but has not as yet reached its full potential. At a mine, long range information is needed concerning the extent and value of an ore body. Selective mining is of more immediate interest. Rapid assay information allows ore/waste sorting. Concentrations of element(s) of interest below a certain economically established level define waste material; concentrations above this cut-off level define ore material. X-Ray analysis is a feasible approach to this decision making.

Waste material may also be economically valuable. In copper production, for example, copper in waste rock is recovered by acid/bacterial leaching. The copper is precipitated by reaction of leach solutions with metallic iron (scrap). Progress and efficiency of this process can be monitored with X-ray instrumentation.

After crushing and grinding, ore materials enter the concentrating phase of the process. Control measurements are needed on the incoming heading material, the outgoing tailing (waste) material, the product concentrate, and on a range of intermediates. The control of this process is so important that it has justified development, installation, and utilization of slurry analyzers for rapid control.

At smelters, analytical information is needed on plant input (concentrate, flux) and on intermediate and final products (matte, slag, metal). X-Ray applications in electrolytic refineries are scarce and mainly related to electrolyte composition.

Elemental concentrations of interest to the metallurgist and to management range from the low metal values in waste rock and in tailings to the high concentrations desired in concentrates. Some typical ranges from nonferrous industries are listed in Tables 13.3 and 13.4. Each sample type listed may represent a number of subtypes; e.g., the slags for reverberatory slag and converter slag. Not every subtype may be analyzed for the element indicated.

TABLE 13.3

Examples of Representative Elemental Ranges
Determined by X-Ray Analysis in Copper Mining and Ore Processing

Element	Sample type	Range (%)
Copper (Cu)	Mine samples	Nil-20
	Headings	0.4-1.0
	Tailings	0.030-0.15
	Copper concentrates	18-38
	Molybdenum concentrates	0.04-1.5
	Slags	0.2 10.0
	Fluxes	0.05-26
	Matte	30-40
Molybdenum (Mo)	Mine samples	0.01-2.0
	Headings	0.02-0.08
	Tailings	0.004-0.025
	Copper concentrates	0.2-3.0
	Molybdenum concentrates	45-57
	Slags	0.05-1.0
Iron (Fe)	Mine samples	0.8-15
	Headings	1.8-4.5
	Tailings	1.8-4.5
	Copper concentrates	11-35
	Molybdenum concentrates	0.4-1.5
	Slags	25-55
	Fluxes	1.0-21
Sulfur (S)	Copper concentrates	25-35
	Fluxes	1.0-8.0
Silicon (Si)	Copper concentrates	1.9-9.5
	Slags	7-21
	Fluxes	18-45
	Lime sands	1.5-6.0
Aluminum (Al)	Copper concentrates	0.4-4.0
	Slags	Nil-7.0
	Fluxes	0.5-13
Calcium (Ca)	Slags	0.7-6.5
	Fluxes	0.01-2.5
	Lime sands	30-38

TABLE 13.4

Examples of Representative Elemental Ranges
Determined by X-Ray Analysis in Lead-Zinc
Mining and Ore Processing

Element	Sample type	Range (%)
Lead (Pb)	Flotation feed	2.5-7.5
	Lead concentrate	60-82
	Zinc feed	0.1-0.3
	Zinc concentrate	0.2-5
	Final tails	0.08-0.3
Zinc (Zn)	Flotation feed	0.3-5
	Lead concentrate	0.5-4.5
	Zinc feed	0.3-5.2
	Zinc concentrate	42-58
	Final tails	0.05-0.2
Copper (Cu)	Heads, concentrates, tails	0.01-3

13.2 IN-PLANT ANALYSIS

On-stream or in-stream analysis is the most significant application of X-ray technology to the ore-processing industry. Eventually, it will become an integral part of fully automated plant control. Real-time analysis became possible with the development of slurry analyzers starting in the late 1950s. Though not all installations have proved cost effective and/or technologically feasible, the approach persisted and, today, can be considered successful.

In some instances, rapid sample transport and handling allows control assays in the service laboratory. Where such transport is difficult or impracticable, batch samples have to be analyzed in the plant. This type of control analysis usually employs rapid chemical techniques with relatively low precision. Now small X-ray units, especially energy-dispersive (EDX), can be placed in the plant to replace these chemical methods.

With improvement in plant technology, the demand for timely assay data has increased steadily. This demand combined with difficulties in sampling led to the development of slurry analyzers.

In most applications, the instrument purchased is of the fixed channel type, with spectrometers for preselected elements and with multistream capabilities. Modern units are usually computerized and provide periodic printouts in percent concentration for each element in each stream.

Design differences resulted in basically two types of X-ray slurry analyzers. One type uses a moving spectrometer head with stationary stream cells, as shown in Fig. 13.1, the other type uses moving flow cells with a stationary spectrometer assembly, as illustrated in Fig. 13.2. Slurry streams such as headings, tailings, and concentrates are brought to the installation via a system of samplers, pumps, pipes, and head tanks. Where a single common head tank is used for several streams, memory effects and residual flotation may negate any results obtained.

At an early stage of planning for an on-stream installation

FIG. 13.1 Wavelength-dispersive slurry analyzer: moving head. (Courtesy Outokumpu Oy, Helsinki, Finland.)

FIG. 13.2 Wavelength-dispersive slurry analyzer; moving flow cells. (Courtesy Applied Research Laboratories, Sunland, California.)

the following must be considered:

  Materials (streams) to be analyzed
  Elements to be determined
  Elemental concentrations expected
  Frequency of analysis and turn-around time
  Precision and accuracy required

Decisions must be made with regard to the following:

  Location and ancillary facilities
  X-Ray equipment and sophistication
  Availability of funds
  Personnel requirements
  Selection of vendor

Since for large, centralized installations the cost of sample transport may equal that of the X-ray and computer units, a minimum number of streams is necessary to make this approach economically feasible. At current costs this number lies between 6 and 10 streams.

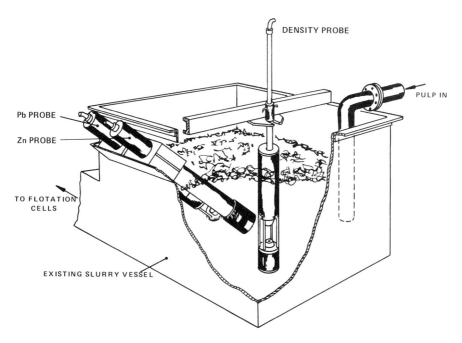

FIG. 13.3 Australian slurry probe. (Reproduced with permission of Philips Industries Holdings Limited, Australia, Copyright 1971.)

Where fewer streams are needed, individual on-site installations become desirable. If expansion in the number of streams is foreseen, however, then a centralized installation might still be preferred. The on-site approach has developed only recently as a result of the availability of EDX instruments with radioisotope excitation.

In a typical on-site on-stream installation, a primary sampling point is selected in an area of high slurry turbulence. The sample goes to a head tank in which the slurry is deairated and screened for the removal of foreign objects. The tank provides a constant-head feed to the X-ray flow cell. After analysis, the sample returns to the process stream. Excitation of the sample is by radiation from a radioisotope source. The characteristic X rays from the slurry pass to the detector. From this point on, the signal can be processed in various degrees of sophistication. Safety

MINING AND ORE PROCESSING   365

circuits are included for protection against equipment failure, especially flow-cell window breakage.

A most significant development in slurry analysis had its beginning in Australia [2-4]. The analytical device is placed directly into the slurry stream, without the necessity for external plumbing. This is illustrated in Fig. 13.3. The system uses radioisotope excitation and scintillation detection in combination with shielding, filters, composite radiators, secondary targets, and collimators to achieve the desired elemental specificity. Additionally, a γ-ray absorption probe is included to determine solids concentrations. At the end of 1975, installations had been made in seven Australian plants, with one more committed.

In the United States during 1975, energy-dispersive equipment with radioisotope excitation and Si(Li) detection was utilized for the first time in fast-flowing mineral slurries [5]. The assembly consisting of a source-detector probe with double Mylar windows and leak detector, preamplifier, liquid nitrogen reservoir, splash shield, and protective cage was positioned so that the probe was in the main slurry stream. The prototype assembly, inserted in a heading slurry, is shown in Fig. 13.4. Copper and iron were determined in copper concentrator heading and tailing streams. The approach was so successful that further development is warranted.

Slurry analysis is affected by sampling, solids matrix, and solids concentration. Sample transport and presentation have caused difficulties. In the past, the sampling problem often was considered separately from the X-ray analysis instrumentation. Suppliers provided X-ray equipment with final head tank(s); but all preceding sampling devices were developed and built by the purchaser. Over the years suppliers have acquired expertise in all phases of installation. In one case a metallurgical company became a major X-ray vendor. Now "turnkey" systems can be purchased; nevertheless, many users prefer to assemble their own sampling and data-processing equipment, purchasing only the spectrometer.

Matrix changes in the solid component of the slurry are compensated for mathematically as is the case with solid samples (see

FIG. 13.4 Si(Li) in-stream probe in high volume high-velocity launder. (Photo by Don Green, Kennecott Copper Corporation.)

Chaps. 2 and 3). Since measurements taken in slurry analysis, however, are usually limited to few elements and a background reading, vigorous mathematical treatment is problematic. In practice, empirical approaches such as multiple regression analysis, have proved fruitful most often. Extensive sampling is done during X-ray counting periods, preferably from the discharge side of the X-ray flow cell. Maximum variations in percent solids and elemental composition should be covered. Element concentrations in the solid phase are measured by independent chemical and/or instrumental means. The resulting values are applied to generate mathematical relationships and error data and to establish calibration curves for manual or computer usage. Corrections for variations in the solids content of the slurry must be included. The solids content ranges generally from 25% to 40%, but can lie outside these values on occasion. While it is possible to establish calibration curves for each level of solids, and this is being done, such procedure is involved and time consuming. Mathematical correction is to be preferred. Percent solids can be measured independently, usually by means of radioisotope density gauges. Another approach, which is quite common, makes use of X-ray background data at a preselected wavelength or of measurements of primary peaks from an X-ray target or a radioisotope source. With increasing aqueous dilution, there is more scatter, and background readings rise. With decreased dilution they drop. This method, however, is badly affected by compositional changes in the solids. For instance, the introduction or increased concentration of heavy elements in a sample may lower background or primary peaks. Hence, an increase in the water phase may be compensated for by compositional changes in the solid. This approach, nevertheless, has great practical value. That it works sufficiently is due to restricted compositional changes in most plant situations. Extreme fluctuations can be recognized easily.

In-plant process control is very sensitive to diminished quality of data and to instrumental down time. Obviously, a hostile plant environment adds to the need for competent regular maintenance. Centralized on-stream assemblies are usually found in protective rooms. Maintenance, therefore, is not different from usual laboratory maintenance, except for routine and emergency changes of flow-cell windows. For these installations the major effort goes to maintenance of sample transport systems outside the instrument room. For on-site batch type, for on-site on-stream, and for in-stream analyzers, however, industrialization of all components is essential, and without it, dust, moisture, vibration, noise, etc., will quickly take their toll.

13.3 SPECIMEN PREPARATION

Details and theoretical aspects of sample handling are provided in Chaps. 10-12. As Bernstein [6] has pointed out, it is difficult to obtain successful X-ray control of mining material without some understanding of the effects of the two most important variables influencing X-ray intensities in powder samples, namely, particle size and mineralogy. The variations in hardness of the materials under consideration often make grinding control a problem. The wide variety of matrices encountered in dealing with mining and ore processing samples necessitates either applying theoretical concepts, or normalizing these matrices, or utilizing calibration curves specifically for the type of matrix in question.

13.3.1 Grinding

We must distinguish between what is needed for high precision and accuracy and what is feasible in the reality of the mining and ore-processing world. The attainable at the research institution might be far from practical in an industrial environment. Though it can be shown that decreased particle size will enhance the X-ray intensity of a metallurgically important component element (e.g., through the unlocking of coated mineral particulates), for most applications consistency in sample treatment is of greater importance than a theoretically best grind. The user must standardize preparation and presentation of standard and unknown samples. The more he can adhere to this requirement, the better the precision and accuracy obtained. The most critical and time consuming work is required for accounting analyses. Here the utmost in quality of results is desirable. For the bulk of work, such as encountered in process control, the requirements are less stringent. Sometimes precision rather than accuracy is the important factor, since the detection of changes in a process may be more significant than absolute values.

Though the instrument operator controls sample treatment in the service laboratory, he has little influence on most in-plant determinations. In slurry analysis, one must assume that particle differences will average out over the counting periods.

In the copper industry, typical conditions for laboratory work are particle sizes of less than 75 μm (200 mesh) and pelletizing pressures of 138,000 kPa (20,000 psi)* for headings, tailings, and copper concentrates. Because of their mineralogical habit, molybdenum concentrates must be briquetted at lower pressure, i.e., 28,000-55,000 kPa (4,000 to 8,000 psi).

13.3.2 Heterogeneities

Gross errors in judgment by process operators can occur if heterogeneities in samples are not recognized. In the nonferrous industries flotation is the major concentrating step. Samples are usually well prepared through several steps of crushing, grinding, and milling. Further grinding and mixing is done in sampling departments prior to analysis. Simple tests for precision aid in the estimation of errors due to residual nonuniformity in the samples. In general, grinding, dilution, dissolution, fusion, or other techniques should be applied so that the contribution of heterogeneities to the total error is minimized. Again, economic factors and actual need dictate how far to go. Effects of sample heterogeneity in slurry streams are reduced by the aqueous phase.

The strength of X-ray analysis in mining and ore-processing lies mainly in its speed. Where this advantage is lost, other approaches may gain prominence. If, for example, extensive preparation is required prior to the X-ray determination of one or two elements, perhaps simple dissolution followed by atomic absorption analysis would be preferred.

In summary, for a certain metallurgical starting material, intermediate or final product, the approach to sample and standards preparation and presentation must be standardized. Grinding should be carried to the level of necessity dictated by sample properties, economics, and desired quality of data.

---

*One kPa = kilopascal = 10^3 pascal = 0.145 psi.

13.4 QUANTITATIVE ANALYSIS

13.4.1 Calibration Standards

For most applications X-ray fluorescence spectrometry depends upon standards developed as needed by the individual laboratory. In some laboratories associated with ore processing, considerable effort has been expended on standards development. Test programs have been used to establish values and associated error data for components in many products. In certain cases such programs have been enlarged to intercompany studies. The standards so developed are often of general value for all types of analytical work. They are useful in evaluating performance of a laboratory and help to prevent introduction of biases into analytical operations. On a broader base, the National Bureau of Standards has made available a few standard materials, such as copper head, tail, and concentrate, molybdenum concentrate, zinc concentrate, and phosphate rock. These allow comparison on a national and even international level. Unfortunately, these standards are expensive and, in the case of copper, have certified values for very few elements. There are more alloy standards available than any other type.

The development of standards through chemical analysis is tedious and costly. Ranges of importance must be covered for all elements of interest. In the usual process control case one to four elements are measured. Regression equations are developed with the standards for each element and applied to the problem at hand as the preferred approach to raw data treatment. In routine work on common materials this calibration procedure has to be followed only once. From then on control standards are analyzed with the unknowns to check the equations. Only when the standard materials become exhausted or when significant changes occur in a matrix need detailed calibration be repeated. Standards may also be used in a bracketing technique to obtain results for single samples or small groups of samples.

In nonroutine, nonrepetitive cases where no standards are available the following procedure may be used. The suite of samples submitted is first analyzed by X-ray methods. From the X-ray intensity data, a selection can be made of a few samples that cover adequately the full concentration range. These samples are submitted for independent analysis and form the basis for subsequent calibration. Preliminary estimates can be made using raw X-ray data based on operator experience. This information may aid the submitter in proceeding with his technical program prior to obtaining the final results.

13.4.2 Internal Standards

The manufacturer may include some kind of internal standardization in his instrument design. For instance, a copper internal standard might be utilized to monitor instrument stability and to correct variations. Internal standards of the chemical type (inquarted in the sample) are rarely used in the mining and ore-processing industry in X-ray work, because of mixing problems and time loss. Benefits do not warrant the effort.

13.4.3 Scatter Background

Scatter background is used widely as an analytical tool. It is discussed in Chaps. 2 and 4. Background intensities are utilized as mathematical correctors for the improvement of raw data as well as to estimate percent solids in slurries. The determination of molybdenum shown in Fig. 13.5 is an example of what can occur when such corrections are neglected. In this specific case two concentrator samples are compared. They are of equal molybdenum content but greatly different iron concentration. The higher iron reduces background significantly. On a suite of similar samples with iron ranging from 2% to 21%, simple ratio correction utilizing background improved the standard error of estimate for molybdenum by a factor of 4.

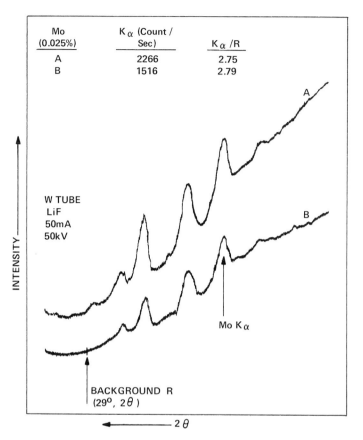

FIG. 13.5 The effect of iron on molybdenum assays (A: 3% Fe, B: 16% Fe).

13.4.4 Mathematical Correction

In the majority of cases, mathematical corrections used in the mineral industries are simple ratio and difference techniques utilizing background measurements. Regression studies determine which correction is best. This empirical approach has proved satisfactory in slurry analysis, on-stream and in-stream, and in the laboratory where the number of elements determined is restricted. In other instances, such as the multielement analysis of environmental samples on filters, a fuller mathematical treatment is necessary.

MINING AND ORE PROCESSING

Manufacturers of multielement analyzers include corrective programs in their software packages such as the Lucas-Tooth-Pyne [7], the Lachance-Traill [8], the Criss-Birks [9], and the Rasberry-Heinrich [10] formalisms.

13.5 SPECIFIC EXAMPLES

In the preceding sections, consideration has been given to necessary precautions in sample preparation and data analysis for application of X-ray methods to mining and ore processing. In this section we will consider specific examples of successful application of X-ray techniques to the analysis of mining and ore-processing samples, starting with exploration techniques and concluding with application to refining processes. Both laboratory and in plant X-ray analytical techniques will be considered.

Table 13.5 is a compilation of important published examples of the utilization of X-ray techniques in mining and ore processing. An attempt has been made to include in the tabulation only cases of proven usage. While some references may have been missed, the extent of present and potential utilization in the industry is evident upon reviewing the information presented. Four distinct techniques are identified in the table: (1) conventional X-ray tube with wavelength dispersion, (2) air-cooled X-ray tube with Si(Li) detector, (3) radioisotope source with Si(Li) detector, (4) radioisotope source with filters and/or radiators and scintillation or proportional detectors.

13.5.1 Exploration

While X-ray fluorescence does not provide the sensitivity to trace elements that can be obtained by optical emission spectrography, it is nevertheless of great value in general survey work where the target elements are present in higher than trace quantities. An important example of this type of study is exploration for economic mineralogical values in deep sea manganese modules. Gordon and

TABLE 13.5

Examples of X-Ray Utilization in Mining and Ore Processing

Ore type	Operation	Assay	Sample type	Sample presentation	X-Ray source	Dispersion or filters	Detector	Application	Refs.
Ag	Exploration, mining	Ag	Field	Powder	^{147}Pm/Al, 0.5 Ci	Mo/Rh	Scintillation	Field assay	[11]
Cu	Exploration	Cu	Core samples	Core	^{238}Pu, 30 mCi	Co/Ni	Scintillation	Core analysis	[11]
	Mining	Cu	Ore pulps and rock faces	In situ and powder	^{3}H/Zr, 12 Ci	Co/Ni	Scintillation	Field assay	[11]
	Mining	Cu, Fe	Drill hole	Down hole and drill chips	^{238}Pu, 90 mCi	Energy	Si(Li)	Feasibility	[12]
	Mining	Cu, Fe, Mo	Mine samples	Briquetted powder, mine water	W-tube water-cooled	Wavelength	Scintillation	Mine control	[13]
	Concentrating	Cu, Fe, Mo	Feed, tail, concentrate	Briquetted powder	W-tube water-cooled	Wavelength	Scintillation	Process control and accounting	[13], Table 13.6
	Concentrating	Cu, Fe	Feed, tail, concentrate	Slurry central	W-tube water-cooled	Wavelength	Scintillation	Process control	[14,15]
	Concentrating	Cu, Fe	Feed	Slurry in-stream	^{238}Pu, 90 mCi	Energy	Si(Li)	Process control	[5]
	Concentrating	Cu, Fe	Feed	Slurry by-pass	^{127}I	Energy	Si(Li)	Process control	[16]
	Concentrating	Cu, Fe	Feed, tail, concentrate	Filter cake	^{238}Pu, 100 mCi	Energy	Si(Li)	Process control	[17]
	Concentrating	Cu	Tail	Briquetted powder	^{244}Cm	Zn/Ni, Zn/Ni/Co/Al	Scintillation	Process control and accounting	[18]
	Concentrating	Mo	Copper concentrate	Slurry central	W-tube water-cooled	Wavelength	Scintillation	Process control	Sec. 13.5.3
	Smelting	Cu, Fe, Mo	Slag	Briquetted powder	W-tube water-cooled	Wavelength	Scintillation	Process control	[13], Table 13.7
	Smelting	Al, Ca, Cu, Fe, S, Si	Concentrate, flux, matte, white metal, slag	Briquetted powder	Au-tube air-cooled	Energy	Si(Li)	Process development	Sec. 13.5.5
	Smelting	Cu	Matte	Briquetted powder	W-tube water-cooled	Wavelength	Scintillation	Process control	[13], Table 13.7

Element	Process	Sample	Sample form	Source	Analysis	Detector	Application	Reference
Cu	Smelting	Precipitates	Oxidized, fluxed, and briquetted	W-tube water-cooled	Wavelength	Scintillation	Process control	[13], Table 13.7
Ca, Cu, Fe, S	Smelting	Flux	Briquetted powder	W-tube water-cooled	Wavelength	Scintillation	Process control	[13], Table 13.7
Cu, Zn, Fe, Si, Al, Pb, Sn, Ni, S	Smelting	Slags	Fused	Cr-tube water-cooled	Wavelength	Proportional scintillation	Process control	[19]
Se, Te	Refining	Copper	Dissolved and co-precipitated on filter	W-tube and Cr-tube water-cooled	Wavelength	Scintillation, proportional	Quality control	[20]
Cu, Ni	Refining	Electrolyte	Liquid	W-tube water-cooled	Wavelength	Scintillation	Process control	Table 13.8
Au	Refining	Copper	On ion exchange resin	W-tube water-cooled	Wavelength	Scintillation	Quality control	[20]
Cu, Zn	Concentrating	Feed, concentrate	Slurry by-pass	^{238}Pu, 30 mCi	Filters	Scintillation	Process control	[21,22]
Pb	Concentrating	Feed, concentrate	Slurry by-pass Slurry in-stream	^{153}Gd, ^{137}Cs	γ-Ray absorption	Ion chamber, scintillation	Process control	[3,21-23]
Cu, Zn	Concentrating	Feed, concentrate, tails	Slurry central	Tube water-cooled	Wavelength		Process control and accounting	[24]
Fe	Mining	Mine	Powder, core, crushed rock	^{238}Pu, 30 mCi; ^{238}Pu, 10 mCi	Cr/Mn	Scintillation	Field assay, mine control	[11]
Cu, Fe, Mn, Ni	Exploration	Deep sea nodules	Powder	^{238}Pu, 20 mCi	Co/Ni, Cr/Mn, V/Cr, Fe/Co	Scintillation	Shipboard assay	[11]
Mn, Fe, Co, Ni, Cu	Exploration	Deep sea nodules	Fused	Tube water-cooled	Wavelength	Geiger	Laboratory assay	[25]
Mo	Mining	Ore, copper concentrate	Powder	^{109}Cd, 2mCi	Y/Zr	Scintillation	Mine control	[11]
Ni	Mining	Ore	Powder	^{109}Cd	Co/Fe	Proportional	Mine control	[26]
Pb	Mining	Ore, drill core	Powder, core	^{238}Pu, 30 mCi	Ga/Ge	Scintillation	Field assay	[11]
Pb	Mining	Ore, mine core	In situ	^{57}Co		Scintillation	Mine control	[11]

TABLE 13.5 (continued)

Ore type	Operation	Assay	Sample type	Sample presentation	X-Ray source	Dispersion or filters	Detector	Application	Refs.
	Mining	Pb	Drill holes	Down hole	^{57}Co	W/Au	Scintillation	Mine control	[27]
Pb, Zn	Concentrating	Pb	Feed, concentrate, tails	Slurry central	W-tube water-cooled	Wavelength	Scintillation	Mill control	[14]
	Concentrating	Zn	Feed	Slurry in-stream	^{238}Pu, 30 mCi	Zn, Al	Scintillation	Mill control	[3,21]
	Concentrating	Zn	Tails	Slurry in-stream	^{238}Pu, 30 mCi	Zn	Scintillation	Mill control	[18,21]
Sn	Exploration, mining	Sn	Ore, drill core, field samples	Powder, rock faces, core	^{147}Pm/Al, 0.5 Ci	Ag/Pd	Scintillation	Field assay	[11,28]
	Exploration, mining, concentrating	Sn	Field samples, ore, concentrate	Powder	W-tube water-cooled	Wavelength	Scintillation	Field assay, mine and concentrator control	[29]
Ti	Exploration, mining, concentrating	Ti	Rock, ore, concentrate	Powder	^{55}Fe, 1 mCi	Sc/Ti or none	Scintillation	Field assay, plant control	[11]
U	Concentrating	Ca	Concentrate	Fusion	Cr-tube water-cooled	Wavelength	Proportional	Process control	[30]
	Exploration, mining	U	Ore	Powder	^{109}Cd, 1 mCi	Y/Zr	Scintillation	Field assay	[11]
V	Mining	V	Ore	Powder	^{55}Fe, 10 mCi	Ti	Scintillation	Mine control	[11]
Zn	Exploration, mining	Zn	Drill core, ore	Core, powder	^{238}Pu, 30 mCi	Ni/Cu	Scintillation	Core and field assay	[11]
	Concentrating	Zn	Feed, concentrate, tails	Slurry central	Tube water-cooled	Wavelength		Mill control	[31]

MINING AND ORE PROCESSING   377

coworkers described such a study in 1960 [25]. More recently, as
mentioned in Table 13.5, EDX equipment using balanced filters has
been utilized on board ship for on-site evaluation of metal contents. The adaptability of the equipment to unfavorable ambient
conditions, and the handling of a large volume of samples in a relatively short time were of importance. Copper, iron, nickel, and
manganese were measured. The radioactive source was ^{238}Pu, 20 mCi,
and filters were varied to fit the element of interest (copper -
Co/Ni, iron - Cr/Mn, manganese - V/Cr, nickel - Fe/Co) [11].

Other techniques of value in exploration depend upon preconcentration by ion exchange, selective precipitation, coprecipitation or chemical reduction [32-34]. Ion exchange with concentration by microfiltration on to glass fiber disks, microporous membrane filters, and ion-exchange resin disks, followed by X-ray
fluorescence measurement was discussed in 1971 by Bertrand and Linn
[20]. The general question of the analysis of rocks by X-ray fluorescence techniques was reviewed in 1957 by Chodos and Branco [35].

13.5.2 Mining

As one moves to mining where the elements of interest are at relatively high concentrations and decisions are to be made regarding
waste vs. ore, X-ray fluorescence techniques become of even greater
importance. Again, EDX instrumentation can be packaged in small
units which can be carried readily to the actual analysis site.
Langheinrich and coworkers [12,36] described an EDX unit with ^{238}Pu
excitation and Si(Li) detector that is small enough to be lowered
down an 8-in. bore hole of the type used in open pit-mining operations. The relative advantages of analyzing drill cuttings as con-
trasted to measuring the integrated content of the bore hole side
walls were considered and it was concluded that because of time factors the measurement of the drill cuttings on-site had distinct advantages. Because of their size and portability EDX units are of
particular value for such measurements.

Marr and Campbell [27] described another drill hole probe which uses a ^{57}Co source, dual scintillation counters, and balanced filters with a pass band for lead Kα radiation. Using 10-sec counting times or less, probe response correlated to lead ore grade at any selected depth in dry uncased drill holes. Other portable X-ray units employing radioisotope sources, and filters for isolating desired energies have been used by a number of workers for mine field assay as shown in Table 13.5. These measurements have been made on rock faces as well as chips, pulps, and briquetted samples.

13.5.3 Concentrating

The advantages of X-ray techniques in mining also pertain to concentrating. In addition, the on-stream capabilities mentioned in Sec. 13.2 are of particular value in providing continuous information for use in automatic or semiautomatic control of concentrator operations. Slurry analyses in concentrators by both wavelength and EDX techniques are listed in Table 13.5.

In 1959, on-stream wavelength-dispersive X-ray units were installed at the Utah Copper Division of Kennecott Copper Corporation to monitor molybdenum losses in copper concentrate from two molybdenite plants. These pioneer, single-stream systems are still in operation having performed well for approximately 17 years with relatively low maintenance costs.

Table 13.6 presents data obtained in a study of the applicability of X-ray fluorescence analysis to the measurement of copper, iron, and molybdenum in typical copper concentrator materials, (A) with and (B) without application of roasting or fusion techniques. All of the sample types noted are now analyzed by X-ray fluorescence for the elements indicated. The sample load has been such that the equipment as originally obtained has been modified to allow computer operation and readout [37]. Currently about 20,000 determinations are made per man month. Sample preparation consists of grinding the sample in a Bleuler mill for 5 min, transferring to a sample cup

TABLE 13.6

Concentrator Samples: Comparative Precision Data

Element	Mean	Repeat precision (2σ)		Chem vs. XRF (2σ)		Samples/man-day		
		Chem	XRF	A	B	Chem	XRF(A)	XRF(B)
Heads								
Cu	0.771	0.010	0.014(A)	0.016	0.100	60	60	200
Fe	2.58	0.11	0.07 (A)	0.10	0.20	60	**[a]	**[a]
Mo	0.03	0.0038	0.003(B)		0.004	60		**[a]
Tailings								
Cu	0.090	0.010	0.009(B)		0.007	60		200
Mo	0.0059	0.002	0.001(B)		0.002	60		**[a]
Fe	If desired, to same precision as Fe in Heads							
Copper concentrate								
Cu	28.55	0.20	0.20 (A)	0.500	2.60	60	60	200
Mo	0.21	0.017	0.012(B)		0.036	60		**[a]
Molybdenite concentrate								
Cu	0.82	0.026	0.030(A)	0.042	0.116	60	60	200
Mo	52.22	0.68	0.41 (A)	0.42	2.10	60	60	**[a]

[a]These analyses are performed by XRF simultaneously with the analysis above, so that the number of samples above times the number of elements read equals total determinations per man-day.

(3.5-cm diameter), and pressing under 15-tonne pressure. Twenty-four samples are placed in a "lazy susan" sample holder and within 18 min results are available in report form.

The determination of iron and copper in concentrator materials, using an automated EDX spectrometer, was reported by Madden [17]. Broad applications of EDX for this type of work were described in 1973 by Langheinrich et al. [36].

X-Ray fluorescence has been applied successfully for continuous-stream analysis (at a central location) and individual sample analysis for lead, zinc, and copper in lead zinc concentrators.

In one case, a wavelength-dispersive X-ray instrument designed for sequential analysis of slurries piped from several points in the plant, was adapted successfully to accept sample powders briquetted in aluminum cups. In this manner the operators were able to make the best possible utilization of the equipment.

Knoke [30] reported on the utilization of X-ray fluorescence analysis in a uranium concentrator. A vacuum-path X-ray spectrometer was used for the determination of calcium in a range of 0.03-1.75% in uranium ore concentrates. The procedure provided a 50% saving in analyst time with no sacrifice of accuracy or precision compared to the chemical method previously used. Because of variation in sample composition and attendant difficulties, sodium tetraborate was utilized to dissolve the samples at 1050°C after which they were case on a polished aluminum plate.

13.5.4 Smelting

XRF techniques have been developed for smelter products that are of comparable precision to chemical methods for copper, iron, and molybdenum at nearly all concentration levels. Considerable work has been done in the analysis of smelter furnace products, particularly reverberatory and converter slag. Assays for copper, iron, and molybdenum in these products may be performed very rapidly with precision that compares favorably with chemical methods. Table 13.7 presents comparative precision data for the analysis of copper smelting materials by both chemical and X-ray fluorescence techniques. In this table, methods are designated as (A) in cases where the instrumental precision is of the same order as the chemical method, and (B) in cases of marginal precision. For an (A) designation it was necessary with some products to develop sample preparation techniques that appreciably reduced the potential output per man per day. For instance, for determining copper in matte with accounting accuracy, the quantitative instrumental method entailed oxidation, fluxing, and pelletizing. These operations reduce the daily sample

TABLE 13.7

Smelter Samples: Comparative Precision Data

Element	Mean	Repeat precision (2σ)		Chem vs. XRF (2σ)		Sample/man-day		
		Chem	XRF(A)	A	B	Chem	XRF(A)	XRF(B)
Reverb slag
Cu	0.43	0.040	0.016	0.034		60	200	
Fe	30.8		0.4	0.8		60	**[a]	
Mo	0.034	0.004	0.003	0.006		60	**[a]	
Reverb matte
Cu	39.5	0.48	0.5	0.5	1.1	60	60	200
Precipitates
Cu	81.57	0.31	0.35	0.45	4.7	60	60	200
Flux
Cu	0.098	0.010	0.010	0.020		90	200	
Fe	2.19	0.12	0.15	0.24		90	**[a]	
Ca	6.56	0.21	0.21	0.71		60	**[a]	
S	0.89	0.14	0.16	0.25		60	100	

[a]These analyses are performed by XRF simultaneously with the analysis above, so that 200 times the number of elements read equals the total number of determinations per man-day.

output to the level of the chemical method. A single sample could be run in about 20 min, however, which could be useful for process control purposes. West and coworkers [19] were successful in the measurement of Al, Cu, Fe, Ni, Pb, S, Si, Sn, and Zn in copper-smelting processes by X-ray fluorescence. They fused the samples in a sodium tetraborate-lanthanum oxide flux with sodium nitrate added to ensure oxidizing conditions in the melt. Several procedures for X-ray fluorescence analysis of slag, including acid solution and fusion methods, have been compared by Carlsson [38]. He gives precisions and accuracies for 11 different combinations and recommends fusion with borax followed by casting of a bead.

Recently, vacuum EDX instrumentation utilizing a low-power X-ray tube source has been applied successfully by Kennecott for copper smelter control purposes. In a typical instance, analyses were run without preparation beyond grinding and pelletizing for Ca, Cu, Fe, and Si in converter slag and Cu, Fe, and S in matte and white metal. Turn-around time for a sample was less than 10 min and 5-7 samples were run per hour around the clock. Air-path applications with radioisotope sources were equally successful for copper and iron.

The importance of flux analyses in proper smelter operation cannot be underestimated. As seen in Table 13.7 results obtainable with X-ray techniques are comparable to wet methods for iron and copper but marginal for calcium and sulfur.

13.5.5 Refining

Comparisons of X-ray and classical analytical methods for refining samples are given in Table 13.8. Copper and nickel in tankhouse electrolyte can be analyzed by XRF with precisions adequate for process control much more rapidly than by chemical methods. The silver method is rapid and could be useful for quick checks of alloying procedures, but does not provide the precision necessary for accounting assays. The XRF method would be useful for antimony, arsenic, bismuth, iron and lead in a number of refinery intermediate products. It is to be stressed that the XRF techniques discussed in Table 13.8 require no preparation other than simple presentation of samples in the X-ray beam.

Bertrand and Linn [20] discussed chemical preconcentration prior to X-ray determination of selenium and tellurium. They considered wirebar, cathode, anode, and fire-refined copper as well as copper concentrates and copper matte. The technique was successful for the determination of gold in anode and cathode copper and in copper concentrates. Both wavelength-dispersion and energy-dispersion X-ray spectrometry were used with these techniques. The separation preconcentration procedure was adapted from methods by

TABLE 13.8

Comparison of X-Ray and Classical
Analytical Methods for Refinery Samples

Copper in tankhouse electrolyte		
	XRF[a]	Chemical
Steps	3	4
Time of single sample	3 min	1 hr
Samples per man-day	150	100
Lower limit of detection	0.03 g/liter	0.01 g/liter
Range	38-52 g/liter	38-52 g/liter
Precision (2σ)	±1.2 g/liter	±0.2 g/liter
Nickel in tankhouse electrolyte		
	XRF[a]	Chemical
Steps	3	4
Time for single sample	3 min	10 min
Samples per man-day	150	100
Lower limit of detection	0.02 g/liter	0.01 g/liter
Range	4.0-5.0 g/liter	4.0-5.0 g/liter
Precision (2σ)	±0.34 g/liter	±0.4 g/liter
Silver in copper alloys		
	XRF	Fire assay
Steps	4	7
Time for single sample	5 min	1 day
Samples per man-day	100	45
Lower limit of detection	100 g/tonne	17 g/tonne
Range	170-1030 g/tonne	170-1030 g/tonne
Precision (2σ)	±70 g/tonne	±10 g/tonne

[a]Copper and nickel can be determined simultaneously with XRF.

Burke and Yanak [39], Luke [40], and Maassen [41], and is based upon the chemical reduction of selenium and tellurium in solution to a finely divided metal. Addition of an arsenic carrier was found necessary to serve as aid in nucleation and to collect the precipitate. It also provided a convenient internal standard for X-ray measurement. For the gold analysis, an ion-exchange resin-loaded paper was used to collect the gold from solution for X-ray spectrographic determination. This method, adapted from Green et. al. [34] and Fukasawa et al. [42] can recover successfully 5-100 µg of gold from 3 g of copper.

13.5.6 Environmental

We would be remiss if we did not mention the role played by X-ray techniques in the diagnosis of environmental problems in and about mining, concentrating, refining, and especially smelting installations. In any instance where airborne particulates are collected on filter media for analysis, X-ray methods are well adapted to handle a majority of the measurements. EDX utilizing either radioisotopes or low-power X-ray tubes is particularly well adapted to this use. An automated vacuum unit can analyze up to 40 filter samples for up to 30 elements in 400 min and provide results in any desired units. Commercially available equipment has computer capability both for control and computation. Various mathematical models are used for computational purposes. Rhodes and coworkers described methods [43,44] for the preparation of standards for these applications.

REFERENCES

1. H. R. Cooper, On-stream X-Ray Analysis, in *Flotation* (M. C. Fuerstenau, ed.), AIME, New York, 1976, pp. 865-894.
2. J. S. Watt, Radioisotope On-Stream Analysis, *Atomic Energy in Australia*, *16*(4), 3 (1973).
3. R. A. Fookes, V. L. Gravitis, D. A. Hinckfuss, N. W. Stump, and J. S. Watt, Plant Trials of Radioisotope Immersion Probes for On-stream Analysis of Mineral Process Streams, *Transactions/Section C of the Institution of Mining and Metallurgy*, *82*, C21 (1973).
4. D. A. Hinckfuss and N. W. Stump, Immersible Probes Eliminate Sampling in On-stream Analysis, Aus. I. M. M., Regional Conference, Adelaide, 1971.

5. A. P. Langheinrich, J. W. Forster, and W. M. Tuddenham, Development and Application of Energy-Dispersive X-Ray (EDX) Instrumentation for In-Stream Slurry Analysis, Kennecott Copper Corporation, Internal Technical Report, 1975.
6. F. Bernstein, Particle Size and Mineralogical Effects in Mining Applications, in *Advances in X-Ray Analysis* (W. M. Mueller and M. J. Fay, eds.), Vol. 6, Plenum Press, New York, 1963, pp. 436-446.
7. H. J. Lucas-Tooth and C. Pyne, The Accurate Determination of Major Constituents by X-Ray Fluorescent Analysis in the Presence of Large Interelement Effects, in *Advances in X-Ray Analysis* (W. M. Mueller, G. R. Mallett, and M. J. Fay, eds.), Vol. 7, Plenum Press, New York, 1964, pp. 523-541.
8. G. R. Lachance and R. J. Traill, Practical Solution to the Matrix Problem in X-Ray Analysis, *Can. Spectry.*, *11*(2), 43 (1966).
9. J. W. Criss and L. S. Birks, Calculation Methods for Fluorescent X-Ray Spectrometry, *Anal. Chem.*, *40*, 1080 (1968).
10. S. D. Rasberry and K. F. J. Heinrich, Calibration for Interelement Effects in X-Ray Fluorescence Analysis *Anal. Chem.*, *46*, 81 (1974).
11. J. R. Rhodes, Application of Portable Radioisotope X-Ray Analyzers in Industry and Mining, *Chemical Engineering Progress Symposium Series*, *Nuclear Engineering - Part XXII*, *66*, 48 (1970).
12. A. P. Langheinrich, J. W. Forster, and T. A. Linn, Jr., Energy Dispersion X-Ray (EDX) Analysis in the Nonferrous Mining Industry, *ISA Transactions*, *10*, 369 (1971).
13. W. M. Tuddenham, A. P. Langheinrich, F. L. Pherson, and J. H. Beyer, Improved Mine, Concentrator, and Smelter Control Through X-Ray Spectrometry, AIME Annual Meeting, New York, 1966; (*preprint No. 2A-V-2*, The Metallurgical Society of AIME).
14. R. O. French, R. W. Vaughn, A. P. Langheinrich, and J. F. Baum, Application of Continuous XRF Analysis to Mill Control, EMD-AIME Symposium on Continuous Process Control, Philadelphia, Pennsylvania, December, 1966.
15. G. R. Marchant, Recent Applications of Computer Control for Copper Beneficiation, 12th Process Automation Symposium, San Diego, California, April, 1970.
16. R. W. Tolmie and L. Urbanoski, On-Stream Analysis in Mine Mills, *Paper No. 116*, 27th Pittsburgh Conference on Analytical Chemistry and Applied Spectroscopy, Cleveland, Ohio, March, 1976.
17. M. L. Madden, New X-Ray Spectrometer for the Mining Industry, *Engineering and Mining Journal*, *174*(12), 84 (1973).

18. J. S. Watt, R. A. Fookes, and V. L. Gravitis, Radioisotope X-Ray Techniques for the On-stream Determination of Low Concentrations of Copper, Zinc, Tin, and Lead in Mineral Slurries, in *Nuclear Techniques in the Basic Metal Industries*, International Atomic Energy Agency, Vienna, 1973, pp. 141-153.

19. N. G. West, G. L. Hendry, and N. T. Bailey, The Analysis of Slags from Primary and Secondary Copper Smelting Processes by X-Ray Fluorescence, *X-Ray Spectrometry*, 3, 78 (1974).

20. C. C. Bertrand and T. A. Linn, Jr., Chemical Preconcentration Technique for the X-Ray Determination of Trace Elements, 26th Northwest Regional Meeting, American Chemical Society, Bozeman, Montana, June, 1971.

21. J. S. Watt and W. J. Howarth, Mineral Processing Methods and On-stream Analysis in Australian Mineral Processing Plants, in *Nuclear Techniques in the Basic Metal Industries*, International Atomic Energy Agency, Vienna, 1973, pp. 105-118.

22. R. A. Fookes, V. L. Gravitis, J. S. Watt, G. J. Wenk, and L. R. Wilkinson, On-stream Analysis for Copper, Zinc, Tin, and Lead in Plant Mineral Slurries Using Radioisotope X-Ray Techniques, in *Symposium on Automatic Control Systems in Mineral Processing Plants: Technical Papers*, Aus. I.M.M., Brisbane (1971), pp. 21-33.

23. D. A. Hinkfuss and B. S. Rawling, The Development and Application of an On-Stream Analysis System for Lead at the Zinc Corporation, Ltd., in *Broken Hill Mines - 1968* (M. Radmanovich and J. T. Woodcock, eds.), Aus. I.M.M., Melbourne, 1968, pp. 475-479.

24. C. L. Lewis, R. A. Hall, J. W. Anderson, and W. H. A. Timm, The On-stream X-Ray Analysis Installation of the Lake Default Mine, *The Canadian Mining and Metallurgical (CIM) Bulletin*, April, 1968, p. 513.

25. G. M. Gordon, D. J. McNely, and J. L. Mero, X-Ray Spectrographic Analysis of Manganese Nodules, in *Advances in X-Ray Analysis* (W. M. Mueller, ed.), Vol. 3, Plenum Press, New York, 1960, pp. 175-184.

26. J. M. Brinkerhoff, B. Sellers, and F. A. Hanser, Measurement of Nickel in Ores Containing Large Amounts of Iron, in *Application of Low-Energy X and Gamma-Rays* (C. A. Ziegler, ed.), Gordon and Breach Science Publisher, New York, 1971, pp. 165-171.

27. H. E. Marr, III and W. J. Campbell, Evaluation of a Radioisotopic X-Ray Drill Hole Probe, *Bureau of Mines Report of Investigations 7611*, U. S. Department of the Interior, 1972.

28. R. Cox, Assessment of Tin Ores in Situ at Cleveland Mine, Tasmania, with a Portable Radioisotope X-Ray Fluorescence Analyzer, *Transactions/Section B of the Institution of Mining and Metallurgy*, 77, B109 (1968).

29. T. R. Sweatman, T. C. Wong, and K. S. Toong, Application of X-Ray Fluorescence Analysis to the Determination of Tin in Ores and Concentrates, *Transactions/Section B of the Institution of Mining and Metallurgy*, 76, B149 (1967).
30. D. R. Knoke and H. F. Waldron, The Determination of Calcium in Uranium Ore Concentrates by X-Ray Fluorescence, in *Advances in X-Ray Analysis* (W. M. Mueller, G. R. Mallett, and M. J. Fay, eds.), Vol. 8, Plenum Press, New York, 1965, pp. 448-455.
31. N. L. Fuller, P. E. McGarry, and J. R. Pellett, X-Ray Assaying and Reagent Control at Friedensville, *Mining Congress Journal*, 53(4), 108 (1967).
32. H. J. Rose, Jr. and F. Cuttita, X-Ray Fluorescence Spectroscopy in the Analysis of Ores, Minerals, and Waters, in *Advances in X-Ray Analysis* (J. B. Newkirk, G. R. Mallett, and H. G. Pfeiffer, eds.), Vol. 11, Plenum Press, New York, 1968, pp. 23-39.
33. C. C. Bertrand and T. A. Linn, Jr., Determination of Traces of Uranium by Radioisotope Energy Dispersive X-Ray (EDX) Analysis, *Anal. Chem.*, 44, 383 (1972).
34. T. E. Green, S. L. Law, and W. J. Campbell, Use of Selective Ion Exchange Paper in X-Ray Spectrography and Neutron Activation-Application to the Determination of Gold, *Anal. Chem.*, 42, 1749 (1970).
35. A. A. Chodos, J. J. R. Branco, and C. G. Engel, Rock Analysis by X-Ray Fluorescence Spectroscopy, in *Proc. 6th Annual Conference on Industrial Applications of X-Ray Analysis* (W. M. Mueller, ed.), Denver Research Institute, University of Denver, Denver, Colo., 1957, pp. 315-327.
36. A. P. Langheinrich, J. W. Forster, and T. A. Linn, Jr., Current Status and Potential of Solid State Detectors in Industrial and Field Applications, in *Nuclear Techniques in the Basic Metal Industries*, International Atomic Energy Agency, Vienna, 1973, pp. 429-471.
37. J. R. Riding and G. R. Marchant, Improved X-Ray Analysis by Application of Real-Time Computers, *Society of Mining Engineers of AIME, Preprint No. 69-B-47*, 1969.
38. G. Carlsson, Comparison of Some Procedures for X-Ray Fluorescence Analysis of Metallurgical Slags, *X-Ray Spectrometry*, 1, 155 (1972).
39. K. E. Burke and N. M. Yanak, Separation and X-Ray Spectrographic Determination of Microgram Quantities of Arsenic in Copper, Iron, and Nickel Based Alloys, *Anal. Chem.*, 41, 963 (1969).
40. C. L. Luke, Determination of Trace Elements in Inorganic and Organic Materials by X-Ray Fluorescence Spectroscopy, *Anal. Chim. Acta*, 41, 237 (1968).

41. G. Maassen, Determination of Low Contents of Selenium in Copper by X-Ray Fluorescence Analysis, *Z. Erzbergbau Metallhuettenw.*, *18*(3), 116 (1965).

42. T. Fukasawa, T. Fujii, and A. Mizuike, X-Ray Fluorescence Determination of Gold and Silver in Copper Using Anion-Exchange Paper and Filter Paper, *Bunseki Kagaku*, *17*(6), 713 (1968).

43. D. C. Camp, T. A. Cooper, and J. R. Rhodes, X-Ray Fluorescence Analysis--Results of a First Round Intercomparison Study, *X-Ray Spectrometry*, *3*, 47 (1974).

44. A. H. Pradzynski and J. R. Rhodes, Standards for Thin Film X-Ray Fluorescence Analysis, *Paper No. 36*, 25th Pittsburgh Conference on Analytical Chemistry and Applied Spectroscopy, Cleveland, Ohio, March, 1974.

ADDITIONAL REFERENCES

1970-1975

A1. G. J. Wenk and L. R. Wilkinson, Rapid Control Assays by Radio-isotope X-Ray Techniques, *Aust. Miner. Develop. Lab. Bull.*, No. 17, 1 (1974).

A2. D. J. Reed, J. L. Dalton, and A. H. Gillieson, The On-Stream Analysis of Hematite Ore Fractions Using Radioisotopes, *X-Ray Spectrometry*, *3*, 15 (1974).

A3. W. J. Howarth, G. J. Wenk, and L. R. Wilkinson, Radioisotope On-Stream Analysis, *The Canadian Mining and Metallurgical (CIM) Bulletin*, Sept. 1973, p. 76.

A4. R. H. Hammerle, R. H. Marsh, K. Rengan, R. D. Giauque, and J. M. Jaklevic, Test of X-Ray Fluorescence Spectrometry as a Method for Analysis of the Elemental Composition of Atmospheric Aerosols, *Anal. Chem.*, *45*, 1939 (1973).

A5. A. S. M. deJesus, A Correction for Interelemental Background Contributions in Energy-Dispersive X-Ray (EDX) Spectroscopy, *X-Ray Spectrometry*, *2*, 179 (1973).

A6. N. H. Clark and R. J. Mitchell, Scattered Primary Radiation as an Internal Standard in X-Ray Emission Spectrometry: Use in the Analysis of Copper Metallurgical Products, *X-Ray Spectrometry*, *2*, 47 (1973).

A7. K. G. Carr-Brion, On-Stream Energy-Dispersive X-Ray Analyzers, *X-Ray Spectrometry*, *2*, 63 (1973).

A8. J. N. Kikkert, The Rapid Multi-Element Analysis of Base Metal Sulphides by X-Ray Fluorescence Spectrometry, *Aust. Miner. Develop. Lab. Bull.*, No. 16, 56 (1973).

A9. D. C. Camp, Applications of Energy Dispersive X-Ray Fluorescence, Winter Meeting, American Nuclear Society, San Francisco, Calif., November, 1973.

A10. D. C-J. Glenn, The Z Cra Probe System, in *Review of On-Stream Analysis Practice*, Australian Mineral Industries Research Association Ltd., Parkville, Victoria, Australia, 1973, pp. 1-14.

A11. S. Kreula, Courier 300 On-stream X-Ray Analysis System, in *Review of On-Stream Analysis Practice*, Australian Mineral Industries Research Association Ltd., Parkville, Victoria, Australia, 1973, pp. 15-34.

A12. T. F. Basinger, Process Control X-Ray Quantometer for High Precision Slurry Stream Analysis, in *Review of On-Stream Analysis Practice*, Australian Mineral Industries Research Association Ltd., Parkville, Victoria, Australia, 1973, pp. 35-46.

A13. G. S. Stacey and W. A. Bolt, Multistream X-Ray Fluorescence Analysis of Mineral Slurries in No. 1 Copper Concentrator at Mount Isa, in *Review of On-Stream Analysis Practice*, Australian Mineral Industries Research Association Ltd., Parkville, Victoria, Australia, 1973, pp. 47-78.

A14. P. Vanninen and S. Kreula, Plant Experience From Courier-300 On-Stream Analysis System, in *Review of On-Stream Analysis Practice*, Australian Mineral Industries Research Association Ltd., Parkville, Victoria, Australia, 1973, pp. 79-98.

A15. N. W. Stump, The Application of Probe Systems for Radioisotope On-Stream Analysis at New Broken Hill Consolidated Limited, in *Review of On-Stream Analysis Practice*, Australian Mineral Industries Research Association Ltd., Parkville, Victoria, Australia, 1973, pp. 99-106.

A16. W. J. Howarth, G. J. Wenk, and L. R. Wilkinson, Installation and Operational Experience with Radioisotope X-Ray On-Stream Analysis Equipment, in *Review of On-Stream Analysis Practice*, Australian Mineral Industries Research Association Ltd., Parkville, Victoria, Australia, 1973, pp. 107-128.

A17. J. S. Watt, Current Status and Potential of Radioisotope X-Ray and Nuclear Techniques for On-Line Analysis of Mineral Process Streams, in *Review of On-Stream Analysis Practice*, Australian Mineral Industries Research Association Ltd., Parkville, Victoria, Australia, 1973, pp. 129-138.

A18. D. J. Reed and A. H. Gillieson, X-Ray Fluorescence Applied to the On-stream Analysis of Sulphide Ore Fractions, *X-Ray Spectrometry*, 1, 69 (1972).

A19. K. G. Carr-Brion, On-stream Application of Semiconductor X-Ray Detectors, *Industrial Measurement and Control by Radiation Techniques, IEE Conference Publication No. 84*, University of Surrey, England, April 1, 1972, pp. 103-107.

A20. J. R. Rhodes, A. H. Pradzynski, and R. D. Sieberg, Energy-Dispersive X-Ray Emission Spectrometry for Multielement Analysis of Air Particulates, ISA Symposium, San Francisco, California, May, 1972.

A21. R. Jenkins and J. L. deVries, Isotope Excitation in On-stream Analysis, *Can. Spectrosc.*, *16*(2), 54 (1971).

A22. P. G. Burkhalter, Radioisotopic X-Ray Analysis of Silver Ores Using Compton Scatter for Matrix Compensation, *Anal. Chem.*, *43*, 10 (1971).

A23. S. P. Kasemsanta, Use of Monoenergetic X-Ray Sources in X-Ray Fluorescence Analysis. Recent Developments and Applications in Mineral Industries, in *Nuclear Techniques for Mineral Exploration and Exploitation*, International Atomic Energy Agency, Vienna, 1971, pp. 81-101.

A24. J. S. Watt, Current and Potential Applications of Radioisotope X-Ray and Neutron Techniques of Analysis in the Mining Industry, *Australas. Inst. Mining Met. Proc.*, *No. 233*, 1970, pp. 69-77.

A25. J. R. Rhodes, A. Pradzynski, R. D. Sieberg, and T. Furata, Application of a Si(Li) Spectrometer to X-Ray Emission Analysis of Thin Specimens, Third Symposium on Low-Energy X-Ray and Gamma-Ray Sources and Applications, Boston College, June, 1970.

A26. S. E. Bramwell and K. G. Carr-Brion, Industrial Application of Lithium Drifted X-Ray Detectors, *Instrument Practice*, *24*(5), 324 (1970).

1960-1969

A27. W. K. Ellis, R. A. Fookes, V. L. Gravitis, and J. S. Watt, Radioisotope X-Ray Techniques for On-stream Analysis of Slurries. Feasibility Studies Using Solid Samples of Mineral Products, *Int. J. App. Rad. Isot.*, *20*, 691 (1969).

A28. J. M. Mathieson, V. L. DaGragnano, and J. P. Hurley, Applications of Energy-Dispersive X-Ray Spectroscopy to the Elemental Analysis of Selected Central and South American Ore Samples, ANS/AEC Topical Conference on the Application of Radioactive Techniques to Latin American Technology, San Juan, Puerto Rico, May, 1969.

A29. S. H. U. Bowie, Portable X-Ray Fluorescence Analyzers in the Mining Industry, *Mining Magazine*, *118*, 230 (1968).

A30. R. Rotter, Multichannel X-Ray Fluorescent Analyzer for Automatic Control of Elementary Composition of Beneficiation Products, VIII International Mineral Processing Congress, Leningrad, USSR, 1968.

A31. A. P. Langheinrich and J. W. Forster, The Application of Radioisotope Nondispersive X-Ray Spectrometry to the Determination of Molybdenum, in *Advances in X-Ray Analysis* (J. B. Newkirk, G. R. Mallett, and H. G. Pfeiffer, eds.), Vol. 11, Plenum Press, New York, 1968, pp. 275-285.

A32. K. G. Carr-Brion, The Effect of Particle Size on Back-Scattered X-Ray Correction Methods in On-Stream X-Ray Fluorescence Analysis, *Analyst*, *91*, 289 (1966).

A33. K. G. Carr-Brion, The Determination of Tin in Powder Samples by X-Ray Fluorescence Analysis, *Analyst*, *90*, 9 (1965).

A34. K. G. Carr-Brion and D. A. Jenkinson, An X-Ray Fluorescence Slurry Presenter Which is Insensitive to Solids Concentration, *J. Sci. Instrum.*, *42*, 817 (1965).

A35. A. H. Smallbone, Briquetting, X-Ray Techniques Refine Onstream Analysis, *Rock Products*, *68*(12), 60 (1965).

A36. K. G. Carr-Brion and J. R. Rhodes, On-stream X-Ray Fluorescence Analysis of Ore Slurries with a Radioisotope X-Ray Source, *Instrument Practice*, *19*, 1007 (1965).

A37. K. G. Carr-Brion, The X-Ray Fluorescence Determination of Zinc in Samples of Unknown Composition, *Analyst*, *89*, 233 (1964).

A38. G. J. Sundkvist, F. O. Lundgren, and L. J. Lidstrom, Automatic X-Ray Determination of Lead and Zinc in the Tailings of an Ore Dressing Plant, *Anal. Chem.*, *36*, 2091 (1964).

Chapter 14

MICROANALYSIS AND TRACE ANALYSIS

John V. Gilfrich

X-Ray Optics Branch
U. S. Naval Research Laboratory
Washington, D. C.

14.1	Introduction	393
14.2	Sample Preparation	395
14.3	Qualitative and Quantitative Analysis	397
14.4	Examples: Detection Limit	404
14.5	Conclusion	409
References		410

14.1 INTRODUCTION

In dealing with microanalysis or trace analysis, we are presented with a different kind of problem than that which has been treated specifically up to this point. The measurement of trace concentrations in bulk samples has been mentioned in previous chapters, but here we shall discuss in detail the measurement of small amounts of the elements whether present as trace quantities in bulk materials or as the constituents of a specimen where the total available material is small, i.e., a microsample.

In order to put this subject in its proper historical perspective and to illustrate that the measurement of small quantities of material is not a recent development, it is appropriate to call attention to the fact that, in his landmark work in 1913 relating the wavelength of X-ray emission to the atomic number of the element [1], Moseley was surprised at how strongly the low-concentration elements appeared in the spectra. Not very long after, in the 1930s, Eddy and Laby in Australia were measuring ppm concentrations in bulk samples [2]. Since that time, X-ray analysts have been striving to improve detection limits to enable the determination of smaller and smaller quantities.

In recent years, some new techniques have been developed for trace analysis. These techniques have been directed primarily at the separation of the elements of interest from the matrix to form a microsample consisting of a chemically concentrated amount of material. Thus, the selective precipitation of certain trace elements (sometimes with a carrier element) [3] and the collection of that precipitate on a filter produces a readily measurable X-ray specimen. The use of ion-exchange resin-impregnated filter paper provides a unique capability [4] for preparing a similarly suitable X-ray sample; resins are available which are general in their ability to collect a wide variety of elements at the same time; other resins can be used under conditions limiting the collection to one or a few specific elements. Bulk ion-exchange resins are also used to concentrate elements of interest but this does not create a microsample, but rather the substitution of a different but constant matrix (the resin) for the original one, coupled to the larger quantity of the elements of interest.

Most recently a rather unique challenge has been presented to the analytical chemistry community. One of the concerns for the environment deals with the elemental composition of particulate matter present in the atmosphere or of trace elements in water. The collection of this air particulate material presents the first example in history of a very high sample-load situation where the material to be analyzed has already been isolated from its matrix. The magnitude of the problem can be appreciated by considering that a single one of the states in the United States might have several hundred sampling sites. If a sample was collected once a day (and there are strong arguments that this is not frequent enough) the sample load would be about a 100,000/year. It is not unexpected that each sample should be analyzed for 20 or more elements and that this must be done on a sample consisting of only a few milligrams.

Another area to which X-ray analysis is being applied with increasing frequency is that of biomedicine [5]. Some reference will be made to specific applications in this field during the discussion to follow.

14.2 SAMPLE PREPARATION

The best method of preparing bulk samples for the measurement of trace concentration is not any different than that necessary for analyzing major constituents. Thus the measurement of trace levels requires that the sample be prepared in standard fashion, e.g., bulk solids should be polished to a flat surface with irregularities small enough so that the emerging radiation is not absorbed in an anomalous manner. Dilution of the sample to eliminate particle size effects or to minimize matrix absorption and enhancement should only be done with appropriate forethought since this operation cannot help but decrease the X-ray intensity emitted by the trace elements, perhaps lowering it below the detection limit.

The measurement of traces in situ can be a difficult problem. An extreme case might be represented by the measurement of lead in ambient air. Let us suppose that there is a need to measure particulate lead in urban air at $1 \mu g/m^3$ (a typical value [6]). An X-ray analyzer might be able to measure a 10-cm^3 volume, that is, $10^{-5} m^3$. This would require a detection limit of $1 \times 10^{-6} g/m^3 \times 10^{-5} m^3 = 1 \times 10^{-11} g = 0.01$ ng, a value which is out of reach with present X-ray instrumentation. By approaching this problem as one of microanalysis, we might filter the particulate material out of 1 m^3 of air onto a 10 cm^2 filter producing a sample having 0.1 µg Pb/cm^2 of filter, a concentration readily measurable.

Microanalysis has many advantages when trace quantities of material are of concern, provided that a microsample can be produced which can be related to the original bulk matrix in a reasonable manner. When one first thinks about the air pollution particulate problem, it is tempting to think of it in the classic microanalytical sense, that is, the measurement of the elemental constituents of a total sample of only one or a few milligrams. It is however, as said previously, a problem in trace analysis because it is the elemental concentration per cubic meter of air which is the important parameter. The same is true of trace elements dissolved in a unit volume of liquid. What must be done is to extract the constituents of interest from the matrix in order to prepare a sample which can be analyzed more conveniently. It happens that particulate material can be easily filtered out of a fluid to prepare a sample which can be measured directly with no further preparation. A small volume of solution of relatively high concentration can be allowed to dry on a thin substrate and analyzed in the same way. Other types of samples require more extensive treatment to prepare them as deposits on thin substrates. As mentioned earlier, this might involve the collection of precipitated elements on a filter or the extraction of the elements of interest by ion-exchange resins.

A technique of sample collection and preparation which does not seem to have been used to any great extent but which may have some application for elements in liquid solution, or for gases, consists of passing the fluid through a medium containing a chemical reagent which removes the element of interest from the fluid. This might consist of a filter paper saturated with the reagent (as the resin in the ion-exchange case) through which the liquid (or gas) could be passed. Or the medium could be a chemical reaction vessel containing the reagent through which a gas could be bubbled, the reaction products forming either a precipitate or a

14.3 QUALITATIVE AND QUANTITATIVE ANALYSIS

component soluble in the reagent solution. Further preparation of the sample after the reaction would depend on the reaction being used. The basic requirement is that the sample should be relatively low mass on a thin substrate.

Qualitative analysis can be performed on these microsamples using either wavelength-dispersion or energy-dispersion equipment as described in Chaps. 1 and 2. Generally, however, quantitative analysis is the objective. Here also either type of instrumentation may be used, and many of the same facts apply. There are some differences in data interpretation but the details of sample preparation previously given and most of the following comments hold equally well for either type of instrument.

The major advantage of the microsample technique is the limited mass presented to the instrument. For the conventional type of X-ray fluorescence analysis using X-ray excitation the background is mostly due to primary radiation scattered into the detector. The magnitude of this background is a direct function of the mass of material which does the scattering. For the microsample case, a small amount of material deposited on a thin substrate, the scattering is decreased and the background is low. Special care with the choice of specimen chamber geometry can reduce the scattering even further; the primary radiation should not illuminate any part of the sample holder which is visible to the detector, and that part of the primary beam which is transmitted through the sample (most of it, in many cases) should be trapped so that it requires several scatterings before it can reach the detector. One additional step helps to minimize the background: the specimen chamber should be evacuated whether the wavelength being measured requires a vacuum or not. Air scattering can contribute significantly to the background. Figure 14.1 shows one possible geometry to accomplish all of this in a crystal spectrometer. For other types of excitation,

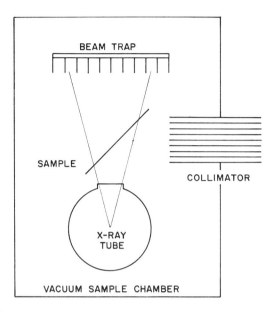

FIG. 14.1 Crystal spectrometer geometry to minimize the background.

e.g., electrons or other high-energy particles, most of the above still holds. The major source of the background may be due to a different mechanism--directly excited bremsstrahlung for electrons or bremsstrahlung due to "knock-on" electrons for proton excitation (see Chaps. 6 and 7)--but the use of a sample which is thin to the exciting quanta improves the detection limit if the transmitted beam cannot "see" the detector. All this concern for the background is related to the criterion which is used to define the detection limit, i.e., the amount of material which gives a signal equal to three times the standard deviation of the background.

It is desirable that the substrate on which the sample is collected (or mounted) be as pure as possible. When filters were first being used to collect air particulate pollution, there was a tendency to consider glass fiber filters very desirable because of their strength and resistance to elevated temperatures. Unfortunately these filters contained many impurities with a high degree of inconsistency from sample to sample. This made the blank correction both large and uncomfortably uncertain. And, of course, it was impossible to analyze such samples for silicon. The application of the X-ray techniques to these samples quickly demonstrated the desirability of pure substrates. Whatman, Millipore, and Nuclepore filters all showed acceptable low impurity levels and mass thicknesses; Mylar, polypropylene, and nitrocellulose films were appropriate for those applications which did not require the substrate to be porous. The small quantities of impurity elements present in these materials are relatively easy to correct for.

One property of the type of sample represented by the particulate material filtered out of air or water is the wide range of concentrations at which the various elements are present. This fact introduces some complication into the quantitative interpretation of the data. The presence of one X-ray peak having an intensity two or three orders of magnitude higher than a neighboring peak places a severe strain on the techniques used to deduce the correct intensity for the smaller one. Ordinarily we think of the resolution as defining whether two peaks can be separated. This is a reasonable expectation for peaks which are the same or nearly the same intensity. As the intensity ratio changes, the quantitative measure of the intensity present in the smaller peak becomes more difficult. Thus, for this situation the improved resolution of a crystal spectrometer compared to the solid state detector becomes a distinct advantage. However, even the superior resolution of a crystal spectrometer cannot completely separate peaks resulting from widely different concentrations. A classic, but not very extreme, example of this effect (not taken from the area of trace analysis but one which should be familiar) can be shown by Fig. 14.2, which illustrates that even a very fine resolution crystal spectrometer (FWHM $\cong$ 20 eV at Mn Kα) cannot completely separate the Cr Kβ from Mn Kα (48 eV apart) in a stainless steel where the Cr to Mn ratio is only 10:1 (NBS SRM 1151). The spectrum of these two peaks recorded with a solid state detector (FWHM = 150 eV at Mn Kα) would look like Fig. 14.3. Fortunately, there are mathematical unfolding techniques

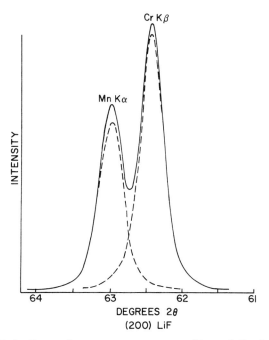

FIG. 14.2 Crystal spectrometer recording of Mn Kα-Cr Kβ.

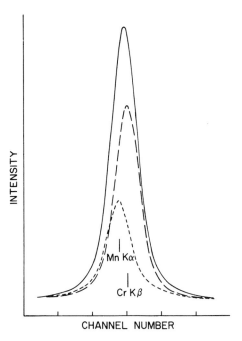

FIG. 14.3 Solid state detector recording of Mn Kα-Cr Kβ.

(see Chaps. 3 and 6) which make it possible to generate values for the two peak intensities but the accuracy with which a small number can be known when it results from the subtraction of two large numbers, as in Fig. 14.3, is quite suspect.

To return to the more positive side of microanalysis, the use of "thin" samples minimizes the usual matrix corrections. By definition it is unnecessary to correct for matrix absorption or secondary fluorescence as is required for bulk samples except for long-wavelength X rays in moderately heavy loadings. This fact makes the preparation of calibration standards very straightforward. A small amount of material deposited uniformly on a thin substrate (a "microstandard," so to speak) will establish the sensitivity in (count/sec)/(μg/cm^2), the slope of the calibration curve, which is linear up to concentrations of about a milligram per square centimeter for moderate energy X rays, even in the presence of elements which are highly absorbing for the X rays being measured [7] as shown in Fig. 14.4.

There is, however, a matrix effect of sorts, which must be considered when dealing with particulate material. This so-called

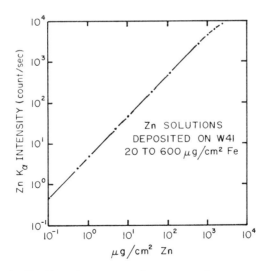

FIG. 14.4 Calibration curve for Zn in the presence of Fe.

particle-size effect results from X-ray absorption (of both primary and fluorescent radiation) within the individual particles, decreases the measured intensity from particles relative to the intensity which would be measured from the same mass of infinitely small particles. In general, for particles significantly smaller than 1 μm, the decrease in sensitivity is negligible. For larger particles the magnitude of the effect is a function of both the size of the particle and the X-ray absorption characteristics of the other elements in the particle. This is a problem common to bulk and microanalysis when dealing with powdered samples; it has been treated with varying degrees of sophistication by several authors [8]. Pollution particulate samples collected on thin substrates represent the simplest kind of sample to correct because they can be approximated as single layers of particles with no significant shadowing. A correction procedure, complete with an analysis of its uncertainties, has been described recently [9] for this specific situation. The technique involves the correction of the mass of the unknown element by a factor $(1 + \bar{b}\bar{a})^2$, where $\bar{a}$ is the particle size in μm and $\bar{b}$ is a parameter dependent on the absorption characteristics of the elements present in the particle. It is assumed that this correction factor would be applied to measurements made on an instrument which had been calibrated using standards not suffering from the particle-size effect (thin films or very small particles). The $\bar{b}$ parameter required for this correction is most accurately calculated from a treatment of X-ray transport which is beyond the scope of this chapter. However, it is sufficient to use the approximate form:

$$\bar{b} \approx \frac{\mu_1 + \mu_2}{5} \text{ cm}^{-1}$$

(for use with $\bar{a}$ in cm), where $\bar{\mu}_1$ is the effective linear absorption coefficient for primary radiation and $\bar{\mu}_2$ is effective linear absorption coefficient for fluorescent radiation. Table 14.1 lists a few $\bar{b}$ values for different compounds of three elements where the primary radiation was the output from standard X-ray tubes having either a W or a Cr target. Figure 14.5 shows the effect of different compounds on the calibration curve for Pb.

TABLE 14.1

Selected $\bar{b}$ Values (from Ref. 9)

X-Ray line	Compound	Using Cr tube (μm^{-1}) (40-50 kV)	Using W tube (μm^{-1}) (40-50 kV)
Pb Lα	PbS	0.024	0.023
	PbCl$_2$	0.017	0.016
	PbSO$_4$	0.016	0.015
K Kα	KBr	0.046	0.034
	KCl	0.033	0.023
	K$_2$O	0.020	0.011
S Kα	PbS	0.20	0.13
	BaSO$_4$	0.12	0.098
	(NH$_4$)$_2$SO$_4$	0.013	0.010

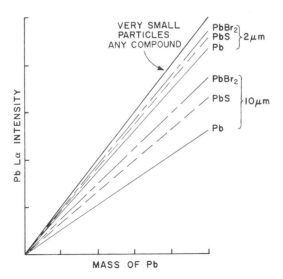

FIG. 14.5 Calibration curves for Pb, showing effects of different particle sizes and different compounds.

14.4 EXAMPLES: DETECTION LIMIT

Throughout this chapter there have been numerous references to the kind of microsample resulting from the collection of pollution particulates on thin substrates. This is certainly not a new application of X-ray fluorescence analysis as publications can be cited [10] from as long ago as 20 years. There has been a substantial increase in emphasis over the last few years, however, to such an extent that it would be impossible to make reference to all of the papers which have been published on the subject since ca. 1970. The primary reason for this increased activity is the concern arising throughout the world for the degradation of environmental conditions. Much of the effort put forth in the area of particle-excited X-ray analysis can be ascribed to the decrease in financial support for basic nuclear physics which has made available the use of high-energy accelerators. The proton or α-particle excitation proponents have used only energy-dispersion techniques [11], as have the analysts depending on radioisotope primary radiation [7]. In fact, a whole community of X-ray analysts measuring particulate pollution using energy dispersion has arisen [12]. Low-power X-ray tubes [13] and higher power X-ray tubes exciting secondary fluorescers [14] are used as primary sources in addition to the high-energy particles and radioisotopes already mentioned. All of the material covered in Chap. 2 applies to this type of analysis in addition to the desirability of trapping the transmitted primary radiation and maintaining the sample chamber in vacuum to minimize the scattering. But, of course, wavelength dispersion is still alive and well, and has been applied to the pollution particulate problem also [15]. An earlier paper [16] attempted to compare the various X-ray methods for the pollution application. Our opinion, as of this writing is the same as expressed in Ref. 16, i.e., that each technique has its place. Photon excitation (X-ray tubes, fluorescers, or radioisotopes) with energy-dispersion detection is a valuable tool for surveying samples if the elements expected are not known. Quantitative analyses under these conditions are possible

TABLE 14.2

3σ-Detection Limits for X-Ray Tube Excited Samples (ng/cm^2)

Element	Wavelength-dispersion Cr tube, 2500 W 100 sec (Ref. 17)	Energy-dispersion, Ag tube, 10 min (Ref. 18)	
		Ag filter (1.8 W)	Ag + V filters (22 W)
Mg	2	80	
Al	3	40	
Si	3	20	
S	9	12	
Cl	9	16	
Ca	2	28	
Fe	18		140[a]
Zn	7		30
Br	28		20
Pb	30		50

[a]Fe impurity in Be window.

if the line overlaps are not too severe and computer data processing is employed. On the other hand, if the elements of interest are known, the multiple spectrometer, wavelength dispersive instruments can provide more quantitative information with less data manipulation due to the superior resolution. The detection limits of these two techniques are comparable to one another when conditions are optimized. A typical example of the detection limits achievable with a state-of-the-art multicrystal simultaneous X-ray analyzer is shown by Table 14.2 [17]. It would not be unreasonable to expect a modern sequential instrument to achieve similar values. Also included in Table 14.2 are the detection limits for a current energy-dispersive system [18]. It must be stressed that these listings are single-element detection limits; the presence of interfering elements will degrade the limits somewhat, more seriously in the energy-dispersive case than for wavelength dispersion.

TABLE 14.3

Detection Limits for Ion Excitation

Sample mount	Ions	Energy	Current or charge	Time	Element	Detection limit	Detection limit criterion	Ref.
1 mg/cm^2 VYNS	α	50 MeV	1 nA	400 sec	Cu Sn Pb	1.9×10^{-12} g 3.2 5.5	P/B = 0.1	[11b]
10-20 µg/cm^2 Carbon or nitrocellulose	p$^+$	5 MeV	5 µC	100-200 sec	K Cu Br Au	1×10^{-9} g/cm^2 2 1 5	3σ Bgd.	[16]
10-20 µg/cm^2 Carbon or nitrocellulose	α	5 MeV	5 µC	100-200 sec	K Br Au	1 10 20	3σ Bgd.	
40 µg/cm^2 Carbon	p$^+$	1.5 MeV	5 µA	30 min	Ca Cu Ba Pb	0.3×10^{-12} g 1 20 10	100 counts above bgd. (of 150 count/keV)	[19a]
4 µm Mylar	p$^+$	1 MeV	10 µC	500 sec	Ca Zn Zr Pb	7×10^{-9} g/cm^2 18 300 90	3σ Bgd.	[19b][a]
4 µm Mylar	p$^+$	3 MeV	10 µC	500 sec	Ca Zn Zr Pb	3 5 30 23	3σ Bgd.	

[a] No tables of values were given in this reference. The values reported here were taken from Fig. 14.3.

High-energy particle excitation (particularly protons) has a detection limit advantage for the unique situation where the sample can be mounted on a substrate of sufficient thinness so that the background is virtually eliminated (e.g., a 20-µg/cm^2 nitrocellulose film rather than the 5-mg/cm^2 millipore filter). The cost of operating these high-energy accelerators is not a trivial consideration when evaluating their use. Some reported detection limits are listed in Table 14.3 [11b,16,19].

In the problem of measuring trace concentrations in bulk matrices, the limit of detection depends on the wavelength being measured and the absorption characteristics of the matrix. Low-concentration elements in metals or alloys present perhaps the most unfavorable situation. With the exception of aluminum, metallic materials represent highly absorbing media for most analytical X-ray lines, particularly those of the low-atomic number elements. Table 14.4 lists detection limits taken from a variety of sources [20].

The other extreme is represented by low-atomic number matrices such as some geologicals, biologicals, aqueous solutions or borate fusions, and ion-exchange resins. Aqueous solutions or borate fusions are techniques used to minimize or eliminate the matrix effect by providing a standard bulk in which the elements of interest are dissolved. In this case, standards can be prepared by dissolving reagent chemicals in the same matrix. Care must be taken to insure that the dilution factor is not so large that the elements of interest fall below the detection limit, nor so small that the attempt to minimize matrix effects fails. Remember also, that there are significant difficulties in measuring long-wavelength X rays in aqueous solution in a vacuum spectrometer. Ion-exchange resins, on the other hand, are useful in concentrating the elements of interest out of solution into a reproducible, low-atomic number matrix (here we are describing bulk resin, pressed into pellets, rather than the resin-loaded filter papers discussed previously). Geological samples may represent low-atomic number matrices (silicates, particularly) but they are plagued by the particle size problem. Frequently, these analyses are carried out using the fusion technique [21],

TABLE 14.4

3σ-Detection Limits for Bulk Samples (ppm)[a]

Element	Iron and steel W target, 2240 W, 10 min (Ref. 20a)	Fe and Ni base alloys, W target, 2025 W, 100 sec (Ref. 20b)	Mineral oil, 1000 W, 10 min (Ref. 20c)	
			Cr Target	W Target
Si	170	4		
P		35	5	
S		8	2	
Cl			0.7	
Ca			0.15	
Ti	1.0			
V	1.9			0.2
Cr	4.0	1		
Mn	1.4	5		
Ni	5.4			0.1
Cu	8.5	12		
Zn				0.1
As	6.8			
Zr	4.6			
Mo	4.5	22		
Sn	3.9			
Ba			0.4	
Pb				0.5

[a] All measurements are by wavelength dispersion.

but bulk ion-exchange resins have been shown to have useful application [22] because of their ability to concentrate the elements of interest. Biological specimens have many elements present in them at concentrations readily measurable; blood was analyzed for Fe, Ca, and Br many years ago [23]. Table 14.5 lists detection limits for various trace elements in whole blood as determined more recently using an energy-dispersive instrument [24]. Concentrations of elements like Cu, Zn, Br, Rb, Sr, and Pb have been determined at levels

TABLE 14.5

Ten Minutes, 3σ-Detection Limits for
Trace Elements in Whole Blood (from Ref. 24)

Element	Detection limit (ppm)
Cr	2.55
As	1.65
Sr	4.25
Ag	148[a]
Cs	8.6
W	5.1
Hg	3.5
Pb	4.8

[a] Potassium K lines interfere with measurement of silver L lines.

below 1 ppm in urine by the ion-exchange collection [25]. It is important to recognize that the detection limit which can be achieved using these concentration techniques is a function of how large a sample can be handled. If an analyst is willing to start with a liter (1 kg) of solution, it should be possible to measure 1 ppb for many elements since detection limits are frequently better than 1 μg (100 ng/cm^2 over an area 10 cm^2, for example).

14.5 CONCLUSION

The most important point to be made in concluding this chapter is that there are certain situations where concentration techniques to produce a microsample make it possible to measure lower level constituents than otherwise. This is not a startling result but one which is reinforced by the improved detection limit achievable by the reduction in background which microsamples permit. Since the physical integrity of the *original* sample is not retained, the nondestructive nature of X-ray analysis is lost. The microsample itself can be retained, however, for confirmatory analysis by another technique, if necessary. Sample preparation may be more extensive than ordinarily considered necessary but data interpretation can be considerably simplified.

REFERENCES

1. H. G. J. Moseley, *Phil. Mag.*, 26, 1024 (1913).
2. C. E. Eddy and T. H. Laby, *Proc. Roy. Soc.*, 127A, 20 (1930).
3. C. L. Luke, *Anal. Chim. Acta*, 41, 237 (1968).
4. W. J. Campbell, E. F. Spano, and T. E. Green, *Anal. Chem.*, 38, 987 (1966).
5a. L. Beitz, *G. I. T. Fachzeit. für Lab.*, 5, 445 (1973).
5b. J. T. Purdham, O. P. Strasz, and K. I. Strausz, *Anal. Chem.*, 47, 2030 (1975).
5c. V. Valkovic, D. Rendic, and G. C. Phillips *Environ. Sci. Technol.*, 9, 1150 (1975).
6. J. A. Cooper, *Nucl. Instrum. Methods*, 106, 525 (1973).
7. J. R. Rhodes, A. H. Pradzynski, C. B. Hunter, and J. L. Lindgren, *Environ. Sci. Technol.*, 6, 525 (1973).
8a. P. F. Berry, T. Furata, and J. R. Rhodes, *Adv. X-Ray Anal.*, 12, 612 (1969).
8b. C. B. Hunter and J. R. Rhodes, *X-Ray Spectrom.*, 1, 107 (1972).
8c. J. R. Rhodes and C. B. Hunter, *X-Ray Spectrom.*, 1, 113 (1972).
8d. R. P. Gardner, D. Betel, and K. Verghese, *Int. J. Appl. Rad. Isotopes*, 24, 135 (1975).
9. J. W. Criss, *Anal. Chem.*, 48, 179 (1976).
10. R. C. Hirt, W. R. Doughman, and J. B. Gisclard, *Anal. Chem.*, 28, 1649 (1956).
11a. T. B. Johansson, R. Akselsson, and S. A. E. Johansson, Lund Institute of Technology, Nuclear Physics Report LUNP 7109, August, 1971.
11b. R. L. Watson, J. R. Sjurseth, and R. W. Howard, *Nucl. Instrum. Methods*, 93, 69 (1971).
12a. D. C. Camp, J. A. Cooper, and J. R. Rhodes, *X-Ray Spectrom.*, 3, 47 (1974).
12b. D. C. Camp, A. L. Van Lehn, J. R. Rhodes, and A. H. Pradzynski, *X-Ray Spectrom.*, 4, 123 (1975).
13. R. D. Giaugue, F. S. Goulding, J. M. Jaklevic, and R. H. Pehl, *Anal. Chem.*, 45, 671 (1973).

14. D. E. Porter, *X-Ray Spectrom.*, *2*, 85 (1973).
15. C. L. Luke, T. Y. Kometani, J. E. Kessler, T. C. Loomis, J. L. Bove, and B. Nathanson, *Environ. Sci. Technol.*, *6*, 1105 (1972).
16. J. V. Gilfrich, P. G. Burkhalter, and L. S. Birks, *Anal. Chem.*, *45*, 2002 (1973).
17. J. Wagman, R. L. Bennett, and K. T. Knapp, Environmental Protection Agency Report EPA-600/2-76-033, March, 1976.
18. R. D. Giauque, R. B. Garrett, L. Y. Goda, J. M. Jaklevic, and D. F. Malone, Lawrence Berkeley Laboratory Report LBL-4402, September 12, 1975.
19a. T. B. Johansson, R. Akselsson, and S. A. E. Johansson, *Nucl. Instrum. Methods*, *84*, 141 (1970).
19b. C. J. Umbarger, R. C. Bearse, D. A. Close, and J. J. Malanify, *Advan. X-Ray Anal.*, *16*, 102 (1973).
20a. W. J. Campbell and J. W. Thatcher, Bureau of Mines Report of Investigations 5966, 1962.
20b. M. C. Peckerar, U. S. Naval Research Laboratory (private communication), 1975.
20c. R. Louis, *Z. Anal. Chem.*, *208*, 34 (1965).
21. A. J. Hebert and K. Street, Jr., *Anal. Chem.*, *46*, 203 (1974).
22. C. W. Blount, D. E. Leyden, T. L. Thomas, and S. M. Guill, *Anal. Chem.*, *45*, 1045 (1973).
23. J. C. Mathies and P. K. Lund, *Norelco Reporter*, 7, 127 (1960); 130 (1960); 134 (1960).
24. M. F. Lubozynski, R. J. Baglan, G. R. Dyer, and A. B. Brill, *Int. J. Appl. Rad. Isotopes*, *23*, 487 (1972).
25. M. Agarwal, R. B. Bennett, I. G. Stump, and J. M. D'Auria, *Anal. Chem.*, *47*, 924 (1975).

Chapter 15

MUSEUM OBJECTS

Victor F. Hanson

Analytical Laboratory
The Henry Francis du Pont Winterthur Museum
Winterthur, Delaware

15.1	Introduction	414
15.2	General Features	415
	15.2.1 Importance of Minor Elements	415
	15.2.2 Effect of Shallow Measuring Depth	417
	15.2.3 X-Ray Fluorescence of Elements Excited by Isotopes	420
	15.2.4 Baseline Compensation	421
	15.2.5 Computation of Weight Percent of Elements from Spectral Count Data	421
	15.2.6 Distribution of Elements Throughout Castings	423
15.3	Analysis Results	424
	15.3.1 Copper-Based Alloys	424
	15.3.2 Copper and Brass	424
	15.3.3 Distribution of Elements in Bronze Castings	425
	15.3.4 Results on Prepared Samples and Art Objects	432
	15.3.5 Analysis of the Liberty Bell	440
15.4	Analysis of Silver	444
	15.4.1 Compositional Features	446
15.5	Analysis of Pewter	454

	15.5.1 Variations of Composition Within a Casting	455
15.6	Analysis of Glass and Ceramics	460
	15.6.1 Analysis Procedure	461
	15.6.2 Analysis of Objects	462
	15.6.3 Analysis Results	465
	15.6.4 Analysis of Near East Ceramic Sherds	478
Acknowledgments		481
References		481

15.1 INTRODUCTION

The Henry Francis du Pont Winterthur Museum, which has a world-famous collection of American antiques housed in over 200 rooms and display areas covering the period of 1640-1840, added a research and library wing in 1969 to aid the curatorial and conservation staff in characterizing and caring for its collection of priceless art objects. The continued increase in the number of fake antique pieces appearing on the market and the recognition of the responsibility of museums to provide the best scientific approaches to the selection and care of the collection led the staff to persuade Mr. du Pont to include a modern research laboratory to develop methods to cope with these problems.

The energy-dispersive X-ray analyzer which had become available commercially at that time was selected as the primary instrument for the laboratory. The fact that it showed promise of being capable of quantitative determinations of extremes in elemental concentrations in all types of art objects without defacing, or sampling the objects, and with accuracies well within the needs of the curatorial experts, has justified its designation as the "Curator's Dream Instrument."

15.2 GENERAL FEATURES

A block diagram of the equipment used to develop the techniques and measurements described in this chapter is shown on Fig. 15.1. A photograph of the laboratory setup is illustrated in Fig. 15.2.

The interactions of various elements, the matrix effects, non-linearities of count vs. wt % of the elements described in other chapters of this book were soon recognized. A practical comparison technique based on a linear relationship of counts in a closely matched standard to the count in an "unknown" minimizes these adverse effects.

The principal element of the alloy is used as internal reference standard and counts from all the elements present are normalized to this element. This provides a first-order correction to geometric factors and matrix effects. The accuracy of the results over a wide range of concentrations can be judged from tables of measured values (M) vs. the theoretical values (T) of the elements added to make up the test specimens.

15.2.1 Importance of Minor Elements

Minor or trace elements in art objects are generally impurities introduced with raw materials.

Traces of gold occur with most silver in nature. Early refiners were unable to recover all the gold so that its absence is a positive indication that the object is less than 100 years old.

Likewise, silver occurs in nature with copper and traces are found in copper-based alloys made before about 1880. This breakpoint in time was confirmed by analyzing hundreds of copper coins covering the period 1790 to 1940 [1].

Old pewter and Britannia ware has a history of many remeltings due to breakage, style changes, etc., so that a wide variety of trace elements are always found. The absence of such elements is an indication of recent fabrication. Artful fakers, however, sometimes add old pewter or "floor sweepings" to make the pieces appear old.

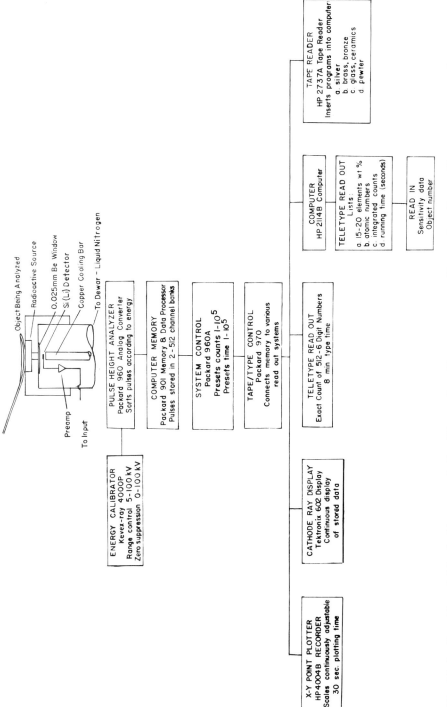

FIG. 15.1 Functional block diagram of the Winterthur Museum energy-dispersive X-ray fluorescence analyzer.

FIG. 15.2 Energy-dispersive X-ray fluorescence analyzer: At the right, the bottom of the tankard being analyzed is resting on the Cd-109 X-ray-emitting radiation source. The liquid nitrogen-cooled detector and preamplifier below the source convert the X rays emitted from the various elements in the tankard bottom to electrical pulses proportional to the X-ray energy. These are then stored in the pulse height analyzer where they are displayed on an oscilloscope and retained for further processing.

A number of elements such as iron, titanium, rubidium, strontium, yttrium, zirconium, and barium appear in clays and sands in varying proportions and concentrations providing valuable and unique "fingerprints" in glass and clay-based objects.

15.2.2 Effect of Shallow Measuring Depth

The fact that the relatively "soft" (i.e., unpenetrating) X rays employed in measurement analyze only the material at the surface has the following advantages over other analysis methods:

1. The eye, being the principal receptor in appreciating art objects, "sees" the same materials that the instrument "sees."
2. Alterations to the surface which may be invisible to the eye can sometimes be detected.
    a. Silver objects. Acid dipping or pickling of silver objects can be detected by the variations in composition at various spots on the silver indicating that repairs or alterations have been made. The presence of high copper and zinc concentrations in silver is due to residual silver solder, indicating that repairs or alterations have been made.
    High, but invisible, gold concentrations indicate that the object has been gilded at one time.
    The presence of mercury with the gold indicates that the gilt was applied by the ancient amalgam process.
    The presence of visible gold and absence of mercury indicates modern electroplating of the gold.
    Surface silver contents of 97-99% indicate that electroplate has been applied to disguise alterations or repairs.
    b. Brass and bronze. Natural patinas being thin oxide or sulfide layers of the substrate metals are essentially transparent to the measuring X rays and reveal the components present in the substrate with reasonable fidelity.
    Artificial patinas produce a diffuse spectrum and show a number of elements not present in the substrate.
    Vestiges of silver or nickel electroplate not visible by eye can be detected.
    When heated to 400°-500°C, bronzes containing substantial amounts of lead cause the lead to diffuse to the surface and appear as an abnormally high lead content,

varying considerably from point to point. The lead-rich surface can be removed to normal by carefully scraping the surface to be analyzed. To demonstrate the magnitude of this effect a bronze test casting was analyzed after filing 0.3 mm of the surface layer, was heated to 400°C for an hour, and reanalyzed, with the following results:

Treatment	Copper (%)	Lead (%)	Tin(%)
Filed surface	77.3	18.6	4.0
After heating to 400°-500°C for 1 hr	69.8	25.3	4.0
After filing 0.3 mm from surface	79.0	17.0	4.0

  c. Pewter and Britannia. Analysis through the black tarnish on pewter introduces slight errors in the substrate analysis. The tin readings made through thick corrosion layers are low, indicating that tin has been lost in the corrosion process.
  High copper, nickel, and silver contents indicate that the piece has been electroplated.
  d. Glass and ceramics. The pigments in paints, enamel, and other overlays can be identified independently of the substrate. The type of glaze (lead, tin, or salt) can be identified.

The shallow measuring depth does not provide an "average composition" such as is reported by many museum analytical laboratories, so the results are sometimes questioned by curators. The actual variations of composition throughout various types of castings in Figs. 15.13 and 15.14 (pages 452 and 453) show that such discrepancies are trivial, after 0.15 mm of the original cast surface has been removed.

15.2.3 X-Ray Fluorescence of Elements Excited by Isotopes

In order to identify the all-important trace elements in antique and archaeological objects, samples of high-purity elements were irradiated with Cd 109, Am 241, and Fe 55 sources for 100 sec. Counts per second vs. energy and channel location for all the peaks observed were tabulated according to element [2]. Semilog plots of counts per second vs. energy were made for each of the excitation sources.

These tables and graphs provide the following critical information:

1. The identification of the most probable elements responsible for the peaks in the spectrum.
2. The choice of the most likely peak from an element which will not overlap another element which is likely to be present.
3. The minimum concentration of a minor element which can be determined in the presence of nearby major element peaks.
4. Methods for determining both elements in overlapping peaks. For example, arsenic and lead are present in many pewter, bronze, glass, and ceramic pieces. Both elements appear at 10.6 keV. The lead contribution to this peak can be determined by subtracting a fraction of the 12.7-keV lead peak from the 10.6-keV peak. The arsenic can be determined from this difference. Cobalt, iron, and manganese which have overlapping peaks appear in glass and ceramics. First-order corrections for the overlap errors can be made from these tables.
5. Normalizing factors for elements not present in multielement reference standards can be clearly estimated. For example, reference standard no. 47.4 for sterling silver contains 92.5% silver, 7% copper, and 0.5% gold. All the thousands of antique silver objects analyzed to date contain varying amounts of lead and gold, which are characteristic of their provenance. The lead normalizing factor

is based on the gold content of the standard in the analyzer. The arsenic normalizing factor of the example of item 4 would be derived from that of copper. With the Cd 109 source it would be the same as for copper and 50% with the Am 241 source.

15.2.4 Baseline Compensation

The need for source-scatter or base-line compensation was recognized. A satisfactory but not perfect solution in metal analysis was to employ the spectrum of the principal element in the object as a baseline which would be subtracted from the object spectrum.

This was a fortuitous choice since it also provides adequate compensation for the overlap of zinc Kα on copper Kβ in brass, bronze, and German silver. It also compensates for antimony Kα on tin Kβ in pewter. These important minor elements are difficult to determine by other base-line compensation techniques.

15.2.5 Computation of Weight Percent of Elements from Spectral Count Data

Spectral normalizing factors of Table 15.1 are derived as follows:

1. Spectral count data and peak channel locations are measured from a prepared multielement reference standard for a predetermined count of the principal element. In routine operation the spectrum of the principal element is stored in the computer. Counts from five channels centered at the peak for each element are taken so that the peak is always included in case of instrumental "drifts."
2. A base-line spectrum of the peak principal element is made for the same peak count.
3. Add the counts from the peak and two adjoining channels; call this sum the total peak counts M_a.
4. Add the counts from the corresponding channels in the baseline; call these the baseline M_b.

TABLE 15.1

Normalizing Factors

Base-line	Internal standard	Silver Sterling solder	Silver Solder	Metals Analysis Copper Brass	Copper Bronze	Chinese silver, copper German silver	Chinese silver, copper Paktong	Tin Pewter
Winterthur		47.4	47.4 F*	29.8	29.8	29.9	29.9	50.1
Source exciter		Cd 109	Cd 109	Cd 109	Am 241	Cd 109	Am 241	Cd 109
Peak counts - Base line element		5000	5000	10000	10000	10000	10000	5000
Range (keV)		4-29	4-29	4-16.5	16.5-29	4-16.5	16.5-29	4-29

Item	El	Z	keV	Channel	K_n		K_n		K_n		K_n
1	Mn	25	5.9	48	0.1		1.6		1.6		0.01
2	Fe	26	6.4	57	0.1		0.9		0.9		0.01
3	Co	27	6.9	68	0.2		0.05		0.05		0.01
4	Ni	28	7.5	80	0.2		1.25		1.1		0.01
5	Cu	29	8.1	90	0.15		1.0	1.0	1.0	1.0	0.056
6	Zn	30	8.6	102	0.12		0.85		1.58		0.03
7	As	33	10.6	140	0.06		0.2		0.2		0.03
8	Au	79	11.6	158	0.11		2.5		2.5		0.01
9	Hg	80	11.9	166	0.11		1.6		1.6		0.01
10	Pb	82	12.6	181	0.11		1.92		1.92		0.044
11	Bi	83	13.1	188	0.08		0.08		0.08		0.029
12	Ag	47	22.1	368	1.0			0.16		0.11	
13	Cd	48	23.3	390				0.08		0.08	
14	Sn	50	25.3	429				0.11		0.11	1.0
15	Sb	51	26.3	450				0.08		0.08	1.15

*Silver solder: Standardize with 50-mg filings from silver for base line and 47.4 for analysis. Use silver factors.

5. Subtracting the base-line count total from the corresponding peak count gives the net peak count for each element $M_a - M_b = M_c$.
6. Divide net peak counts by the weight percent of the element in the reference sample. This gives M_d, count/wt %, Mc/wt % = M_d.
7. The normalizing factors are the counts per weight percent of the principal element divided by those of each of the other elements y in the sample

$$K_n = \frac{M_{d}(\text{principal element})}{M_{d}(\text{element y})}$$

8. Calculation of "unknown."
   a. Process the count data from an "unknown" in steps 1 through 5.
   b. Multiply net counts M_c for each element by its K_N factor which gives normalized counts M_n for each element.
   c. Add the normalized counts M_n to obtain normalized total counts of all elements, M_t.
   d. Weight percent of each element is 100 M_n/M_t.

   This same procedure can be computerized.

15.2.6 Distribution of Elements Throughout Castings

Since the penetration of the exciting incident radiation and the escape depth of the lower energy fluorescence radiation is only a few micrometers, the following experiments were conducted to determine how representative of the casting as a whole the "surface" measurements are. Furthermore, it has long been known that some elements in an alloy solidify before others when they contact a chilled mold. While it would be logical to assume that the higher melting elements would be the first to solidify and the lower melting elements would concentrate in the last part to solidify, this is not always the case and is called inverse segregation.

Various carefully prepared alloys were induction melted and cast in chilled copper molds. A relatively flat, smooth surface was produced on the bottom of the casting and a crater-like surface appeared on the top.

These two surfaces were analyzed after which the billet was chucked in a lathe and thin slices were removed from each face. The new faces were analyzed as well as the turnings. This process was repeated until the center of the casting was reached. The composition as a function of the distance from the chilled cast face was then plotted. Results of these experiments are given in the following detailed sections on specific alloys.

15.3 ANALYSIS RESULTS

15.3.1 Copper-Based Alloys

Copper, brass, and bronze are the most important base metals employed in making art objects. Where protected from the elements they last for thousands of years. In making a melt for castings, foundrymen start with low-melting scrap brass or bronze and add copper and other metals to make an alloy having desired properties for the end product. Unknown elements are introduced with the scrap as well as impurities with the alloying elements. These impurities and the major elements provide clues for detecting whether art objects are genuine or fake.

15.3.2 Copper and Brass

The energy-dispersive X-ray analyzer is ideally suited for analyzing these objects since all the elements present appear simultaneously. A clean spectrum free of impurities is strong evidence that the object is made from modern high-impurity ingredients. Oriental "antique" bronzes having a high zinc content (note Run 2004 of Table 15.2) instead of the tin found in genuine ancient objects, are easily detected. Traces of silver present in copper ores were not completely recoverable by refining methods employed prior to about 1875. The complete absence of silver is a strong indication that the piece was made after that date. The presence of traces of silver is no guarantee that the object was made before that date since old scrap combining many impurities can be used as part of the deception.

15.3.3 Distribution of Elements in Bronze Castings

The classical methods of analyzing brass and bronze museum objects has been to drill holes in obscure parts of the object and to determine the composition of the drillings by wet chemical analysis and emission spectroscopy. The drillings are assumed to be representative of the object as a whole.

In analyzing some 15th-century bronze figures by X-ray fluorescence at a number of points, significant variations in composition were observed (see Table 15.3).

Rechecks of the same measured points gave reproducible results indicating that the variations were real. A group of bronze castings of a similar carefully controlled composition were made in rectangular copper molds having a 13 × 40-mm base and 50-mm height to determine the relevance of surface-made X-ray fluorescence measurements to the composition of the casting as a whole. Figures 15.3 and 15.4 show the results of measurements made on a section across a casting of a similar composition as the 15th-century Italian bronzes. The cast faces were analyzed followed by removing thin layers and reanalyzing the new faces. The turnings removed were also analyzed. Since the soft fluorescent X rays can escape only from a shallow depth, a series of experiments (Fig. 15.3) were made on a number of specially prepared alloys to determine how representative of the whole are the measurements made at the surface. In Fig. 15.3 the original cast surface was analyzed after which thin layers were removed and the new surface and the "turnings" were analyzed. The curves show the anomalous copper-rich surface

TABLE 15.2

Miscellaneous Brass and Bronzes Analyzed in 1973 (Dates Unknown)

Locality	Run	Description	Cu	Mn	Fe	Ni	Zn	As	Pb	Au	Bi	Ag	Sn	Sb
South-east Asia	2003	Han coin	75.66	0.03	1.04	0	0	0	21.77	0	0.02	0.44	0.03	0.95
	2004	Han coin fake	65.76	0.11	2.52	0.25	29.90	0.02	0.76	0	0	0	0.32	0.12
	2005	Han knife money fake	61.39	0.13	1.35	0.22	32.06	0.14	3.05	0	0	0	0.89	0.52
	2006	Chou bow finial	39.72	0.90	2.92	1.00	1.61	0	25.38	0.303	0.09	0.48	32.99	0.44
	2007	Han mirror	35.24	0.73	2.33	1.03	1.75	0.32	25.50	0.448	0.11	0.44	25.81	1.04
	2010	Chou bow finial	36.23	0.72	2.15	1.00	1.49	0.11	27.09	0.394	0.10	0.50	24.97	0.54
	2017	Thai Ban Chieng collar	51.36	0.33	2.07	0.53	0.47	0	38.28	0.08	0.08	0.12	5.04	0.27
	29T3	Thai bracelet	41.11	0.34	2.74	0.17	2.37	0	36.84	0.14	0.18	0.11	13.75	0
	29T4	Thai large ring	82.19	0.27	1.91	0.13	1.63	0.02	2.49	0.09	0.04	0.04	7.61	0
	29T5	Thai large ring	83.78	0.18	1.30	0.08	1.23	0	7.54	0.06	0.03	0.03	3.81	0.03

MUSEUM OBJECTS

Egypt	E7124	Egyptian brass	82.70	0	0	0	15.19	0	0.60	0.08	0	0.02	1.95	0
	E3100	Egyptian bronze	80.06	0.93	0.32	0.02	0.84	0	7.66	0.09	0.03	0.17	9.16	0.16
	E1118	Egyptian bronze	87.13	0.46	0	0	0.75	0.02	1.10	0.05	0.02	0.05	9.97	0
Ashenti, West Africa	557	Brass gold weights	76.28	0.50	0.50	0.001	20.58	0	0.70	0.01	0	0.09	1.13	0.04
	514	Brass gold weights	65.06	0.65	0.54	0.03	31.60	0	1.62	0.02	0	0.09	0.18	0.01
	90	Brass gold weights	85.75	0.30	0.10	0	13.08	0	0.08		0	0.02	0.63	0.01
	708	Brass gold weights	69.09	0.40	0.26	0.01	29.04	0	1.27	0	0	0.004	0.04	0
	542	Brass gold weights	69.52	0.49	0.44	0.01	25.60	0	1.28	0.02	0.01	0.045	2.53	0.05
	577	Brass gold weights	70.88	0.62	0.69	0.03	21.95	0	1.22	0.02	0.01	0.07	4.27	0.14

TABLE 15.3
15th-Century Italian Bronzes (Louvre and National Gallery)

Run number	National Gallery accession number	Attributed to	Description	Element (%)															
				Cu	Mn	Fe	Co	Ni	Zn	As	Pb	Au	Bi	Ag	Ba	In	Sn	Sb	Cd
991	Louvre	Bellano, B.	St. Christopher	81.70	0.39	1.30	0.67	0.53	12.59	0	0.2	0.14	0.06	0.07		0.04	1.27	0.40	
2104-1	A206.44C	Bellano, B.	Christ Child	81.75	0.30	0.97	0.54	0.51	13.12	0.02	0.88	0.11	0.05	0.08		0.04	1.38	0.23	0.02
2104-14	A909.172A	Bellano, B.	Ant. Roselli	81.42	0.08	0.71	0.25	0.39	13.54	0.03	1.26	0.08	0.04	0.22	0.002	0.04	1.78	0.13	
2104-3	A292.158	Bellano, B.	Dead Christ	89.22	0.08	0.49	0.22	0.20	5.56	0.14	0.49	0.07	0.04	0.46		0.08	2.62	0.29	0.02
2104-2	A-97	Bellano, B.	David	81.67	1.20	2.37	1.54	1.45	4.09	0.19	1.44	0.21	0.12	0.21		0.15	4.79	0.52	0.06
2104-13	A172.10C	Riccio, A.	Arion on a Shell	85.04	0.23	1.32	0.51	0.64	6.61	0.07	1.34	0.10	0.06	0.18		0.09	3.03	0.78	0.003
2104-12	A227.68C	Riccio, A.	Sea Monster	84.34	0.39	1.64	0.60	0.60	9.66	0.28	1.15	0.08	0.04	0.11		0.02	0.84	0.24	0.002
2104-10	A228.69C	Riccio, A.	Sea Monster	78.20	0.52	1.87	0.71	0.78	8.42	0.24	3.26	0.15	0.11	0.65	0.003	0.11	4.08	0.90	
2104-7	A168.6C	Verrocchio, A.	Faun	77.68	0.61	1.95	0.92	0.98	15.16	0.20	0.79	0.16	0.05	0.22	0.02	0.04	1.09	0.15	
2104-8	A214.55C	Paduan	Casket	85.06	0.23	0.80	0.46	0.45	3.05	0.27	2.26	0.10	0.10	0.49	0.002	0.14	5.38	1.18	
2104-8B	A214.55C	Paduan	Casket	83.16	0.28	0.77	0.40	0.44	3.50		1.39	0.12	0.10	0.42	0.001	0.21	8.63	0.58	
2104-9	A217-58C	Paduan	Δ Box, center lid	82.85	0.20	0.85	0.40	0.37	5.05		3.58	0.09	0.13	0.19	0.001	0.15	5.75	0.48	
2104-9B	A217.58C	Paduan	Δ Box, side	86.64	0.32	1.12	0.62	0.75	4.19	0.0006	2.06	0.10	0.08	0.19	0.004	0.08	3.24	0.59	
2104-11	A218.59C	Paduan	Box top	83.25	0.18	0.87	0.32	0.29	4.66		5.68	0.08	0.12	0.25	0.002	0.12	4.34	0.28	
2104-16A	A215.56C	Paduan	Box top	87.32	0.13	0.78	0.33	0.28	2.40		1.44	0.10	0.08	0.07	0.0007	0.17	6.62	0.30	
2104-16B	A215.56C	Paduan	Box side	86.63	0.11	0.38	0.30	0.29	2.10	0.10	0.62	0.10	0.08	0.12	0.0006	0.22	8.80	0.16	
2104-15A	A219.60C	Paduan	Lamp top	77.45	0.43	1.72	0.76	0.79	3.00	0.10	5.10	0.15	0.17	0.24	0.005	0.16	4.58	5.32	
2104-15B	A219.60C	Paduan	Lamp bottom	79.94	0.40	1.27	0.56	0.66	4.60	0.01	3.69	0.14	0.12	0.24	0.005	0.12	3.86	4.29	

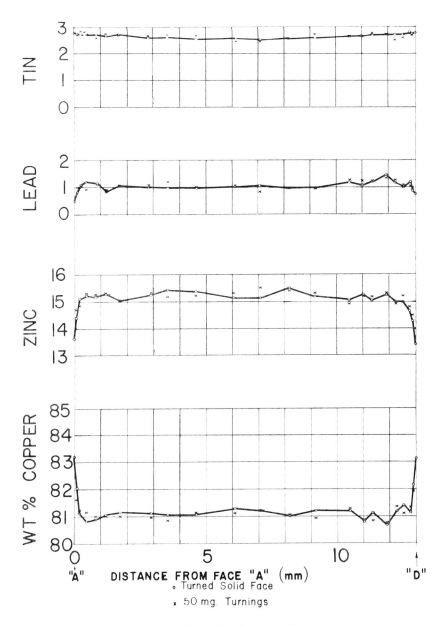

FIG. 15.3 Element distribution in a bronze casting.

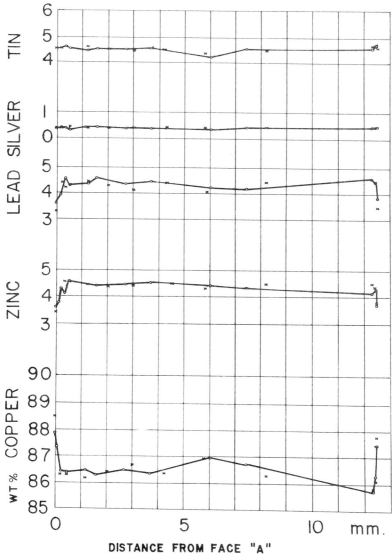

FIG. 15.4 Element distribution in a bronze casting. The effect of gravitational separation of lead during solidification of the melt. The upper half of the casting has a slightly lower lead content than the lower half indicating that a separation occurred.

12 POSITION AVERAGE

EL.	AS CAST %	AS CAST S	MINUS 0.13 mm %	MINUS 0.13 mm S	TURNINGS %
Cu	79.90	0.37	79.55	0.26	79.83
Zn	4.85	0.21	5.45	0.04	5.24
Pb	4.95	0.31	4.95	0.01	5.06
Sn	9.12	0.13	9.22	0.09	9.30

S = STANDARD DEVIATION

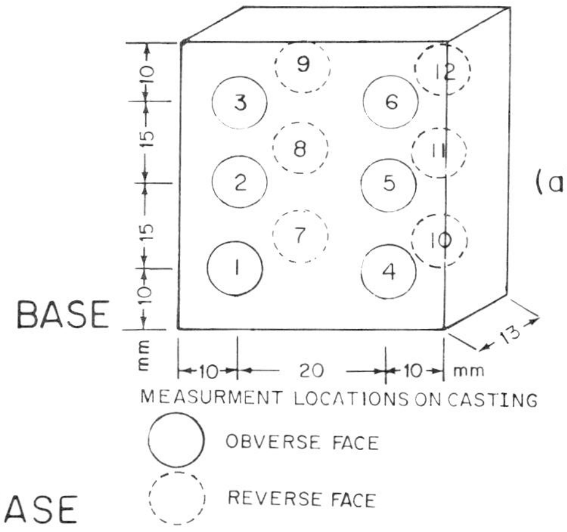

(a)

PARAMETER - DISTANCE FROM CASTING BASE

mm	AS CAST 10	AS CAST 25	AS CAST 40	MINUS 0.13 mm 10	MINUS 0.13 mm 25	MINUS 0.13 mm 40
EL.	%	%	%	%	%	%
Cu	79.44	79.96	80.31	79.60	79.56	79.50
Zn	4.99	4.87	4.68	5.38	5.46	5.51
Pb	5.02	4.93	4.85	4.87	5.00	4.97
Sn	9.13	9.15	9.09	9.27	9.22	9.19

(b)

FIG. 15.5 Element distribution in a 0.13 mm bronze casting surface layer. There is a slight copper and tin enrichment in the surface layer. The surface layer is richer in copper in the upper part of the original cast base than in the lower part. These differences are imperceptible after removing a 0.13-mm thick surface layer. Ordinary finishing of bronze art objects removes this surface-rich layer.

layer which is about 0.15 mm thick. This layer is generally removed during the finishing of art objects, leaving a surface composition close to that of the average of the whole piece.

The chilled cast surface is high in copper and low in zinc indicating normal segregation considering the large difference in melting points. Normal average body composition is attained after removing a 0.15 mm layer. The smooth surface of the Italian figures indicated that at least 0.15 mm had been removed so that the surface measurements were truly representative of the casting as a whole and that the surface variation observed was an indication of poor mixing of the ingredients in the melt.

Figure 15.5(a) shows the 12-point position average analysis on cast block 29.3 before and after machining off a 0.13 mm layer. There is a significant increase in uniformity after removing the surface layer. Figure 15.5(b) shows the distribution of elements as a function of distance from the base before and after removing a 0.13-mm surface layer. Segregation due to the element density differences is barely detectable.

Figure 15.4 shows the distribution of elements through two sections of 29.2 bronze casting. Here again, the differences in composition as a function of distance from the base is undetectable. The casting is extremely homogeneous except for the outer 0.15 mm layer which would normally be removed in the finishing process.

15.3.4 Results on Prepared Samples and Art Objects

15.3.4.1 Prepared Samples

A series of copper-based alloys was prepared by Handy and Harman of Fairfield, Connecticut by carefully weighing the ingredients, induction-melting the mixture to insure intensive mixing, and casting into chilled copper molds. The surface layer was removed incrementally after chucking in a lathe.

15.3.4.2 Asiatic, Egyptian, and West African Brasses and Bronzes

Table 15.2 lists a group of copper-based alloys that were analyzed during 1973 to indicate the wide range in composition found in museum objects.

The two fake Han coins of Runs 2004 and 2005 of Table 15.2 are high in zinc and low in lead. The absence of silver indicates that they are 20th-century pieces. The Egyptian brass piece (Run 7124) is unusual since zinc was not known in ancient Egypt. The composition of this piece resembles modern free machining brass.

The West African gold weights are intricate brass castings used by the natives for weighing gold since about 1800. Since they have become collectors' items a thriving business has developed to supply tourists with fake reproductions of these rare pieces. The absence of silver is a positive indication of modern brass. About 100 of these African pieces in a private collection were analyzed.

The computer which is set up to perform the arithmetic steps of Sec. 15.2.5, items 1-8, used the normalizing factors K_y for alloy 29.8 in Table 15.1. Results of the analysis of alloys 29.1 through 29.8 and alloys 29.21 through 29.31 employing these factors are on Tables 15.4 and 15.5. The nickel, zinc, and tin factors were modified for bell-metal, German silver, and copper-nickel alloys shown on Table 15.6. While it has not been determined if the discrepancies are due to sample preparations or measurement errors, they are small and the accuracies are well within curatorial needs. Table 15.7 shows compositions for a number of paktong or German silver objects. Paktong and German silver were made in England in an effort to duplicate alloys of copper, nickel, and zinc (no silver) which had been used in China for many years and later in Germany. It is a strong durable alloy now commonly used as the base metal of plated silverware.

Table 15.7 shows the range in composition found in a few oriental objects of unknown origin analyzed at Winterthur in 1973. British-made paktong is very similar in composition. The nickel imparts the silver-white color rather than the zinc which is present in a

TABLE 15.4

Brass Standards

			Weight percent theoretical (T) vs. measured (M) values														
Standard No.			29.8[a]		29.21		29.22		29.23		29.24		29.25		29.26		29.27
Element	Z	T	M	T	M	T	M	T	M	T	M	T	M	T	M	T	M
Cu	29	80	80.24	80	78.81	70	69.93	60	60.81	75	74.57	75	74.50	75	75.00	75	74.07
Mn	25	1	1.15														
Fe	26	1	0.85														
Zn	30	5	4.97	20	20.87	30	29.77	40	38.70	25	25.12	24.9	25.00	24.8	24.49	24.5	25.06
Pb	82	5	5.14														
Au	79	1	1.15														
Ag	47	1	0.98									0.1	0.11	0.2	0.19	0.5	0.50
Sn	50	5	4.93														
Sb	51	1	1.05														

[a] Reference standard.

TABLE 15.5

Bronze Standards

Standard No.		29.8[a]		29.1		29.2		29.3		29.4		29.5		29.6		29.7		29.8	
Element	Z	T	M	T	M	T	M	T	M	T	M	T	M	T	M	T	M	T	M
Cu	29	80	79.95	85	86.27	85	85.30	80	79.42	80	80.02	80	77.74	80	80.26	80	78.90	80	78.92
Mn	25	1	0.98															1	0.94
Fe	26	1	0.96															1	0.84
Zn	30	5	5.02	5	5.03	4.5	4.10	5	5.02	5	4.53					15	16.29	5	4.90
Pb	82	5	5.01	5	4.60	5	4.76	5	4.80	5	4.92	5	5.14	15	14.14	2	1.56	5	5.04
Au	79	1	0.96															1	1.14
Ag	47	1	1.00							1	0.92							1	0.96
Sn	50	5	4.99	5	4.70	0.5	0.43	10	10.10	8	8.32	15	16.37	5	4.52	3	2.79	5	6.14
Sb	51	1	1.01							1	0.84							1	1.02

Weight percent theoretical (T) vs. measured (M) values

[a] Reference standard.

TABLE 15.6

Bell Metal, German Silver, Paktong, and Copper-Nickel Standards

Weight percent theoretical (T) vs. measured (M) values

Standard No.		29.31		29.32[a]		29.33		29.9[a]		29.10		29.34		29.35	
Element	Z	T	M	T	M	T	M	T	M	T	M	T	M	T	M
Cu	29	78	77.26	73.1	71.67	65.75	64.19	55	54.24	33	34.85	90	88.48	85	83.67
Fe	26		0.17		0.23				0.18		0.29		1.50		0.14
Ni	28							20	19.87	33	33.12	10	9.93	15	15.69
Zn	30							25	24.74	33	31.11				
Ag	47	2.5	2.48	2.5	2.87	0.25	0.55								
Sn	50	19.5	19.10	24.4	24.13	0.34	33.24								

[a]Reference standard.

TABLE 15.7
Paktong and German Silver

Object	Cu	Ni	Zn	Pb	Ag
Covered box	36	8	53	2.0	0
Small brazier	36	9	50	2.0	0.01
Large brazier	36	10	52	1.0	0
Cosmetics case	35	11	52	0.4	0.04
Plate	34	11	54	0.25	0
Candlestick (1)	33	7	57	0	0
Candlestick (2)	40	6	50	0.4	0.04
Tea caddy	48	10	42	0	0
Candlestick (2)	50	12	37	0.4	0.01
Candlestick (2)	52	12	32	0.2	0.01
Covered cup	57	13	30	0	0
Heated bowl	56	6	35	6.0	0.03
Candlestick (2)	58	10	25	0	0

much higher concentration. The analysis results are listed to the second decimal place for convenience in transcribing from the computer readout and not as a claim for the implied accuracy.

15.3.4.3 Fifteenth-Century Italian Bronzes

At the request of Dr. Robert Feller and Dr. Douglas Lewis of the National Gallery of Art, a piece of 15th-century Italian bronze from the Louvre and a piece at the National Gallery thought to belong together were brought to the Winterthur Laboratory and analyzed. Analysis of the two figures showed an identical compositional pattern (within experimental limits), giving unequivocal proof that the two pieces were cast from the same melt, and were therefore both made by the same artist. These are shown assembled in Fig. 15.6a. Thus, the Louvre's "Atlas" and the National Gallery's "Boy With A Ball" became "St. Christopher carrying the Christ Child" after being separated for over 100 years, and the results were reported in the

FIG. 15.6a Confirmation of identity of Renaissance bronzes. Bertrans Jestaz, curator of art objects at the Louvre, during a visit at the National Gallery of Art, speculated that the Renaissance bronze statue of "The Boy with a Ball" (supported by the raised hand) belonged on the hand of the Louvre's standing bronze figure listed as "Atlas." The tenon in the boy's rump exactly fitted the hole in the hand of Atlas and the stylistic similarity of the two pieces was evident. The joining of the two bronzes represents the original artist's concept of "St. Christopher carrying the Christ Child astride the World."

FIG. 15.6b The medallion used to identify the sculptor of the bronze figures shown in 15.6a.

Annual Report of 1971 by the National Gallery of Art [3]. After this spectacular success the National Gallery brought 16 additional pieces of 15th-century Italian bronze for analysis with the hope that the sculptor of these two pieces might be identified. A medallion of Pope Antonnis Roselli (A909.172A), Table 15.3, Fig. 15.6b, signed by Bartolommeo Bellano had the same composition and had to have been poured from the same melt as St. Christopher and the Christ Child (206.44C). Thus an alert curator at the Louvre, a curator and research scientist at the National Gallery, and The Winterthur Museum research laboratory were able to piece together an attribution puzzle against almost infinite odds. The analysis results proved beyond doubt that both pieces had been made from the same melt and the identical composition of the signed metallion identified the artist. The wide variation in composition of the other pieces adds credence to this confirmation.

Equally dramatic is the fact that all these analyses were made within three hours. Higher accuracies could have been achieved with longer measuring times or with duplicate measurements but these would not have changed the dramatic conclusions that were drawn from these short-period measurements.

15.3.5 Analysis of the Liberty Bell

In support of a metallurgical study to assess the potential hazard involved in moving the Liberty Bell to a new location for the Bicentennial celebration, a semiportable instrument was employed to analyze 10 points on the rim of the Liberty Bell. The instrument was also taken to the belfry of Old Christ Church which has eight bells cast by the Whitechapel Foundry which cast the original Liberty Bell two years earlier (1752).

Figure 15.7 shows the instrument setup in making the measurements and Fig. 15.8 shows the composition at 10 points around the rim of the Liberty Bell. Also shown are the results from the Christ Church bell.

FIG. 15.7 Equipment setup for analyzing the rim of the Liberty Bell. Mrs. Karen Papouchado, on the left, is keeping notes, Mr. Victor Hanson, center, is holding the radioactive source and liquid nitrogen-cooled detector in contact with the bell rim, and Mrs. Janice Carlson is operating the instrument controls.

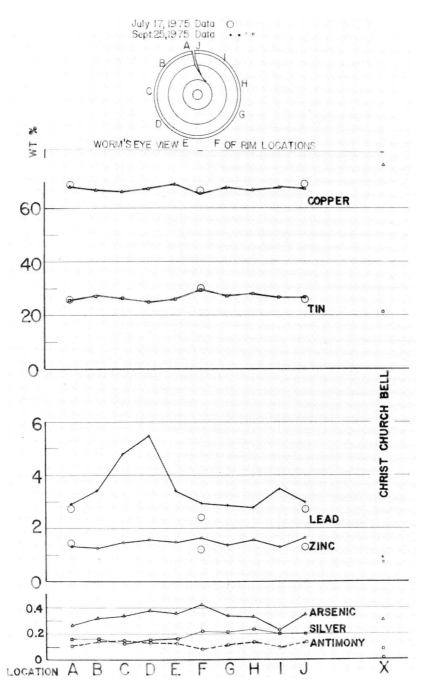

FIG. 15.8 Analysis of the White Chapel foundry bells. The surprisingly wide variation in the elements found in the rim of the Liberty Bell is compared with a bell in Christ Church, made in the same foundry two years after the original bell was cast in London. The present Liberty Bell is the second recasting made in a small foundry in Philadelphia under conditions described by the author and his associates in *American Scientist*, "The Liberty Bell: Composition of the Famous Failure," November-December 1976.

15.3.5.1 Results

In the following discussion the original London-made Whitechapel bell which cracked on its first trial will be designated as No. 1, the first recast by Pass and Stow as No. 2, and the second recast (the present Liberty Bell) as No. 3. The following observations, speculations, and conclusions are based on the surprising analysis results displayed graphically in Fig. 15-8.

1. The tin content of the Liberty Bell is 4-8% higher than the handbook values for "bell metal" resulting in an extremely brittle alloy which caused the Liberty Bell to fail.

2. The variable tin content (25-30%), as well as variation in the other elements shown in Fig. 15.8 and Table 15.8, is probably a result of the limited melting facilities of the Pass and Stow brass foundry. They probably melted their alloy in several crucibles using pewter to increase the tin content resulting in a nonuniform melt. Pewter of the period had the following composition range:

Tin	55-95	Lead	2-45
Copper	0.5-2	Arsenic	0-0.2
Antimony	0-2	Zinc	0-0.1

   The foundrymen probably thought the pewter to be practically pure tin so that the alloys made in the various pots differed with the lead content of the pewter.

3. The surprisingly high (2.2-5.5%) lead in the Liberty Bell compared to 0.84% in the Christ Church bell probably came from the low-quality pewter added to one of the melting crucibles to elevate the tin content of bell No. 3. The presence of 0.14% antimony also suggests the use of pewter in the No. 3 bell.

4. The presence of 1.2-1.6% zinc indicates that some brass may have been mistakenly added as copper.

5. The arsenic contents of bell No. 3 (0.32%) and the Christ Church bell (0.31%) are the same within experimental limits of error.

TABLE 15.8

Analysis of Whitechapel Foundry Bell[a]

								Liberty Bell							Std.[c]	
	A	Ao[b]	B	C	D	E	F	Fo[b]	G	H	I	J	Jo[b]	x	Dev.	X[d]
Iron	0.06			0.87	0.19	0.12	0.11		0.57	0.47	0.53	0.46		0.37	0.26	0.19
Copper	67.87	67.65	66.72	66.06	66.88	68.57	64.95	65.46	67.38	66.61	67.64	67.18	69.12	67.08	1.17	76.77
Zinc	1.30	1.32	1.25	1.44	1.54	1.48	1.61	1.23	1.35	1.55	1.26	1.65	1.17	1.40	0.16	0.70
Arsenic	0.26	0.22	0.32	0.34	0.38	0.36	0.42	0.38	0.34	0.33	0.19	0.35	0.26	0.32	0.07	0.31
Gold	0.02	0.03	0.03	0.03	0.05	0.06	0.06	0.04	0.02	0.02	0.02	0.03	0.03	0.03	0.02	0
Lead	2.90	2.65	3.40	4.83	5.47	3.27	2.91	2.68	2.84	2.73	3.48	2.95	2.20	3.25	0.92	0.84
Silver	0.17	0.21	0.16	0.14	0.16	0.17	0.24	0.26	0.20	0.18	0.14	0.20	0.26	0.19	0.04	0.10
Tin	25.67	25.26	27.67	26.22	25.20	25.83	29.61	30.16	27.14	27.96	26.61	26.81	25.42	26.89	1.60	21.08
Antimony	0.11	0.18	0.16	0.15	0.14	0.13	0.08	0.17	0.12	0.14	0.10	0.15	0.14	0.14	0.03	0.01

[a] Weight % of elements at rim location; analysis dates: September 27, 1975 and July 17, 1975.
[b] Columns Ao, Fo, and Jo are the results of the July 17 measurements at points A, F, and J, respectively.
[c] Standard deviation.
[d] Christ Church

6. The high silver content of the Liberty Bell (0.19% as compared to 0.1% for the Christ Church bell and other bronzes of the period) suggests that silver was deliberately added to enhance the tone of the Liberty Bell.
7. The Whitechapel bell analyzed at Christ Church has tin at the normal bell metal level of 21%. This is one of eight bells that have been in service at Christ Church for over 230 years without failure.

15.4 ANALYSIS OF SILVER

Sterling silver made by early American silversmiths commands prices many times higher than equivalent pieces of European origin. This has provided a strong incentive for some dealers to repair damaged pieces and apply fraudulent marks, to disguise the identification marks on pieces in good condition, and in some cases create copies complete with their favorite makers, marks. Some of these forgeries are so cleverly done that they deceive many of the truly great experts. The average collector is often easily deceived.

In 1970, the Winterthur Museum started its program to provide a library of analysis data on the thousands of silver objects in the collection. This study has been extended to silver from other collections. The Baltimore Museum of Art has included compositional data in their catalogue on silver published in 1975 [4]. The energy-dispersive X-ray fluorescence analysis was ideally suited to this task for the following reasons:

a. The object is not defaced in any way.
b. Up to 15 elements in a wide concentration range can be analyzed simultaneously.
c. It is rapid; it takes less than 5 min to obtain a teletype readout of the weight percent of all the elements present.
d. All the parts of the piece can be readily analyzed instead of depending on spectrographic analyses of scrapings taken from obscure parts of the piece and assuming that this is representative of the piece as a whole.

TABLE 15.9

Analysis Results of Six Paul Revere Tankards:
Winterthur Museum; Creamer: Art Institute of Chicago, Massachusetts[a]

Accession number	Bottom of body				Side of body				Base rim				Lid			
	Ag	Cu	Au	Pb	Ag	Cu	Au	Pb	Ag	Cu	Au	Pb	Ag	Cu	Au	Pb
57.859.1	93.8	5.4	0.19	0.26	94.0	5.5	0.15	0.22[b]	94.5	4.7	0.09	0.20	94.6	4.5	0.13	0.28
57.859.2	93.6	5.4	0.04	0.26	92.9	6.0	0.04	0.31	92.9	6.5	0.15	0.26	90.7	8.5	0.15	0.31
57.859.3	93.3	6.0	0.18	0.33	86.5	12.7	0.12	0.30	93.3	6.2	0.16	0.21	93.4	6.1	0.14	0.22
57.859.4	93.9	5.4	0.08	0.20	92.7	6.6	0.09	0.17	93.1	6.3	0.17	0.23	94.3	3.8	0.16	0.20
57.859.5	93.4	6.0	0.13	0.22	91.4	8.0	0.17	0.24	92.8	6.9	0.12	0.18	94.0	5.4	0.16	0.23[b]
57.859.6	93.1	6.4	0.14	0.32	88.7	10.5	0.10	0.37	93.3	5.5	0.13	0.22[b]	93.8	5.6	0.17	0.27[b]
Creamer					91.3	7.4	0.15	0.4	92.3	6.4	0.15	0.40				

Accession number	Handle face				Handle side				Finial				Hinge			
	Ag	Cu	Au	Pb	Ag	Cu	Au	Pb	Ag	Cu	Au	Pb	Ag	Cu	Au	Pb
57.859.1	93.4	6.1	0.15	0.20	90.0	9.2	0.13	0.27	94.2	5.2	0.16	0.29	92.7	6.8	0.03	0.14
57.859.2	94.3	5.2	0.16	0.29	93.8	5.7	0.17	0.20[b]	92.1	6.5	0.32	0.35	85.4	12.0	0.11	0.27
57.859.3	91.2	7.4	0.13	0.21	90.5	8.5	0.18	0.28	94.5	4.9	0.13	0.10	91.4	8.1	0.07	0.14
57.859.4	94.1	5.4	0.19	0.19[b]	91.0	8.1	0.20	0.16	93.8	5.7	0.19	0.24[b]	90.1	8.7	0.05	0.17
57.859.5	93.1	6.4	0.18	0.17	93.5	5.7	0.21	0.24[b]	93.9	5.7	0.15	0.11	90.2	9.2	0.15	0.19
57.859.6	94.4	5.2	0.16	0.21	89.2	9.5	0.09	0.22	91.3	7.0	0.28	0.35	92.5	6.8	0.08	0.30
Creamer					91.4	7.3	0.12	0.32								

[a] About 1760.
[b] Parts having identical composition, indicating that they were probably cast at the same time or from the same melt.

e. The measurements are made at the surface of the piece in such a fashion that alterations or repairs are readily detected.

f. The composition patterns of the various parts of pieces from various countries are well enough characterized and documented that most pieces can be judged to be American or foreign with confidence by comparing them to the data on file. Table 15.9 shows the analysis data of Paul Revere tankards in the Winterthur laboratory file.

15.4.1 Compositional Features

a. Most of the American silver made before 1750 is very similar to British silver of the period indicating the materials and even partially completed objects were obtained from England and finished and hall marked in America.

b. Over 90% of the flat silver in the Winterthur collections has under 90% silver indicating that it was probably made from coins. All British silver had to be analyzed in an assay office after which it was "hall marked" if it contained over 92.5% silver. No such control was required of American silversmiths.

c. The composition of the various parts of British-made pieces are very similar. Parts of comparable American pieces vary greatly. Note the silver content of the 6 Paul Revere tankards of Fig. 15.9 and Table 15.9 range from 85% to 96%. This indicates that the Americans melted their silver in smaller batches that were not carefully controlled. A careful examination of the data indicates that several castings were made from the same melt.

d. Gold, which occurs naturally with silver ores, is present in amounts from 0.02% to 1% until late in the 19th century. The absence of gold is a strong indication that the piece is modern. Note the absence of gold in the body of the "Joseph Richardson" false tankard (Fig. 15.10 and Table 15.10).

e. The bar graphs of Fig. 5.11 show the silver and gold content of European silver analyzed by the Winterthur Museum during the

FIG. 15.9 All the parts in these six Paul Revere tankards (about 1780) of the Winterthur Collection were found to have varying compositional features. Parts having similar compositional features were probably made from the same batch of alloy enabling the art historian to deduce the fabrication techniques employed in Mr. Revere's shop.

period 1970-1975. The point "S" indicates the silver content of the sterling standard. It should be noted that only one British object fell below this value. Some of the Continental countries have 80% silver as their standard. The gold content provides a valuable clue as to the minimum age of an object.

    f. The bar graphs of Fig. 15.12 show the silver and gold content of American flatware and holloware made between 1700 and 1950. In contrast with Fig. 15.11 it should be noted that there are many more pieces containing silver below 92.5% than above. Also note the reduction of the gold content after 1850.

Compositional features found in altered or fake pieces:

    a. The absence of gold in a part of a piece indicates that it is modern replacement.

FIG. 15.10 This tankard bearing JR hallmarks on the body near the handle (Joseph Richardson was a leading Philadelphia silversmith), has compositional features of 18th century British silver except for the body which is of 20th century alloy. Probably, a previous owner, in an attempt to upgrade a British-made piece to bring a higher price in the American market, ruined the body while attempting to remove the evidence of its true origin. While the hallmarks were good copies of true Richardson pieces, the compositional features positively disclosed the chicanery of the forger.

TABLE 15.10

Comparison of Fake and Authentic Joseph Richardson Tankard Compositions

Part	"J. Richardson" fake tankard (Accession No. 64.52)								Joseph Richardson tankard (Philadelphia, 1739; Run 5078)							
	Ag	Mn	Fe	Cu	Pb	Zn	Au		Ag	Mn	Fe	Cu	Pb	Zn	Au	
Lid	92.8	0	0.03	6.5	0.32	0.07	0.13		92.7	0.01	0.02	6.54	0.25	0.16	0.10	
Base rim	93.8	0.03	0.02	5.5	0.31	0.07	0.11		91.9	0.02	0.02	7.42	0.28	0.17	0.10	
Body	92.5	0.02	0.02	7.4	0.01	0.02	0.00		91.8	0.01	0.04	7.62	0.27	0.11	0.07	
Handle	91.7	0.06	0.03	7.4	0.37	0.14	0.16		89.6	0	0.03	9.68	0.31	0.20	0.09	
Hinge top	94.7	0.02	0.07	4.6	0.20	0.09	0.12		91.5	0.01	0.04	7.52	0.29	0.11	0.07	
Hinge bottom	94.6	0.03	0.13	4.6	0.24	0.09	0.13		91.6	0.03	0.04	7.50	0.26	0.13	0.08	
Underside	74.7	0.08	0.06	4.5	0.31	0.07	0.14		93.0	0.01	0.02	6.30	0.26	0.13	0.10	

Solders	Ag	Mn	Fe	Cu	Pb	Zn	Au
Bottom to rim	84.7	0.04	0.06	11.6	0.43	3.0	0.09
Top of handle to side	77.8	0.11	0.10	13.8	0.17	7.6	0.02
Bottom of handle to side	79.3	0.14	0.14	16.3	0	4.1	0

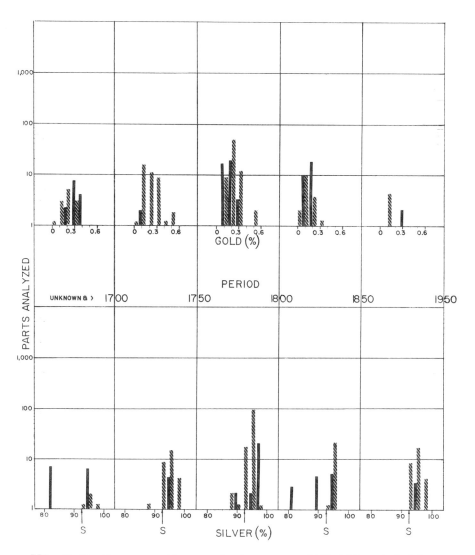

FIG. 15.11 European silver analysis summary 1650-1950: British (broken bars); Continental (solid bars). The rigid control of the assay office in Great Britain is very much in evidence from these graphs. The early British silversmiths lacking precise assay facilities tended to keep their silver content well above the 92.5% requirement to avoid severe penalties.

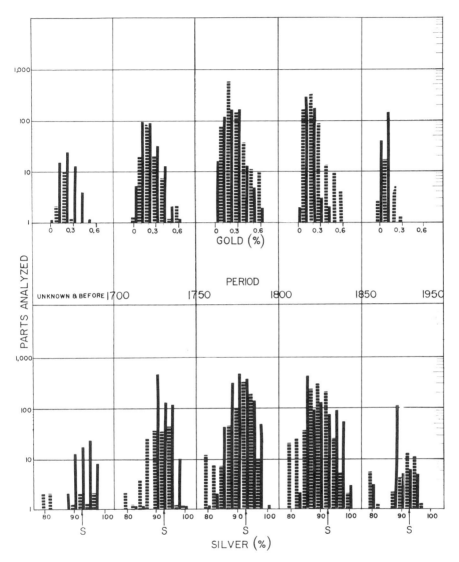

FIG. 15.12 American silver analysis summary 1650-1950: holloware (broken bars); flatware (solid bars). American makers were not troubled by consumer advocates at that time and tended to use materials on the basis of availability rather than purity.

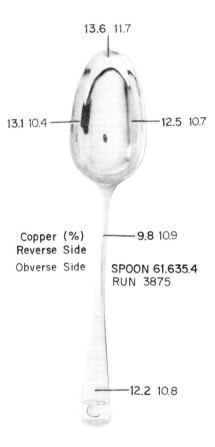

FIG. 15.13 Effect of the segregation of copper and silver during solidification from a "sterling silver" melt. The copper content on the reverse side is shown in bold lettering and the obverse side in narrow lettering.

b. The presence of abnormally high amounts of copper and zinc is an indication of silver solder. The object should be examined for evidence of repairs or alteration.

c. Abnormally high silver content (98-99% silver) indicates that the piece has been electroplated. The electroplate is applied to disguise evidence of repairs or alteration. Some base metal objects have been electroplated and hallmarked without being detected by experts.

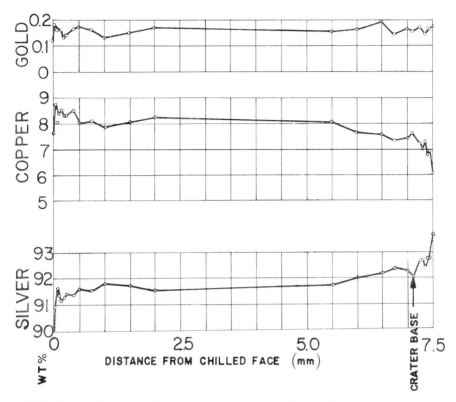

FIG. 15.14 Element distribution in a Sterling silver casting. Copper solidifies preferentially over silver when contacting a cold mold surface. Silver concentrates is the last part to solidify. Note how this effect appears on the spoon of Fig. 15.13.

d. Wide variations in silver on a flatware indicates that it has recently been pickled in acid to remove discoloration caused by heating for repairs. The acid treatment not only removes the dark scale but also the copper at the surface. A few polishings abrade through parts of the silver-rich surface, thus exposing the bulk metal of the piece. The composition could appear to differ by 4-8% at nearby spots at this stage. Repeated polishings reduce these variations to a very small amount. Close examination might reveal that repairs had been made which required the pickeling treatment. Note the spoon, Fig. 15.13, whose repair was detected by wide variation in silver content at the surface. The higher melting copper solidifies preferentially on a chilled mold leaving a copper-rich skin and a silver-rich crater on the top. (Note Fig. 15.14.) These compositional differences remain after the billet is rolled or forged into a sheet or bar. The copper content on the two sides of this spoon differ by 2%. Similar differences are observed on the two sides of silver coins. The absence of these differences might indicate that the object is a cast forgery.

e. Lead, which is not deliberately added to silver solder, is found in quantities as high as 1 percent in old solders. The joint between the handle and the modern body of the antique fake tankard of Fig. 15.10 was made with 20th-century silver solder.

15.5 ANALYSIS OF PEWTER

Pewter is a tin-based alloy containing a few percent of copper, antimony, and sometimes arsenic to improve its ductility and other working properties. "High-quality" pewter generally contains 90% tin and a small amount of lead. Lead content of 20-50% is found in "low-quality" pieces.

Flatware is cast in two-piece brass or bronze molds and finished by filing, scraping, and buffing. Holloware pieces are cast in several parts and assembled by fusing or soldering.

"Britannia-grade" pewter containing from 5% to 10% antimony was introduced in the late 18th century. It is much more ductile than the cast pewter so that it can be rolled into a sheet and formed by spinning or other sheet-forming techniques.

Hollow handles and spouts are "slush cast" by pouring the molten metal into split brass molds and draining the mold after the metal in contact with the mold has solidified.

The tin content of the various parts of coffee pots, tea pots, and tankards made before 1820 was typically above 92%. Measures contained 60-80% tin. The lead content after this period increased gradually either as a result of using scrap pewter or by deliberate adulteration.

At a given period the tin content of British-made pewter was 5-10% higher than that found in similar American-made objects. Compositional data by X-ray fluorescence techniques on about 100 Winterthur objects is given in Charles Montgomery's book [5]. Detailed data on the composition of over 2,000 American and European pewter objects analyzed at the Winterthur Museum by Janice Carlson will be published in *Winterthur Portfolio* in 1977 [6].

15.5.1 Variations of Composition Within a Casting

The wide difference in melting points of the elements used in pewter results in segregation of the elements during solidification. Some of the high-melting copper appears to have disappeared between the melting and the casting since the analyzed copper content is generally about half that prescribed in pewter formulas of the period. Furthermore, it was suspected that a different distribution of

Element	Melting point °C	Typical wt %
Tin	232	60-95
Copper	1083	0-2
Lead	327	0-40
Antimony	630	0.5-10

elements would occur if the melt was cast into a hot mold than into a cold mold. A pewter billet, made up from 86% tin, 1% copper, 10% lead, and 3% antimony, was cast under a zinc chloride flux that was supplied by Colonial Williamsburg.

The composition at various parts of part of the billet is given in Table 15.11. The billet was then melted and a cylinder was cast into a 30°C copper mold and another cylinder was cast into a 300°C mold. Dross from the first remelt was analyzed to determine in what order the elements oxidized in the melt. About 15% zinc appeared in the dross from the zinc chloride flux used in the first melting. The greatest loss due to oxidation were copper and lead. This ex-

TABLE 15.11

Element Distribution of Williamsburg Pewter (Average Values)

	Tin		Copper		Lead		Antimony		Number of analyses in average
	%^a	S	%^a	S	%^a	S	%^a	S	
Ingredients added	86.0		1.00		10.00		3.00		
Original billet									
Bottom (chilled face)	87.30	0.61	1.97	0.30	7.48	1.03	2.91	0.31	6
Cross section	82.71	1.15	1.02	0.07	12.76	0.84	2.47	0.25	6
Top	87.66	1.06	1.21	0.27	7.80	0.81	2.64	0.23	6
Recast billet (midsection)	84.07	0.30	1.01	0.03	11.46	0.27	2.87	0.12	12
Dross (first melting)	70.39		1.91		10.21		2.47		1
Dross (second melting)	83.93		0.88		11.71		2.66		1

^a% = Average weight percent; S = standard deviation.

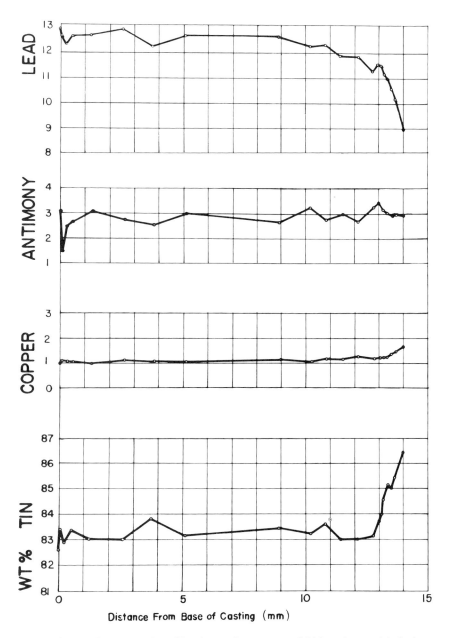

FIG. 15.15 Element Distribution of a pewter billet in a chilled copper mold. Compositional variations are frequently found on different parts of the same piece of pewter. This might result from removing varying amounts of the heterogeneous "skin" in the finishing operation.

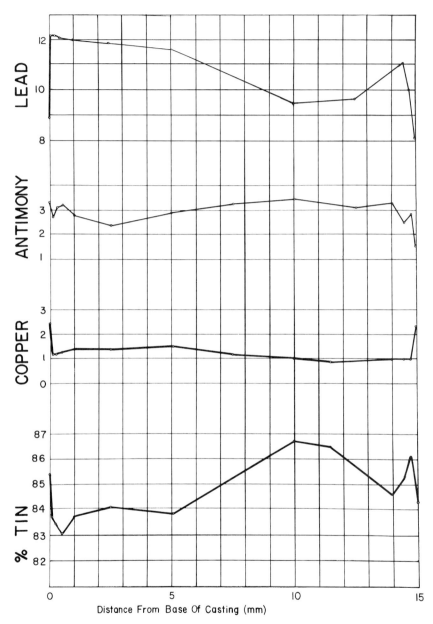

FIG. 15.16 Element distribution of a pewter billet cast in a hot copper mold. These experiments indicate that greater compositional differences occur at the cast surface layer when the melt solidifies slower in a hot mold than that shown on Fig. 15.15, in which a chilled mold was employed.

TABLE 15.12

Composition of Pewter Objects

Date	Object	Acc. no.	Maker	Location	Sn	Cu	Pb	Sb
1690	Multireed charger		John Cave	Bristol	95.84	1.10	2.14	0
1690	Multireed charger		John Cave	Bristol	96.40	1.92	0.41	0.14
1725	Multiple reeded dish		Joseph Gidding	London	98.43	0.92	0.33	0.05
1770	Single reeded plate		William Hunter	Edinburgh	95.15	0.86	2.95	0.37
1689	Multireeded plate		John Trout[a]	London	80.83	1.20	15.03	1.34
1689	Multireeded plate		John Trout[a]	London	84.48	1.12	12.01	1.44
1810	Plate	58.639	George Lightner	Baltimore	82.67	0.82	13.60	2.58
1810	Charger	56.59.19	George Lightner[a]	Baltimore	90.68	2.51		6.81
1825	Plate	65.1454	Thomas Boardman	Hartford	80.48	1.47	15.01	2.59
1840	Deep dish	68.617	Thomas Boardman[b]	Hartford	96.42	0.86	1.82	0.74
1820	Porringer body	65.1532	T. D. and S. Boardman	Hartford	79.55	1.02	18.02	0.70
1820	Porringer handle	65.1532	T. D. and S. Boardman	Hartford	66.19	0.78	31.42	0.44
1820	Porringer body	53.37.5	T. D. and S. Boardman	Hartford	91.11	1.50	1.10	6.20
1820	Porringer handle	53.37.5	T. D. and S. Boardman	Hartford	82.20	1.51	10.58	4.88
1970	Porringer body	70.426	John Stauffer	Lancaster	92.34	5.13	2.23	9.95

[a] Known fake pieces.
[b] British-made plate (18th century), upgraded by the removal of British marks and the application of an American maker's mark.

plains in part why the analyzed values for copper are always lower than one might expect from the published formulas for pewter.

The faces of the cast cylinders were analyzed as a function of distance from the cast face by machining layers away and reanalyzing. The composition as a function of the distance from the cast face is shown in Figs. 15.15 and 15.16. It is significant to note that the surface-rich layer is less than 0.2 mm thick. Since more metal than this is ordinarily removed in finishing a pewter piece, this experiment provides assurance that the surface analysis is indeed representative of the body of the piece. The extreme variations at the right end of the curve is a result of segregation since this is the last part to solidify. Such variations occur in the sprues and risers in the mold and are not a part of the final casting.

Analysis results of a few pieces of American and British-made pewter objects are given in Table 15.12. A few known fake pieces are included in this table as well as a modern reproduction.

Analysis information on about 2,000 pieces of American, British, and continental European pewter will be published in 1977 in the *Winterthur Portfolio* [6]. Other American pieces are included in Montgomery [5].

15.6 ANALYSIS OF GLASS AND CERAMICS

What has been said before about silver and pewter is true again: energy-dispersive X-ray fluorescence analysis is an exceptionally powerful tool for comparing a wide range of museum and archeological glass and ceramic objects, because 30 or more elements over a wide range in concentration can be determined simultaneously without sampling or defacing the object.

The National Bureau of Standard (NBS) prepared a series of test glasses designated 610, 612, 614, and 616, containing 61 elements in the 500, 50, 1, and 0.02 ppm concentration range. Dr. Robert H. Brill of the Corning Museum of Glass prepared a series of four synthetic ancient glasses, designated Brill A, B, C, and D, containing as many as 28 elements in concentrations from a few ppm for trace elements to 66% for the major elements. The Brill and

NBS glasses have been analyzed by many prominent museums, academic, and industrial laboratories throughout the world, and after a statistical review the most probable values were reported [7]. Dr. Brill also made up a special series of test glasses for the Winterthur Museum as analogs of glass made since the 17th century (Fig. 15.18).

These test glasses have made it possible to obtain quantative analyses of glass objects. A project is now under way to build up a library of analysis test data to make possible the assignment of the proper attribution to pieces of questioned origin.

The test glasses are also very useful in comparing ceramic bodies that are free of lead glazes. The absolute accuracy obtainable using these prepared glasses as standards for analyzing ceramics will be the subject of research in the near future.

15.6.1 Analysis Procedure

There are about 30 elements which have been found in ancient and modern glasses in amounts measurable by the techniques to be described. This covers X-ray energies up to 40 keV from the K or L electron orbits.

  a. Elements K, Ca, Ti, and V are excited by the 5.9-keV Mn K radiation from a Fe 55 10-mCi isotope source for 200 sec. The analog-digital converter rejects data above 5.3 keV.
  b. Elements Cr, Mn, Fe, Co, Ni, Cu, Zn, Ga, As, Rb, Sr, Y, Zr, Pb, and Bi are excited by the 22.1 keV Ag K X ray from a 50-mCi Cd 109 isotope source for 500 sec. The analog-digital converter does not accept data below 5.2 keV and above 18 keV during this period.
  c. Elements Ag, Cd, Sn, Sb, La, Cs, and Ce, are then excited by the 26.4 keV and 59.6 keV gamma radiation from a 200-mCi Am 241 source for 300 sec. The analog-digital converter rejects data below 18 keV during this period.

This procedure produces a continuous spectrum free from the back scatter of the exciting sources.

A computer program has been composed to perform the following steps and functions.

15.6.1.1 Make Ready for Analysis

a. Command 60. The atomic number, channel peak location, and weight percent of the elements in the reference standard are inserted through the teletypewriter and stored in the memory.
b. Command 10. A base-line spectrum from a 5 mm thickness of high-purity silica is placed in memory.
c. Command 20. The spectrum of the reference standard is inserted.
d. Command 50. The base-line spectrum is subtracted from the spectrum of the standard providing a net count for each channel in the memory.
e. Command 80. The ratio of weight percent of each element to the sum of the counts in the assigned peak channel plus the two adjacent channels is automatically calculated and stored in the memory.

The data required for calculating the weight percent of the elements in any number of unknowns are now stored in the computer.

15.6.2 Analysis of Objects

The object is excited by the three sources according to steps a, b, and c in Sec. 5.6.1 which provide a continuous spectrum of photon counts in 500 channels covering the energy range from 0 to 40 keV.

a. The data are reproduced graphically on an x-y point plotter. Typical plots are illustrated throughout this chapter.
b. Command 20. Digital count data from the analyzer memory are transferred to the computer.
c. Command 50. The stored base-line spectra are subtracted from the object spectra channel by channel.

d. Command 70. The following computations are made:

   a'. The net counts in each assigned channel are added to those of the two adjoining channels.

   b'. Arithmetic calculations are made for those elements having overlapping spectral peaks such as As and Pb at 10.6 keV and Co and Fe at 6.9 keV.

   c'. Normalizing factors calculated from Sec. 15.2.5 are applied to each element.

   d'. The object description number is inserted after which the atomic number, net counts, and weight percent of each element are typed out. The total analysis time is about 20 min.

   Twenty-five to thirty-five elements are determined according to the nature of the object and the purpose of the analysis. A punched tape of the analysis data is kept for future comparison with other glass by computer methods. A format is being prepared to permit comparison with analysis data obtained at other institutions.

The procedure just described assumes the following:

a. The concentration of the elements in the reference standard is within a factor of 5-10 of that in the unknown so that adverse matrix effects are minimized and there is a linear-relationship between counts accumulated and weight percent of that element.

b. The object approaches "infinite" thickness for the incident and fluorescent radiation.

c. The object size "fills" the measuring aperture of the instrument.

d. The object is within the same geometric distance from the source and detector as the reference standard was.

These effects are evaluated in Fig. 15.17 in which glass of known composition was sliced into thicknesses and widths from 1.2 to 40 mm

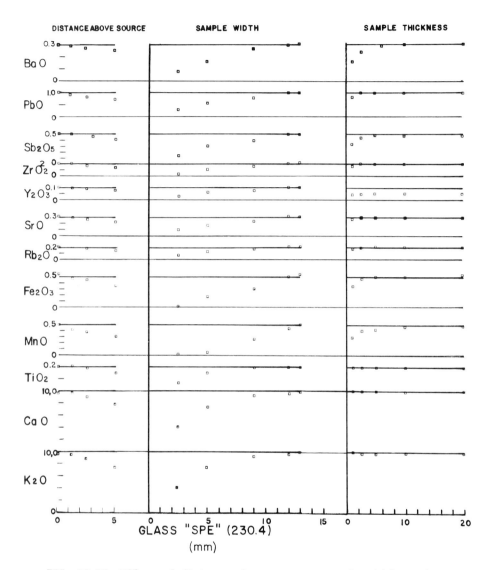

FIG. 15.17 Effect of distance above source, sample width, and sample thickness when analyzing glass objects. Such effects are minimized when analyzing alloy art objects by employing the principal element as an internal standard. This procedure is not applicable to glass and ceramic objects so that such measuring conditions affect the result. These results provide correction factors that may be applied to the various elements when small or odd shapes are analyzed.

and analyzed at various distances from the sample aperture. These curves show that ideally the sample should be at least 13 mm in diameter, 5 mm thick, and in the plane of the aperture.

First-order corrections based on these curves can be applied in nonideal cases.

15.6.2.1 Instrument Stability

In order to separate instrument drift from other measurement vagaries, a large group of glass objects was analyzed over a period of two weeks with standardization checks at least twice per day but without readjusting the instrument during the period (Table 15.13). The experiment was repeated for another week at a later date. The variations in measurement of the standard sample are shown on Fig. 15.18 (see also Fig. 15.19 and Table 15.14).

Such slight instrument drift is reduced by checking the current channel location of calcium at the low end of the scale and barium at the high end and making electrical adjustments to bring them to their assigned location. This action places the glass standard sample in control of the instrument calibration and independent of instrumental drift. Drift in sensitivity due to natural radioactive source decay is automatically compensated for when the instrument is standardized by the routine described in Sec. 15.6.1.

15.6.3 Analysis Results

A group of rare signed glass objects made by a German emigrant, John Amelung near Frederick, Maryland, during the period 1780-1795, when his factory was destroyed by fire, were analyzed to establish characteristic features that distinguish them from objects made at other glass works.

Since these were the first objects to be analyzed by the techniques described above and since they were loaned for the purpose, more detailed measurements were made than were deemed necessary in analyzing subsequent objects. Duplicate measurements made on at

TABLE 15.13

Concentration Range of Elements Found in
Amelung Glass (June-July 1974 Series)

Element	Concentration range (wt % of oxide)		Corning SPE (230.4) Mean (X)
	Low	High	
Potassium	10	15	9.97
Calcium	8	10	10.03
Titanium	0	0.02	0.19
Vanadium	0	0	
Chromium	0	0	
Manganese	0.2	1.2	0.46
Iron	0.1	0.3	0.51
Cobalt	0	0.001	
Nickel	0	0	
Copper	0	0	
Zinc	0	0	
Rubidium	0.005	0.02	0.20
Strontium	0.002	0.02	0.29
Yttrium	0.002	0.04	0.093
Zirconium	0.01	0.08	0.20
Silver	0	0	
Cadmium	0	0	
Tin	0	0	
Lead	0	1.4	0.99
Arsenic	0	0.002	
Antimony	0.02	0.4	0.50
Barium	0.02	0.08	0.29
Lanthanum	0	0.0006	
Bismuth	0	0.001	
Gallium	0	0.001	

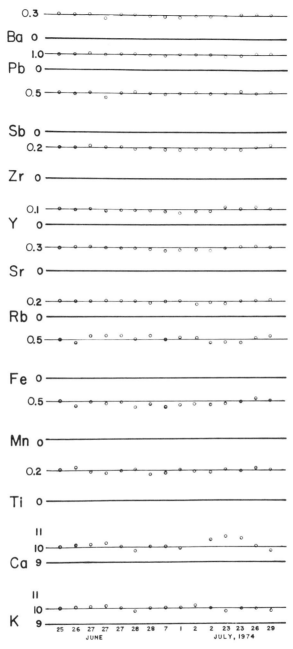

FIG. 15.18 Standard glass checks in which the inherent stability of the analyzer system was determined over a period of one month, during which no changes were made on the instrument settings. The observed 2% to 5% instrument drift was essentially eliminated by resetting the span adjustment so that the peak of a selected element always appeared in the same channel.

TABLE 15.14

Comparison of Test Glasses[f]

Item	EL	Z	keV	Wt% T[b]	CPS[a]	CPS/Wt	Wt%/CPS	Glass number: 230.4 T[b]	M[c]	315.1 T[b]	M[c]	C.L.[e]	315.2 T[b]	M[c]	C.L.[e]
1	K	19	3.3	10	79	7.9	0.127	10	9.68	10	9.79	10	3.1	3.49	3.0
2	Cu	20	3.7	10	10.4	10.4	0.096	10	9.61	10	9.96	15	5.7	6.52	10.0
3	Ti	22	4.5	0.2	15.4	76	0.013	0.2	0.23	0.2	0.20	0.2	0.2	0.24	0.2
4	V	23	4.9	0.1	12.6	126	0.008	0	0.04	0.1	0.10	0.2	0.1	0.13	0.2
5	Cr	24	5.4	0	0			0	0	0	0		0	0	
6	Mn	25	5.9	0.5	3.4	6.8	0.15	0.5	0.36	0.5	0.51	0.05	0.5	0.67	0.05
7	Fe	26	6.4	0.5	4.6	9.2	0.11	0.5	0.49	0.5	0.51	0.50	0.5	0.62	0.40
8	Co	27	6.9	0.1	1.6	16	0.063	0	0.02	0.1	0.11	0.60	0.1	0.14	0.60
9	Ni	28	7.5	0.1	1.8	18	0.056	0	0	0.1	0.11	0.08	0.1	0.13	0.08
10	Cu	29	8.1	0.2	5.5	27.5	0.036	0	0	0.2	0.21	0.15	0.2	0.27	0.15
11	Zn	30	8.6	0.1	4.8	48	0.021	0	0.01	0.1	0.10	0.25	0.1	0.13	0.25
12	Ga	31	9.2	0.1	7.2	72	0.014	0	0.03	0.1	0.10	<0.5	0.1	0.11	<0.5
13	Rb	37	13.3	0.2	54	270	0.0037	0.2	0.21	0.2	0.20	0.1	0.2	0.24	0.1
14	Sr	38	14.1	0.3	90	300	0.0033	0.3	0.33	0.3	0.30	0.25	0.3	0.35	0.25
15	Y	39	14.9	0.1	55	550	0.0018	0.1	0.12	0.1	0.10	0.25	0.1	0.12	0.25
16	Zr	40	15.7	0.2	82	410	0.0024	0.2	0.23	0.2	0.20	0.10	0.2	0.23	0.10
												0.20			0.20
17	Ag	47	22.1	0.1	20.6	206	0.0049	0	0.01	0.1	0.10	0.15	0.1	0.11	0.15
18	Sn	50	25.1	0.2	48	240	0.0042	0	0.01	0.2	0.20	0.20	0.2	0.22	0.20
19	Pb	82	12.7	1.0	78	78	0.013	1	1.04	1.0	1.01	0.20	1.0	1.2	0.60
20	As	33	10.5	0.1	8.7	87	0.012	0	0	0.1	0.09	NF[d]	0.1	0.12	NF[d]
21	Sb	51	26.1	0.5	102	204	0.0049	0.5	0.54	0.5	0.50	0.50	0.5	0.53	0.45
22	La	57	33.5	0.1	21	210	0.0048			0.1	0.10	0.10	0.1	0.10	0.10
23	Ba	56	31.8	0.3	61	203	0.0049	0.3	0.34	0.3	0.30	0.50	0.3	0.32	0.50
24	Bi	82	13.1	0.1	26	260	0.0038	0	0.07	0.1	0.10	0.10	0.1	0.12	0.10

[a] CPS = Counts per second from isotopic sources.
[b] c_T = Theoretical composition.
[c] c_M = Measured composition.
[d] d_{NF} = None found.
[e] C. L. = Commercial laboratory results.
[f] Test Glass 230.4 was the interim standard when the glass analysis program was initiated. Glasses 315.1 and 315.2 were subsequently prepared to cope with additional elements actually found in museum pieces. Glass 315.2 is identical with 315.1 except for its low potassium and calcium content. These glasses were analyzed by a commercial analytical laboratory whose results are tabulated in the C. L. column. With few exceptions, the results are very close to the theoretical values.

The differences in measured values between 315.1 and 315.2 indicates the nonlinearity in the normalizing factors for low concentrations of low atomic numbered elements.

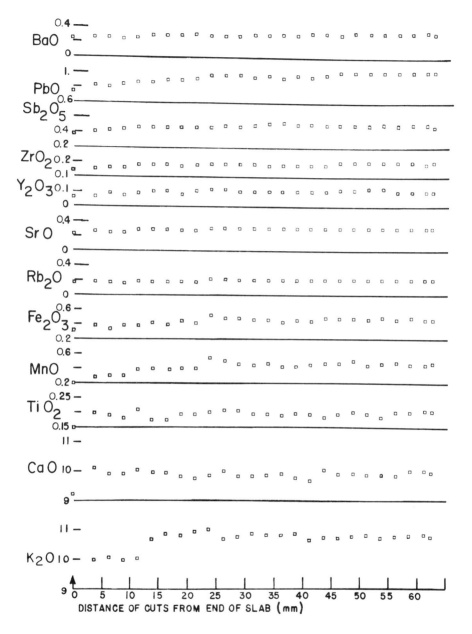

FIG. 15.19 Element distribution in Corning Test Glass (230.4). A section of the cast slab of a specially prepared interim glass standard was sliced into 5-mm thick coupons to determine the uniformity of the element distributions. The composition of these slabs are plotted according to their original location in the "patty." The low values at the left are probably due to the narrow sections near the edge of the cast patty. An arbitrary reference standard was selected from the center of the slab.

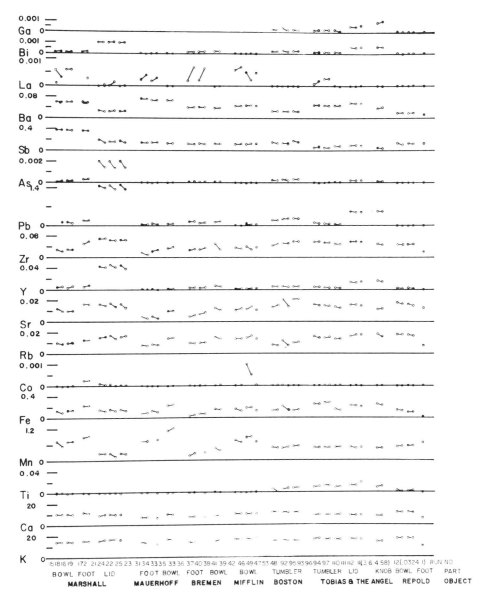

FIG. 15.20 A group of rare, signed and dated tumblers and goblets made by John Frederick Amelung near Frederick, Maryland, during the period of 1780-95, were analyzed and subsequently exhibited at the Corning Museum of Glass as their bicentennial feature in 1976.

TABLE 15.15

Analysis Results of Brill Synthetic Ancient Glasses A, B, and D

Brill base glasses TNH, TNI, and THS[a]

Element	Brill glass A				Brill glass B				Brill glass D			
	P	T	$\bar{X}$	$\bar{S}$	P	T	$\bar{X}$	$\bar{S}$	P	T	$\bar{X}$	$\bar{S}$
K	2.88	3.01	4.04	0.025	1.06	0.30	2.15	0.48	11.56	12.02	12.67	0.47
Ca	5.29	5.01	7.10	0.24	8.69	8.00	10.12	0.43	14.99	15.02	14.57	0.45
Ti	0.80	0.80	0.52	0.01	0.13	0.08	0.075	0.003	0.39	0.40	0.22	0.04
V	0.006	0.006	0.07	0.05	0.03	0.03	0.026	0.021	0.02	0.015	0.03	0.02
Cr	0.001	0.001	0	0	0.005	0.005	0.018	0.031	0.002	0.003	0	0
Mn	1.02	1.22	1.98	0.14	0.25	0.25	0.59	0.22	0.55	0.65	0.67	0.08
Fe	1.07	1.10	1.49	0.08	0.37	0.30	0.56	0.16	0.51	0.50	0.48	0.06
Co	0.15	0.20	0.18	0.01	0.04	0.05	0.18	0.32	0.02	0.02	0.02	0.01
Ni	0.03	0.02	0.20	0.16	0.10	0.10	0.34	0.36	0.06	0.05	0.02	0.01
Cu	1.22	1.20	0.99	0.30	2.68	3.00	3.16	0.32	0.38	0.40	0.20	0.04
Zn	0.04	0.04	0.04	0.01	0.20	0.20	0.16	0.03	0.10	0.10	0.04	0.01
Rb	0.01	0.01	0.01	0	0.001	0.001	0.006	0.005	0.005	0.005	0.004	0.001
Sr	0.10	0.10	0.12	0.05	0.02	0.01	0.025	0.006	0.05	0.05	0.06	0.002
Y			0.003	0.0005			0.014	0.002			0.005	0.0005
Zr	0.005	0.005	0.019	0.002	0.025	0.025	0.03	0.007	0.01	0.013	0.016	0.002
Ag	0.003	0.002	0.002	0.0005	0.01	0.01	0.002	0.001	0.005	0.005	0.002	0.0004
Cd			0.003	0.0007			0.001	0.0003			0.002	0.0005
Sn	0.20	0.20	0.23	0.012	0.03	0.02	0.052	0.005	0.15	0.10	0.13	0.015
Pb	0.05	0.05	0.09	0.006	0.50	0.50	0.68	0.08	0.25	0.20	0.25	0.01
As			0.001	0.0008			0.001	0.004			0.004	0.005

MUSEUM OBJECTS 473

Element	T	P	X̄	S̄	(P) TNH	Wint.	(P) TNI	Wint.	(P) TNK	Wint.	
Sb	1.76	1.81	1.58	0.10	0.45	0.40	0.53	0.02	0.98	1.00	0.94 0.03
Cs		0.10	0.02	0.014			0.003	0.003			0.011 0.01
Ba	0.55		0.38	0.02	0.10		0.098	0.007	0.33	0.25	0.30 0.01
La			0.004	0.002			0.001	0.001			0.001 0.0005
Ce			0	0			0	0			0 0

Element	(P) TNH	Wint.	(P) TNI	Wint.	(P) TNK	Wint.
K	3.10	3.50	1.13	1.80	12.40	13.10
Ca	5.70	6.64	9.26	10.58	16.20	14.60

[a] Corning SPE (230.4) reference, SiO$_2$ baseline.
P = most probable value; T = theoretical value; mean X̄ and standard deviation S̄ are for 7 runs made between September 1974 and February 1975. No corrections were applied for Cd 109 strength loss during this period.

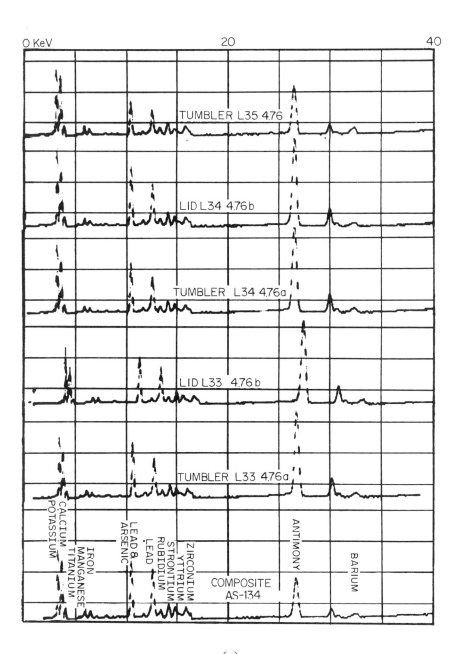

(a)

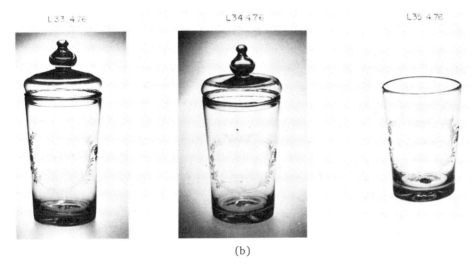

(b)

FIG. 15.21 Amelung glass spectra: (a) The bottom graph is the spectrum of a composite of several hundred colorless archeological fragments found at the Amelung factory site which were subsequently melted and cast into a "patty." This spectrum fits into the middle group of those obtained from signed and dated pieces proving that it is a good average of glass produced during Amelung's operating period of 1780-95.

The upper curves are the spectra of the 5 pieces shown in (b), and were used to confirm the authenticity of this rare group after which they were acquired by the Smithsonian Institution.

least two places on each part of an object are plotted in Fig. 15.20. For example, two measurements were made at each of two points on the bowl of the Marshall stemmed goblet, two measurements on the foot, and two measurements at each of 3 points on the lid. These measurements proved that the base and the bowl were of a different composition although this difference was not discernible to the eye. The lid was of quite a different composition. The circle on the curve measured at the same point on the object is connected by solid lines. The differences shown in lead and arsenic are real and are visible as a swirl when the piece is observed under ultraviolet light.

The following compositional features common to all the signed pieces appear in several nonsigned pieces and provide strong evidence to justify attribution to the Amelung factory [8]:

Ratios
  Potassium:calcium = 3:2
  Rubidium:strontium:yttrium:zirconium = 8:10:2:4
Concentrations
  Barium > 0.02%
  Manganese and iron > 0.05%

A number of glass fragments found at the site of the Amelung factory were melted and cast into a patty by Dr. A. A. Erickson of the Corning Glass Works Research Laboratory. A section sliced from the patty was analyzed. The spectrogram of this composite is shown on Fig. 15.21. It is interesting to note that this composite of archeological fragments has a composition very close to the average of 30 Amelung pieces analyzed.

Figure 15.22 shows the relationship of concentration of various elements found in some of Winterthur Museum's glass collection.

None of the pieces attributed to Amelung are signed nor do they have the chemical pattern of the signed pieces making attribution questionable. On the other hand, flask 59.3131 attributed to Stiegel is probably Amelung.

MUSEUM OBJECTS 477

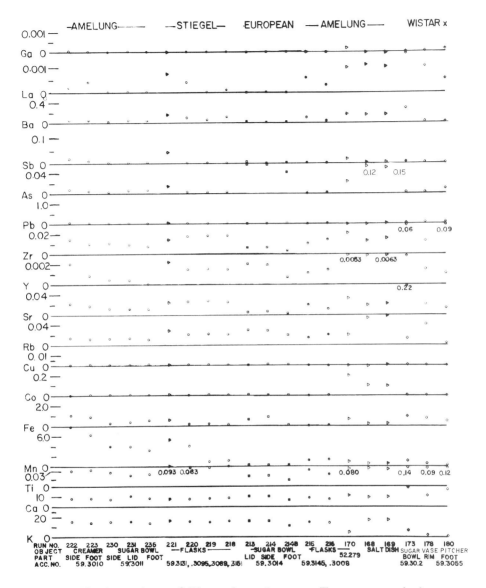

FIG. 15.22 Comparison of Winterthur glasses. These curves depict
the range of composition found in a group of nonlead glass objects
in the Winterthur collection. They provided the basis for a computer program which is currently used in glass objects.

15.6.4 Analysis of Near East Ceramic Sherds

A group of sherds from archeological sites in the Near East were analyzed to determine the feasibility of ascertaining the point of origin based on the composition.

The high concentration of the colorants, iron and calcium, in the decoration obscured the pattern of those elements occurring naturally in the clay. However, the relationship of rubidium, strontium, yttrium, and zirconium is very distinctive and is not affected by the pigments. Figure 15.23 shows the element distribution of four sherds found in these different areas.

Portions of these same sherds were pulverized and the resulting powder analyzed. Figure 15.24 shows the instrument readings obtained on the original painted surface, unpainted surface, powdered samples, and pellets from portions of the sherds shown on Fig. 15.23. This series of experiments designed to establish the optimum sample preparation method for analyzing unglazed sherds gave the following results:

1. Painted surfaces (small triangle) exaggerate the elements in the paint.
2. Paint-free surface provides realistic analysis of the "body."
3. A minimum of 0.5 g should be provided if powdered samples are used.
4. Pellets made from 0.4 g give the same values as the scraped surface of the solid sample.
5. The instrument should be standardized with prepared samples in the form used in the analysis of specimens.

The best results are obtained on 0.4- to 0.6-g pellets. The powder samples and pellets are not affected appreciably by the pigments.

Research is in progress to include data on sodium, magnesium, silicon, and aluminum which cannot be "seen" with Winterthur's isotope excited instrument.

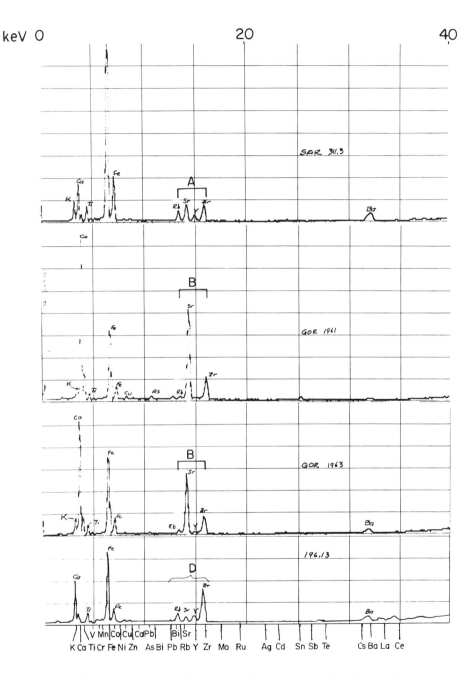

FIG. 15.23 Compositional features of ancient unglazed ceramic sherds from the Near East indicate that element "finger prints" may provide the archeologist with valuable clues regarding the origin of the clays used. Items found in site "B" were probably from the same clay bed whereas those designated "A" and "D" which were found almost 150 km from site "B" are distinctly different.

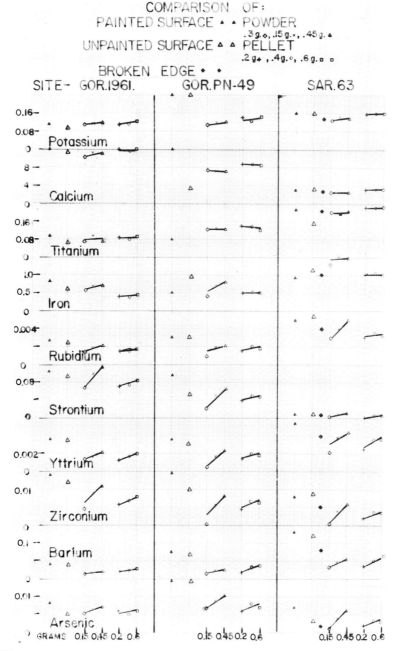

FIG. 15.24 Comparison of painted surface (▲▲), unpainted surface (△△), and a broken edge (♦♦) in unglazed pottery sherds.

ACKNOWLEDGMENTS

I wish to acknowledge the encouragement and constant support that my wife Dot has provided throughout all the phases of this project.

I also wish to acknowledge the valuable contributions of my associates at the Winterthur Museum, especially the following: Karen Anderson, Janice Carlson, Charles Hummel, Karen Papouchado, and George Reilly.

I am most grateful to our consultant Dr. P. H. Gaither for setting up the computer program and procedures which have made it possible for us to make hundreds of determinations per day.

REFERENCES

1. Karen Papouchado, Trace Element Analysis of U. S. Coins, paper presented at Winterthur Scientific Advisory Meeting, November, 1974.
2. V. F. Hanson, Quantitative Elemental Analysis of Art Objects by Energy-Dispersive X-ray Fluorescence Spectroscopy, *Appl. Spectrosc.*, *27* (5), September (1973).
3. Annual Report of the National Gallery of Art, 1971.
4. J. Goldsborough, *Maryland Silver in the Collection of the Baltimore Museum of Art*, 1975, pp. 23-28.
5. Charles Montgomery, *A History of American Pewter*, Praeger Publishers, New York, 1973.
6. Janice Carlson, X-Ray Fluorescence Analysis of British and American Pewter, (will appear) in *Winterthur Portfolio 12* (to be released in) 1977.
7. Robert H. Brill, A Chemical-Analytical Round-Robin on Four Synthetic Ancient Glasses, Corning Museum of Glass.
8. Corning Museum of Glass, *Journal of Glass Studies*, *18* (1976).

Chapter 16

X-RAY ASTRONOMY AND OTHER EXOTIC APPLICATIONS

H. K. Herglotz

Engineering Department
E. I. du Pont de Nemours and Company
Wilmington, Delaware

16.1	Introduction	483
16.2	X-Ray Astronomy	484
16.3	Plasma Diagnostics	488
16.4	Nuclear Structure	494
References		495

16.1 INTRODUCTION

In Chaps. 10 to 15 of this book the reader is introduced to most of the practical applications of X-ray spectrometry. He can benefit from the experience of the authors who are willing to share the fruits of their efforts.

For the sake of thoroughness, a few fields of application will now be described briefly which occupy the interest of few people, do not have the economic impact of other applications, do not attract much attention, but which are on the frontier of science, and are pregnant with discovery and future impact.

In contrast to all other chapters of the book, the author of this chapter cannot resort to personal experience in the field; he dares to act as referee because of his background as an X-ray physicist and astronomer and because of his fascination with novel applications. Those readers with more than a casual interest in any of these fields will find more profound sources of information in the references.

16.2 X-RAY ASTRONOMY

The earth's atmosphere acts as an effective shield against cosmic X rays of all wavelengths, and ground-based astronomical observatories, therefore, have not concerned themselves with this wavelength range of the electromagnetic spectrum. The first information about X-ray sources in the sky was obtained from instruments carried by rockets and high-flying balloons which made their observations from above 99% of the atmosphere [1,2]. The first X-ray source was identified conclusively in 1962. It was named Scorpius X-1, because of its location in the constellation of the same name. Since then, X-ray astronomy has made enormous contributions to our knowledge of the universe by revealing unsuspected objects and processes. Recently, it has been instrumental in increasing confidence in the existence of black holes (Cygnus X-1), objects that cannot emit light and manifest themselves only indirectly [3].

The X-ray astronomer requires the same instrumental features as the classical astronomer. He would like to have instruments of good imaging quality, i.e., with high spatial resolution (<1 sec of arc) and true, undistorted images of celestial objects, and with adequate dispersion to provide a high-resolution spectrum of the radiation emitted by the celestial objects.

While the inhomogeneous and turbulent atmosphere limits the resolution of ground-based optical telescopes, the X-ray observer is hampered only by the limits of his instruments. There are neither lenses nor conventional mirrors for X rays; he always has weight restrictions in his rocket-borne or orbiting instruments; and the need for remote control and data transfer add to the complexity of his instruments.

Chapter 16

X-RAY ASTRONOMY AND OTHER EXOTIC APPLICATIONS

H. K. Herglotz

Engineering Department
E. I. du Pont de Nemours and Company
Wilmington, Delaware

16.1	Introduction	483
16.2	X-Ray Astronomy	484
16.3	Plasma Diagnostics	488
16.4	Nuclear Structure	494
References		495

16.1 INTRODUCTION

In Chaps. 10 to 15 of this book the reader is introduced to most of the practical applications of X-ray spectrometry. He can benefit from the experience of the authors who are willing to share the fruits of their efforts.

For the sake of thoroughness, a few fields of application will now be described briefly which occupy the interest of few people, do not have the economic impact of other applications, do not attract much attention, but which are on the frontier of science, and are pregnant with discovery and future impact.

In contrast to all other chapters of the book, the author of this chapter cannot resort to personal experience in the field; he dares to act as referee because of his background as an X-ray physicist and astronomer and because of his fascination with novel applications. Those readers with more than a casual interest in any of these fields will find more profound sources of information in the references.

16.2 X-RAY ASTRONOMY

The earth's atmosphere acts as an effective shield against cosmic X rays of all wavelengths, and ground-based astronomical observatories, therefore, have not concerned themselves with this wavelength range of the electromagnetic spectrum. The first information about X-ray sources in the sky was obtained from instruments carried by rockets and high-flying balloons which made their observations from above 99% of the atmosphere [1,2]. The first X-ray source was identified conclusively in 1962. It was named Scorpius X-1, because of its location in the constellation of the same name. Since then, X-ray astronomy has made enormous contributions to our knowledge of the universe by revealing unsuspected objects and processes. Recently, it has been instrumental in increasing confidence in the existence of black holes (Cygnus X-1), objects that cannot emit light and manifest themselves only indirectly [3].

The X-ray astronomer requires the same instrumental features as the classical astronomer. He would like to have instruments of good imaging quality, i.e., with high spatial resolution (<1 sec of arc) and true, undistorted images of celestial objects, and with adequate dispersion to provide a high-resolution spectrum of the radiation emitted by the celestial objects.

While the inhomogeneous and turbulent atmosphere limits the resolution of ground-based optical telescopes, the X-ray observer is hampered only by the limits of his instruments. There are neither lenses nor conventional mirrors for X rays; he always has weight restrictions in his rocket-borne or orbiting instruments; and the need for remote control and data transfer add to the complexity of his instruments.

The first observation of an X-ray-emitting object in the sky was an accomplishment in itself, even though neither the exact location and shape of the object nor the wavelength distribution of the emitted X rays was established. Since then, more than a decade has elapsed and tremendous advances have been made in all fields. Several satellites (Uhuru, Copernicus, Ariel V, Netherlands Astronomy Satellite), which are not limited in time like previous short rocket flights, are equipped with X-ray telescopes and spectrographs. Skylab, the manned orbiting laboratory, carried instrumentation for X-ray observations of the sun. All these missions have amassed a huge amount of data: more than 100 X-ray objects are known and the number is growing; spatial resolution of 2 sec arc has been achieved [4]; however, the spectral resolution is still limited by the X-ray detector (for example, 30% at 5.9 keV of the flow proportional counter).

The X-ray spectroscopist who enters the field of celestial X-ray analysis will be astonished to find instrumentation that does not even remotely resemble his earth-based tools [2,5]. Figure 16.1 shows a typical early "X-ray telescope," which is simply a

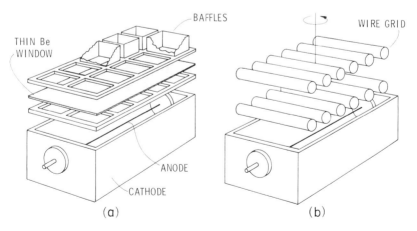

FIG. 16.1 Typical large-area proportional counter for celestial X-ray spectroscopy: (a) with baffles and (b) with grids and rotation for directional selectivity.

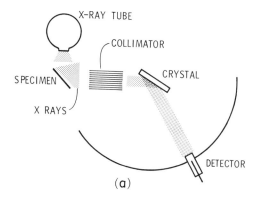

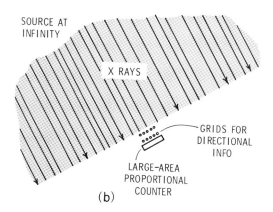

FIG. 16.2 Geometric relations between source and detector in (a) laboratory X-ray fluorescence and (b) X-ray astronomy.

large-area proportional counter. There are many reasons why the physical arrangements are so different from those of laboratory spectroscopy. Perhaps this will be clear if we take a side-by-side look at the two arrangements (Fig. 16.2). In laboratory X-ray fluorescence analysis we keep the distance short between the exciting source and the emitting sample and between the sample and detector (see also Fig. 1-2) to intercept as much radiation as possible from the tube and the emitting sample. In X-ray astronomy, the emitting source is at infinite distance; the radiation from point sources is parallel, or only minutes to few degrees divergent if the source is an extended area object, e.g., the sun. The latter is roughly 0.5° in diameter, which is also the divergence of the X rays that it emits. Because of its relative proximity, namely "only" 1 AU (astronomical unit, or $\sim 1.5 \times 10^{11}$ m $\approx 93 \times 10^6$ miles) the sun is a strong X-ray source. It is ironic, but without significance, that the X-ray astronomer studying the sun measures X rays of order of magnitude 1-Å (ångström, 10^{-10} m) wavelength, which come from an object 1 AU (astronomical unit, $\sim 1.5 \times 10^{11}$ m) away.

The X-ray astronomer-spectroscopist works with very small signals superimposed over an omnidirectional background. Therefore, his prime concern is to collect a maximum of the parallel radiation of Fig. 16.2; the determination of the wavelength distribution is of secondary importance. The use of crystals for wavelength dispersion is not yet directly possible; use of cryogenic energy-dispersive devices is in its infancy [9]. The large-area collimated proportional counter of Fig. 16.1 remains the workhorse of the X-ray astronomer. The shape of the collimating baffles or grids and the scanning motions of the device provide some information about the location of the X-ray source [2].

After the existence of interesting X-ray objects was established beyond doubt, the instrumentalist among the X-ray astronomers faced his greatest challenge: to increase the collecting power of instruments and to build imaging X-ray telescopes. Nearly all of these instruments are based on total reflection, the effect discussed in Chap. 8. As described there, X rays can be totally reflected at grazing incidence on solid surfaces. In Chap. 8, use of this effect for the dispersion of X rays in a grating spectrograph was discussed. However, in X-ray astronomy total reflection is used merely for obtaining a mirror, although a very limited one (Fig. 16.3). These parabolic mirrors form a useful, though stigmatic or comatic, image of the object at the focus of the mirror where it can be recorded [4-7]. Advances in the development of these X-ray telescopes might bring them to the state where they can be coupled with a wavelength-

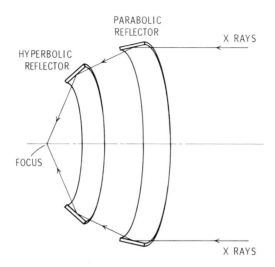

FIG. 16.3 Example of an X-ray telescope using a combination of parabolic and hyperbolic total reflectors.

dispersive spectrometer. So far, proportional or scintillation counters, sometimes aided by filters, are used as energy-dispersive devices, and the spectroscopist has to be content with their poor spectral resolution. The information value of the resolved lines from elements, e.g., in a gas cloud like the Crab nebula, excited by a central pulsar would be worth the effort.

Most recent and thorough reviews of X-ray astronomy are found in Refs. 8 and 9.

16.3 PLASMA DIAGNOSTICS

When high-velocity electrons hit a solid target, a small part of their kinetic energy is converted into electromagnetic radiation while most of it generates thermal energy. Every X-ray spectroscopist recognizes this effect by the bremsspectrum which is generated in the target of his X-ray tube, and knows that he has to remove a large amount of heat from the target by effective cooling. The wavelength distribution of the bremsspectrum gives information about the energy of the electrons. Particularly, the shortest wavelength of the bremsspectrum, λ_o, is a measure of the highest kinetic energy carried by the electrons.

$$eV = h\nu = \frac{hc}{\lambda_o} \tag{16.1}$$

where e = charge of electron, V = tube voltage, h = Planck's constant, ν = frequency, c = velocity of light, and λ_o = shortest wavelength of bremsspectrum. In the early days of X-ray physics when gas-filled tubes were operated by inductors instead of transformers, it was used to measure the peak tube voltage V by the "λ_o spectrograph" [10].

Similarly, the lines of the Cu K series (e.g., $K\alpha_1, \alpha_2$ Kβ) do not appear in the radiation from a copper target, unless the electrons strike the target with an energy of at least 8.96 keV. Absence of the lines indicates that eV < 8.96 keV; their presence means that eV > 8.96 keV.

Both these concepts, with variations, find use today in measuring the temperatures of plasmas. *Plasmas* are often called "the fourth state of matter," in addition to the solid, liquid, and gaseous states. They are electrically neutral and consist of equal numbers of positive and negative free charges. From the standpoint of the universe, it is the most prevalent state of matter, since all active stars consist of plasmas. Plasma research has been catapulted to the frontier of science and to prominence in the search for new energy sources, of which nuclear fusion would be one of the more desirables. Nuclear fusion in the laboratory, as well as in stars, requires a certain "temperature," alias kinetic energy of the participating particles, and a certain density, i.e., number of particles per unit volume.

With the intensive research in this field, X-ray spectroscopy has been an effective way to measure temperature in the 10^6-K (Kelvin) region [11,12]. The wavelength distribution of the radiation emitted by the plasma is measured and compared with that calculated for different temperatures. Similarly, the intensity of characteristic lines is compared with those theoretically expected.

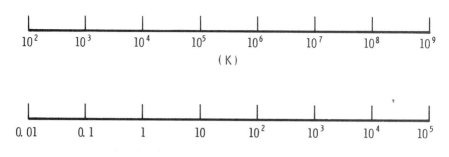

FIG. 16.4 Comparison of temperature scales in Kelvin (K) and electron volts (eV).

The relationship between the kinetic energy of free electrons and temperature is

$$E \approx kT \tag{16.2}$$

where E is energy in eV, k is the Boltzman constant (8.64×10^{-5} eV/K), and T is the temperature in Kelvin. Figure 16.4 shows the relationship between the two scales according to Eq. (16.1); temperature in Kelvin (K) and electron volts (eV). Room temperature, 300 K, corresponds to only 0.02 eV; similarly, the 10 kV electrons of an X-ray tube would correspond to 10^8 K, if they were a thermal distribution instead of a nearly monoenergetic flow of electrons. Or another way to express it: the thermal emission of hot plasmas is in the X-ray range. Figure 16.5 (see also Ref. 11) shows the intensity vs. energy function measured in plasma. The high-energy limit, where the extrapolated thermal line intersects the abscissa, is indicative of the temperature. The "nonthermal" section comes from superheated electrons, which are a byproduct in plasma devices (Stellarator, Tokamak, laser). Figure 16.6 shows the same function plotted against wavelength instead of energy, according to

$$I(\lambda) \, d\lambda = I(E) \, dE$$
$$I(\lambda) = I(E) \frac{dE}{d\lambda}$$

X-RAY ASTRONOMY AND OTHER EXOTIC APPLICATIONS   491

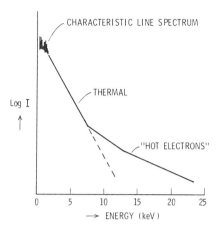

FIG. 16.5 Schematic X-ray spectrum log I vs. energy from a high-temperature plasma.

$$E = h\nu = \frac{hc}{\lambda}$$

$$\frac{dE}{d\lambda} = -\frac{hc}{\lambda^2}$$

Therefore,

$$I(\lambda) = -I(E)\frac{hc}{\lambda^2} \qquad (16.3)$$

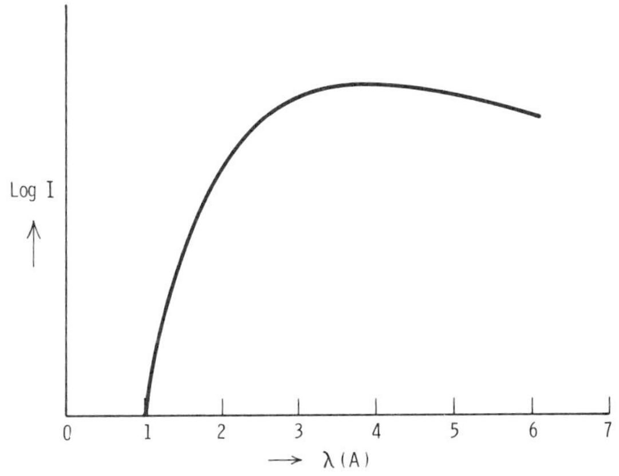

FIG. 16.6 Same as Fig. 16.5 but log I vs. wavelength (λ). Note similarity with bremsspectrum curve of Fig. 1.5.

One recognizes the similarity of the $I(\lambda) = f(\lambda)$ curve with the familiar humped curve of the bremsspectrum. The bremsspectrum process, i.e., decelerated charges, is the most significant, though not the only process in thermal emission [11].

The total intensity of the emitted X rays of Fig. 16.5, i.e., the area under the I versus E function $\int_0^\infty I(E)\, dE$, is a measure of the number N of particles participating in the emission process according to

$$\int_0^\infty I(E)\, dE \propto N^2 \sqrt{T} \qquad (16.4)$$

Thus, the fusion researcher obtains from the plot of Fig. 16.5 two features salient to thermal fusion: the temperature and the number of particles. Since the determination of the number of particles in absolute terms is difficult, the researcher usually settles for relative values [11].

The appearance of characteristic lines of the atoms in the plasma is another temperature indicator. However, these lines are not familiar ones of the K series, emitted by solid targets under electron bombardment. Plasma lines are emitted by highly ionized isolated atoms. The spectroscopist who works with the short-wavelength UV and excites his samples by a condensed spark discharge [13] has been familiar with these lines for a long time.

The instrumental equipment of the plasma diagnostician is similar to that of the classical analytical chemist [14]. His most important problem is to extract the soft radiation from the plasma generator. Today, much of the interest is in laser-induced plasmas, where the energy of an intense laser pulse, focused on a solid, heats and evaporates parts of the solid. Since these laser pulses are short, scanning devices for recording the spectra are impossible. Some of the old-fashioned divergence spectrographs with photographic recording (Fig. 16.7) have found a new application [15]. In a divergence spectrograph, the stationary crystal reflects a variety of wavelengths due to the spread in the angles of incidence. Other situations call for a different type of instrumentation as shown in Fig. 16.8. The fact that industrial firms manufacture such

X-RAY ASTRONOMY AND OTHER EXOTIC APPLICATIONS 493

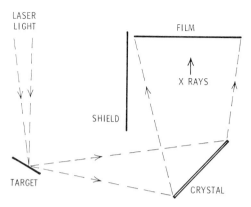

FIG. 16.7 Divergence spectrograph with photographic recording for diagnostics of plasma generated by laser light impacting on solid target.

FIG. 16.8 Cluster of four multiplex solid-state detectors for plasma diagnostics. (Courtesy Keves Corporation.)

devices indicates that plasma diagnostics is an application of X-ray spectroscopy not quite as exotic and rare as the title of this chapter suggests.

16.4 NUCLEAR STRUCTURE

When Moseley discovered the close relationship between the atomic number of an element and the wavelength of the X-ray lines that it emits [Chap. 1, Fig. 1.1, Eq. (1.2)], it became evident that X-ray spectra reveal atomic structure. X-Ray spectroscopy became the atomic physicist's favorite tool for probing the electronic shells of atoms and provided experimental facts for the then novel quantum mechanics. The Nobel Prize to Manne Siegbahn in 1924 acknowledged and rewarded the contributions of X-ray spectroscopists to the advancement of science. Analytical applications of X-ray spectroscopy were rare in those days because the analytical chemist shied away from the use of complicated and difficult apparatus. In later years (after about 1950), analytical chemistry became unthinkable without X-ray spectroscopy, thanks to instrumental development that invited application, but the contributions to atomic structure nearly vanished.

Recently, physicists have returned to X-ray spectroscopy in a search for new knowledge. This time the object of their studies is not the atom's electronic shell but its nucleus. Nuclear and particle physicists have replaced some of the electrons of atoms by short-lived (10^{-6}-10^{-10} sec) muons and kaons. Such muonic or kaonic atoms emit X rays with much higher energy than normal atoms. Their study, done primarily with scintillation counters, reveals the properties of these particles [16].

Another contribution by the X-ray spectroscopist to fundamental nuclear and particle physics is the study of X-ray spectra excited by bombardment of atoms with high-energy ions. The analytical application of this branch is described in Chap. 6. In contrast, the physicist who probes X-ray spectra which are emitted by atoms after excitation by highly energetic ions (e.g., in the 10-100 MeV range) is interested in the short-lived intermediate species that exist while the atom and ion interact. These intermediate species emit X-ray spectra different from those of the atom and are called *quasimolecular*. The X-ray spectra which are observed deliver the evidence for the existence of these species [17]. Another interpretation of these nuclear compounds assumes the existence of superheavy quasiatoms. Bombardment of gold (Z = 79) with 50 MeV iodine ions (Z = 53) has generated a compound entity whose X-ray spectra can be interpreted as the M series of a superheavy atom with Z = 132 [18].

Development of heavy-ion accelerators has been a major effort in modern physics, and the X-ray spectroscopist is again an indispensable member of the team. These heavy-ion accelerators are huge machines; only a few are in existence and they are not a tool for the practical spectroscopist. Why, then, does this book mention them? Because the discovery of X rays happened in the pure search for knowledge and brought innumerable practical applications. It is therefore healthy that X-ray physics never stops this search for novelty because that will in turn spawn new practical applications in the future.

REFERENCES

1. H. Friedman, X-Ray Astronomy, *Advan. X-Ray Anal.*, 13, 289-312 (1970).
2. H. W. Schnopper and J. P. Delvaille, The X-Ray Sky, *Sci. American*, July, 27-37 (1972).
3. H. Friedman, Cosmic X-Ray Sources: A Program Report, *Science*, *181*, 395-407 (1973).
4. A. S. Krieger, G. S. Vaiana, R. Petrasso, and J. K. Silk, The X-Ray Spectrographic Telescope, *Instrumentation in Astronomy II*, 185-205 (1974).
5. W. P. Reidy, Soft X-Ray Instrumentation for Space Experiments, *Advan. X-Ray Anal.*, 13, 313-329 (1970).
6. J. A. Bowles, T. J. Patrick, P. H. Sheather, and A. M. Eiband, Design of an X-Ray Telescope System for an Orbiting Astronomical Observatory, *J. Phys. (E) Scientific Instrum.*, 7, 183 (1974).

7. H. Helava, R. Mitchell, M. C. Weisskopf, and R. S. Wolff, One-Dimensional Focusing X-Ray Telescope for Stellar X-Ray Astronomy, *Rev. Sci. Instrum.*, *46*, 1066-1071 (1975).

8. R. Giacconi and H. Gursky (eds.), *X-Ray Astronomy*, Reidel, Boston, 1974.

9. L. E. Peterson, Instrumental Technique in X-Ray Astronomy, in *Annual Rev. of Astronomy and Astrophysics* (G. R. Burbridge, D. Layzer, and J. G. Phillips, eds.), *13* (1975).

10. F. Regler, *Grundzüge der Röntgenphysik*, Urban Schwarzenberg Berlin-Vienna, 1937, Chap. IX-3.

11. D. J. Nagel, X-Ray Emission from High-Temperature Laboratory Plasmas, *Advan. X-Ray Anal.*, *18*, 1-25 (1975).

12. W. Lochte-Holtgreven, *Plasma Diagnostics*, North Holland, Amsterdam, 1968.

13. M. Siegbahn and T. Magnusson, Zur Spektroskopie der ultraweichen Röntgenstrahlung II, *Zeitschrift für Physik*, *87*, 291 (1934).

14. C. E. Violet, Instrumentation for X-Ray Diagnostics of Plasmas, *Advan. X-Ray Anal.*, *18*, 26-61 (1975).

15. L. S. Birks, X-Ray Spectroscopy and Its Uses, *Siemens Review XLI*, 4-7 (1974); see also, L. S. Birks and J. V. Gilfrich, X-Ray Absorption and Emission, *Anal. Chem.*, *46*, 360R-366R (1974).

16. J. P. Davidson, D. A. Close, and J. J. Malanify, Evidence for Higher Shape Deformations in Muonic X Rays, *Phys. Rev. Lett.*, *32* (7), 337-339 (1974).

17. J. S. Greenberg, C. K. Davis, and P. Vincent, Evidence for Quasi-Molecular K X-Ray Emission in Heavy Ion Collisions from the Observation of the X-Ray Directional Anisotropy, *Phys. Rev. Lett.*, *33* (8), 473-476 (1974).

18. P. H. Mokler, H. J. Stein, and P. Armbruster, X-Rays from Superheavy Quasi-Atoms Transiently Formed During Heavy-Ion-Atom Collisions, *Phys. Rev. Lett.*, *29*, 827-830 (1972).

AUTHOR INDEX

Numbers in parentheses are reference numbers and indicate that an author's work is referred to although his name may not be cited in the text. Underlined numbers give the page on which the complete reference is listed.

A

Abbey, S., 322(146,147,148,149, 150), 347
Adams, F., 318(111), 329(111), 346
Adler, I., 299(9,11), 308(9,11), 309(11), 313(11), 314(11), 320(116), 321(127,129), 329(11), 339(11,231), 342(276,277), 342, 346, 347, 350, 352, 353
Afonin, V. P., 323(175), 348
Agarwal, M., 409(25), 411
Agus, F., 316, 346
Ahrens, L. H., 341(258), 351
Akselsson, R., 124(2), 128(2), 133(9), 134(9), 138(12), 152(16), 159(9), 161, 162, 256(5), 276, 404(11), 406(19a), 410, 411
Albee, A. L., 184, 202, 327, 349
Al-Derzi, N. W., 321(132), 347
Allen, A. L., 311(76), 345
Allen, J. D., 333(222), 350
Alley, B. J., 290(84), 294, 307(35), 323(164,165), 343, 348
Allison, S. K., 5, 15, 213(26), 215, 223
Al-Saudi, N. A., 321(132), 347
Alvarez, R., 290(78), 294
Andermann, G., 321(136), 333(222), 347, 350

Andersen, C. A., 198(84), 204, 340(255), 351
Anderson, C. H., 326(198), 327(198), 349
Anderson, D. H., 322(151), 331(151), 347
Anderson, J. W., 375(24), 386
Angell, G. R., 310(58), 344
Annell, C. S., 341(270,271), 352
Argov, L., 309(52), 327(52), 331(52), 344
Armbruster, P., 495(18), 496
Armstrong, J. T., 194, 204
Arriens, P. A., 325(191), 333(191,219), 341(219), 349, 350
Asaad, W. N., 23(2), 56
Asley, R. W., 290(71), 293
Atherton, M. P., 339(232), 351
Atkins, L., 312(80), 329(80), 345
Aucott, J. W., 324(181), 349
Austen, C. E., 312(82), 314(82), 321(82), 323(82), 329(82), 345
Averbach, B. L., 213(25), 223
Axlerod, J. M., 299, 308(9), 321(127,129), 342, 347
Azároff, L. V., 216(33), 223

B

Baborovska, J., 192(52), 203
Backerud, L., 290(80,89), 294

Baer, Y., 208(8), 217(8), 221(8), 222
Baglan, R. J., 408(24), 409(24), 411
Bailey, J. C., 341(265), 352
Bailey, N. T., 375(19), 381(19), 386
Baird, A. K., 299(13), 308(13, 45), 309(45), 310(45), 312(13), 325(197), 328(45), 329(13,209), 342(197), 343, 344, 349, 350
Baker, A. D., 208(9), 217(9), 221(9), 222
Ball, D. F., 328(204), 349
Ballard, D. B., 196(77), 204
Bambynek, W., 49(15), 57, 67(10), 68
Bancroft, F., 302(22), 308(22), 312(22), 314(22), 315(22), 327(22), 329(22), 343
Banks, M. S., 122
Barber, C., 321(126), 333(126), 347
Barnhart, M. W., 35(7), 56
Basinger, T. F., 389
Baskin, A. B., 138(12), 161
Bates, S. R., 323(177), 325(177), 348
Baum, J. F., 374(14), 376(14), 385
Baun, W. L., 208(6), 209(14,15, 16), 216(6), 221, 222, 290(49,50), 293
Bayard, M., 172(24), 202
Beaman, D. R., 60, 68, 192, 203
Bean, L., 308(43), 309(43), 310(43), 344
Bearden, J. A., 209(22), 211(22), 223
Bearse, R. C., 138(12), 162, 406(19b), 407(19), 411
Beattie, H. J., 299, 323(5), 326, 342
Becker, W., 290(69), 293
Beeson, M. H., 340(244,245,246), 351
Beitz, L., 395(5), 410
Belcher, C. B., 314(101), 346
Bell, K., 333(220), 350
Bence, A. E., 184, 202, 341(273), 352

Bennett, J. M., 209(21), 223
Bennett, R. B., 409(25), 411
Bennett, R. L., 405(17), 411
Berg, O., 2(3), 15
Bergengren, J., 206, 216(2), 221
Berger, R., 290(70), 293
Bergmark, T., 208(7,8), 217(7,8), 221(7,8), 222
Bergstol, S., 312(77), 326(77), 328(77), 329(77), 345
Berman, S. S., 333(221), 350
Bernstein, F., 302(18), 307(18, 36,38), 308(18), 329, 343, 350, 368, 385
Berry, P. F., 306(30), 307(33), 343, 402(8), 410
Bertin, E. P., 76(2), 110, 226(2), 239, 282, 285, 291, 300(15), 308(15,44), 309(15, 44), 310(15,44), 320(15), 323(15), 324(15), 343, 344
Bertrand, C. C., 375(20), 377(20, 33), 382, 386, 387
Betel, D., 402(8), 410
Bethe, H. A., 187, 202
Betteridge, D., 208(9), 217(9), 221(9), 222
Beyer, J. H., 374(13), 375(13), 385
Bichsel, H., 138(11), 162
Bingham, E. R., 320, 321(120), 347
Bird, R. J., 271(20), 276
Birks, L. S., 8(7), 16, 22(1), 51(17), 56, 57, 73(1), 96(8), 106(1), 110, 124, 161, 164(6), 193(60), 201, 203, 232(5), 233(8), 239, 240, 299, 323(6), 325(6), 342, 373, 385, 404(16), 406(16), 407(16), 411, 492(15), 496
Birnbaum, H. K., 60(1), 68
Bizouard, H., 299, 342
Blodgett, H., 342(277), 353
Blount, C. W., 311(67,69), 331(69), 345, 408(22), 411
Blum, F., 340(254), 351
Boehm, G., 212(24), 213(24), 223
Bojic, M., 313(87), 314(87), 329(87), 345
Bolon, R. B., 165(15), 192(50), 201, 203

AUTHOR INDEX

Bolt, W. A., 389
Boniforti, R., 323(179), 348
Bonissoni, G., 290(63), 293
Booker, G. R., 166(20), 201
Borggren, J., 125(5), 161
Borovskii, I. B., 164(5), 201
Bourdieu, J. M., 313(87), 314(87), 329(87), 345
Bove, J. L., 404(15), 411
Bowie, S. H. U., 390
Bowles, J. A., 487(6), 495
Bramhall, P. S., 122
Bramwell, S. E., 390
Branco, J. J. R., 377(35), 387
Brandle, J. L., 324(187), 349
Brandt, M. P., 312(78), 314(78), 326(78), 329(78), 340(254), 345, 351
Brenner, I. B., 309(52), 327(52), 331(52), 344
Brett, R., 341(268), 352
Brill, A. B., 408(24), 409(24), 411
Brill, M., 290(96), 294
Brill, R. H., 461(7), 481
Brinkerhoff, J. M., 375(26), 386
Brissey, R. M., 299, 323(5), 326, 342
Brooks, C. K., 321(130), 347
Brooks, E. J., 164(6), 201, 290(94), 294
Brown, C. M., 341(264), 352
Brown, D. B., 8(5,6,7), 15, 16, 67(8), 68
Brown, G. C., 324(180), 327(180), 328(180), 331(180), 348
Brown, H., 341(257), 351
Brown, R., 340(256), 351
Brown, R. W., 341(272), 352
Brunner, W. F., 290(46), 293
Brydon, J. E., 313(94), 314(94), 321(94), 327(94), 329(94), 346
Buehler, A., 282, 284, 291
Buffoni, G., 323(179), 348
Bunch, T. E., 341(269), 352
Burhop, E. H. S., 23(2), 56
Burke, K. E., 383, 387
Burkhalter, P. G., 8(5), 15, 22(1), 56, 232(5), 239, 390, 404(16), 406(16), 407(16), 411

Busch, C. E., 138(12), 161
Buseck, P. R., 194, 204

C

Cahill, T. A., 125, 134(3), 161
Camp, D. C., 35(9), 56, 384(43), 388, 389, 404(12), 410
Campbell, J. L., 133, 142, 161
Campbell, W. J., 102(10), 110, 310(56), 311(56), 344, 376(27), 377(34), 378, 384(34), 386, 387, 394(4), 407(20), 408(20a), 410, 411
Campbell, W. T., 253(2), 275
Carl, H. J., 290(86), 294
Carlson, J., 455, 460(6), 481
Carlsson, G., 381, 387
Carmichael, I. S. E., 324(182), 325(182), 331(182), 333(218), 349, 350
Carnevale, A., 290(85), 294
Carr-Brion, K. G., 388, 389, 390, 391
Carron, M. K., 340(256), 341(263, 270,271), 351, 352
Carter, H. V., 271(18), 276
Castaing, R., 164, 175(3), 184, 185(3), 201, 202
Cate, J. L., Jr., 37(10), 56
Caul, H. J., 290(78,86,101), 294, 295
Cave, H. J., 290(81), 294
Cerquiera, M. I., 324(187), 349
Champness, P. E., 341(265), 352
Chan, K. C., 138(14), 162
Channell, R. E., 311(66,67), 345
Chappell, B. W., 308(49), 312(49), 319(49), 323(49), 325(49,191), 329(49), 333(191,219), 341(219), 344, 349, 350
Chmeleff, J., 313(89,93), 314(89, 93), 315(93), 345
Chodos, A. A., 339(234), 341(257), 351, 377(35), 387
Chopra, K. L., 285(23), 292
Chow, E., 290(64), 293
Christian, R. P., 341(263,270, 271), 352
Christiansen, L. J., 290(46), 293

Christie, O. H. J., 312(77), 326(77), 328(77), 329(77), 345
Chung, F. H., 285, 291
Ciccarelli, M. F., 165(15), 201
Cipolla, S. J., 49(14), 57
Claisse, F., 282(7), 285(18), 291, 299(10,106), 307(10, 32,34), 308(10), 313(10), 314(10), 315, 323(161,162), 342, 343, 346, 348
Clanton, U. S., 342(274,275), 352
Clark, B. C., 325(197), 342(197), 349
Clark, N. H., 388
Clarke, W. E., 122
Cliff, G., 192(53), 203
Cline, J. E., 290(47), 293
Close, D. A., 138(12), 162, 406(19b), 407(19), 411, 494(16), 496
Coats, J. S., 324(181), 349
Cocozza, E. P., 290(64,65), 293
Cohen, B. L., 135, 138(10), 162
Colby, J. W., 192(51), 193(61), 195(70), 203, 204
Colella, C., 323(179), 348
Coleman, J. R., 172(25), 198, 202
Compston, W., 325(191), 333(191, 219), 341(219), 349, 350
Compton, A. H., 5, 15, 213(26), 215, 223
Conley, D. K., 192(51), 203
Cook, W. B., 133(8), 142(8), 161
Cooper, H. R., 356, 384
Cooper, J. A., 395(6), 404(12), 410
Cooper, J. W., 25(3), 56
Cooper, T. A., 384(43), 388
Coster, D., 2(2), 15
Cox, R., 376(28), 386
Craig, J. V., 283(11), 290(88), 291, 294
Cranston, R. W., 270(13), 271(13), 276
Creton, F., 290(57,74), 293, 294
Criss, J. W., 52(21), 57, 64(5), 67(12), 68, 73(1), 106(1), 110, 260(6), 276, 299,

[Criss, J. W.] 323(6), 325(6), 342, 373, 385, 402(9), 403(9), 410
Cullen, T. J., 290(91), 294
Cullity, B. D., 279(2), 291
Cuttitta, F., 320(121), 324(121), 333(121), 339(121,239,240), 340(256), 341(263,270,271), 347, 351, 352, 377(32), 387
Czamanske, G. K., 312(85), 329(85), 339(241), 340(247, 248,249,250), 345, 351

D

DaGragnano, V. L., 390
Dalton, J. L., 388
Daniels, A., 290, 292
Darashkevich, V. R., 310(57), 344
D'Auria, J. M., 409(25), 411
Davidson, E., 290(56), 293
Davidson, J. P., 494(16), 496
Davies, R. C., 340(253), 351
Davis, C. K., 495(17), 496
Davis, E. N., 270(9), 271(9), 276
de Broglie, M., 12, 16
Deceuleneer, P., 290(70), 293
deJesus, A. S. M., 388
DeLaeter, J. R., 341(260), 352
De Laffolie, H., 290(51), 293
de la Roche, H., 319(113), 346
Del Grande, N. K., 279(3), 291
Delvaille, J. P., 484(2), 485(2), 487(2), 495
deNeve, R., 318(111), 329(111), 346
Desborough, G. A., 340(250), 351
Descamps, J., 184, 202
Despujols, J., 299, 342
DeVries, J. L., 76(3), 102(11), 110, 288(29), 290(29), 292, 390
Dewolfs, R., 318(111), 329(111), 346
Dick, J. G., 290(53), 293, 326(199), 333(199), 349
Dobner, W., 290(69), 293
Dodge, F. C. W., 339(238), 351
Doe, B. R., 339(234), 351

Dolbey, W., 122
Dolby, R. M., 165(11,12), 201
Donderer, E., 321(135), 347
Doughman, W. R., 404(10), 410
Drummond, C. H., 314, 346
Drumond, W. E., 232(6), 239
Dryer, H. T., 333(223), 350
Dubois, R. J., 311(71), 345
Duffield, W. A., 340(246), 351
Duncumb, P., 165(9,10), 188, 201, 202
Dunham, A. C., 341(265), 352
Duzevic, D., 290(54), 293
Dwiggins, C. W., 270(15), 276
Dwornik, E. J., 341(263,270,271), 352
Dyer, G. R., 408(24), 409(24), 411
Dzubay, T. G., 35(8), 52(20), 56, 57

E

Eby, G. N., 311(63), 333(63), 344
Echlin, P., 173(26,27), 202
Eddy, C. E., 394, 410
Eiband, A. M., 487(6), 495
Eick, J. D., 290(81), 294
Eldad, H., 309(52), 327(52), 331(52), 344
Eller, E., 342(277), 353
Ellis, W. K., 390
Elsheimer, H. N., 301(16), 302(19), 306(19), 308(16,19), 313(19), 314(19), 324(16,19), 327(19), 331(19), 343
Emeleus, C. H., 341(264), 352
Engel, C. G., 377(35), 387
Engels, J. C., 306(29), 322(29), 343
Ereiser, J. A., 329(207), 349
Erlank, A. J., 341(258), 351
Ermakov, A. N., 311(64), 344
Espos, L. F., 301(16), 308(16), 310(54), 324(16,185), 331(185), 343, 344, 349
Esson, J., 324(180), 327(180), 328(180), 331(180), 341(265), 348, 352

Evans, N., 270(13), 271(13), 276
Everhart, T. E., 165(18), 201

F

Fabbi, B. P., 301(16), 302(19, 20,21), 306(19), 307(39), 308(16,19,20,21,39,42,50,51), 309(42,53), 310(42,54), 313(19,21,53), 314(19,53), 316(50), 320(50,51), 324(16, 19,183,184,185), 326(50,51), 327(19,53), 328(51), 329(21, 53,210,213), 331(19,51,184, 185), 339(238), 343, 344, 349, 350, 351
Fabian, D. J., 216(29), 223
Fadley, C. S., 208(11), 221(11), 222
Faessler, A., 208(5,12), 210, 211(23), 212(24), 213(24), 216(5,30), 221, 222, 223
Fahlman, A., 208(7), 217(7), 221(7), 222
Fairbairn, H. P., 321(140), 322(140), 325(140), 333(140), 347
Fatemi, M., 8(6,7), 16
Faure, S., 322(152), 347
Faye, G. H., 322(144), 347
Fechtig, H., 194(68), 204
Fergason, L., 290(48), 293
Ferguson, A., 290(65), 293
Field, K. M., 122
Fink, R. W., 41(13), 49(16), 56, 57
Fiori, C. E., 62(4), 68, 195(74), 204
Fischer, D. W., 209(14,15), 222, 290(49,50,82), 293, 294
Fitzgerald, R., 165(14), 176(14), 201
Fitzsimons, F. J., 314(97), 346
Flanagan, F. J., 299(11), 308(11), 309(11), 313(11), 314(11), 322(141,142,145), 329(11), 339(11), 342, 347
Flitsch, R., 290(78), 294
Florestan, J., 286, 290(24), 292

Flynn, A. R., 290(4), 292
Folkmann, F., 125, 161
Fookes, R. A., 365(3), 374(18), 375(3,22), 376(3,18), 384, 386, 390
Forster, J. W., 331(214), 350, 365(5), 374(5,12), 377(12, 36), 379(36), 385, 387, 391
Franks, A., 209(21), 223
Franzini, M., 320(124), 323(124), 325(124), 331(124), 347
Fraser, A. R., 290(53), 293
French, R. O., 374(14), 376(14), 385
Freund, H. V., 49(16), 57
Friedman, H., 484(1,3), 495
Fuiii, T., 384(42), 388
Fukasawa, T., 384, 388
Fuller, N. L., 376(31), 387
Furata, T., 307(33), 343, 390, 402(8), 410
Fyffe, W. S., 341(265), 352

G

Gaarde, C., 125(5), 161
Gacesa, T., 290(54), 293
Gallagber, W. J., 49(14), 57
Gardner, R. P., 402(8), 410
Garrett, R. B., 49(22), 52(22), 57, 405(18), 411
Gast, P. W., 341(266), 352
Gates, O. R., 290(94), 294
Gatinskaya, N. G., 311(64), 344
Gedcke, D., 285(19), 291
Gehrke, R. J., 340(253), 351
Gelius, U., 208(8), 217(8), 221(8), 222
Geninasca, R., 314(98), 326(98), 329(98), 346
Gerald, J., 342(277), 353
Gerrard, P. W., 331(216), 350
Giacconi, R., 488(8), 496
Giauque, R. D., 35, 49, 52(22), 56, 57, 285, 291, 310(61), 311(61), 331(61), 344, 388, 404(13), 405(18), 410, 411

Giles, M. A., 175(30), 202
Gilfrich, J. V., 8(5), 15, 22(1), 56, 67(8), 68, 232, 233(8), 239, 240, 290(44), 292, 404(16), 406(16), 407(16), 411, 492(15), 496
Gilliam, E., 290(73), 294
Gillieson, A. H., 322, 347, 388, 389
Gisclard, J. B., 404(10), 410
Glade, G. H., 282(8), 283(8), 291
Glenn, D. C-J., 389
Goda, L. Y., 49(22), 52(22), 57, 405(18), 411
Godijn, E., 339(234), 351
Goehring, M., 208(12), 222
Goff, F. E., 340(250), 351
Goldberg, E. D., 321(128), 347
Goldsborough, J., 444(4), 481
Goldstein, J. I., 195(70), 204
Goodman, R. J., 324(186), 325(186), 331(186), 349
Gordon, G. M., 375(25), 377(25), 386
Goto, H., 290(55,67,72), 293
Gould, R. W., 285(19), 291, 323(177), 325(177), 348
Goulding, F. S., 35(6), 54(23), 56, 57, 160(17), 162, 232(7), 240, 310(61), 311(61), 331(61), 344, 404(13), 410
Govindaraju, K., 310(62), 311(62), 344
Graf von Harrach, H., 166(20), 201
Gravitis, V. L., 365(3), 374(18), 375(3,22), 376(3,18), 384, 386, 390
Green, T. E., 102(10), 110, 253(2), 275, 310(56), 311(56), 344, 377(34), 384, 387, 394(4), 410
Green, T. F., 290(42), 292
Greenberg, J. S., 495(17), 496
Greenland, L. P., 341(270,271), 352
Gregor, P. R., 329(207), 349
Griffiths, J. M., 290(76), 294
Grodjian, D. S. M., 305(27), 343
Groves, D. I., 339(235), 351

Grubb, W. T., 253(1), 275
Guill, S. M., 311(69), 311(69), 345, 408(22), 411
Gulacar, O. F., 311(73), 331(73), 345
Gulson, B. L., 316(227), 350
Gunicheva, T. N., 323(175), 348
Gunn, B. M., 308(48), 321(134), 331(134), 344, 347
Gunn, E. L., 271(17), 276, 307(37), 310(60), 320(117), 323(117), 343, 344, 346
Gursky, H., 488(8), 496
Gutknecht, W. F., 62(2), 68, 127(6), 133(6), 136(6), 161
Guy, A. G., 280, 291
Gwozdz, R., 323(173), 348

H

Haas, D. J., 226(3), 234(3), 235(3,10), 237(3), 238, 239, 239, 240
Hadding, A., 298, 342
Haeda, F., 290(72), 293
Hagström, S. B. M., 208(11), 221(11), 222
Haine, M. E., 164(7), 201
Hakkila, E. A., 290(66,92), 293, 294
Hall, R. A., 375(24), 386
Hall, T. A., 173(27), 193(55,56,57), 202, 203
Hall, W. E., 340(248), 351
Hammerle, R. H., 388
Hamrin, K., 208(7,8), 217(7,8), 221(7,8), 222
Hansen, J. S., 41(13), 49(16), 56, 57
Hansen, M., 280, 291
Hanser, F. A., 375(26), 386
Hanson, V. F., 420(2), 481
Harmon, R. S., 341(268), 352
Harper, T., 54(25), 57
Harris, D. C., 340(251), 351
Harvey, P. K., 302(22), 308(22), 312(22), 314(22), 315, 324(181), 327(22), 329(22), 343, 349

Hattori, H., 313(91), 314(91), 315, 328(91), 329(91), 345
Havens, R. G., 319(112), 346
Hayasaka, T., 290(99), 295
Haycock, R. F., 270(12), 271(12), 276
Haylett, D. W., 122
Heal, H. T., 290(73), 294
Hebert, A. J., 299, 309(4), 314(4), 315, 327(4), 329, 342, 407(21), 411
Hedén, P. F., 208(8), 217(8), 221(8), 222
Hedman, J., 208(7,8), 217(7,8), 221(7,8), 222
Heinrich, K. F. J., 62(4), 67(13,14), 68, 174(29), 175(30), 180(33), 182(34), 186(39), 187(29,41), 189(39,45,46), 192(39,46), 194(65), 195(72,75), 202, 203, 204, 299, 323(7), 327, 328(229,230), 342, 350, 373, 385
Helava, H., 487(7), 496
Helin, E., 341(257), 351
Helz, A. W., 341(270), 352
Hendry, G. L., 324(181), 349, 375(19), 381(19), 386
Hendry, R. D., 302(22), 308(22), 312(22), 314(22), 315(22), 327(22), 329(22), 343
Henke, B. L., 165(13), 201, 209(13), 216, 222, 299, 329(3,208,209), 342, 350
Hénoc, J., 186(39), 189(39,47), 192(39), 202, 203, 328(229), 350
Herglotz, H. K., 208(4,10), 209(19), 214(19), 215(19), 216(4), 217(10), 218(31), 220(31), 221(10), 221, 222
Herman, A. W., 133(8), 142(8), 161
Herrmann, H., 313(93), 314(93), 315(93), 345
Hesp, W. R., 316, 346
Heymann, D., 341(261), 352
Hicho, G. E., 288(31), 292
Hillier, J., 164(2), 175(2), 201
Hinckfuss, D. A., 365(3,4), 375(3,23), 376(3), 384, 386

Hinthorne, J. R., 340(255), 351
Hirokawa, K., 290(52,72), 293
Hirt, R. C., 404(10), 410
Hisano, K., 339(237), 351
Hoffmann, H. J., 194(68), 204
Hohling, H. J., 197(82), 204
Holden, G. R., 323(174), 348
Holland, J. G., 341(264), 352
Holliday, J. E., 209(17), 215(17), 222
Holloway, S. M., 327(200), 328(200), 349
Holt, D. B., 171(23), 202
Hooper, P. R., 312(80), 329(80), 345
Hooser, G. E., 290(87), 294
Hopper, F. N., 323(167), 327(167), 348
Hormann, P. K., 328(206), 339(206), 349
Howard, R. W., 404(11), 406(11b), 407(11b), 410
Howarth, R. J., 324(181), 349
Howarth, W. J., 324(181), 349, 375(21), 376(21), 386, 388, 389
Howell, R. H., 37(10), 56
Hower, J., 312(85), 320, 325, 329(85), 345, 347
Hubbard, N. J., 341(266), 352
Hubbel, J. H., 279(3), 291
Hubbell, J. H., 26(4), 56
Hughes, D. J., 324(180), 327(180), 328(180), 331(180), 348
Humpel, J., 324(182), 325(182), 331(182), 349
Hunt, E. C. R., 290(75), 294
Hunter, C. B., 51(19), 57, 311(74,75), 345, 401(7), 402(8), 404(7), 410
Hunter, D. R., 339(236), 351
Hurley, H., 290(92), 294
Hurley, J. P., 390
Hurley, P. M., 321(140), 322(140), 325(140), 333(140), 347
Hurley, P. W., 276, 283(10), 291
Hutton, C. O., 304(24), 343
Hutton, J. T., 299, 308(12), 312(12,83), 313(12), 314(12, 83), 315, 325(12,83), 327(12), 329(83), 342, 345

Huus, T., 125(5), 161

I

Ilin, N. P., 164(5), 201, 315, 346
Ingamells, C. O., 306(28,29), 314(100), 322(28,29,151), 331(151), 343, 346, 347
Inouye, T., 54(25), 57
Isari, J. A., 60(1), 68, 192, 203

J

Jack, R. N., 324(182), 325(182), 331(182), 349
Jackson, E. D., 340(244,245), 351
Jackson, P. F. S., 322(153), 348
Jacobs, M. H., 192(52), 203
Jakes, P., 316 (228), 350
Jaklevic, J. M., 35(6,8), 49(22), 52(22), 54(23), 56, 57, 160(17), 162, 232(7), 240, 285(21), 291, 310(61), 311(61), 331(61), 344, 388, 404(13), 405(18), 410, 411
Jarrett, B. V., 35(8) 56
Jecko, G., 313(87), 314(87), 329(87), 345
Jeldgaard, A. K., 125(5), 161
Jenkins, R., 76(3), 102(11), 110, 226(3), 233(9), 234, 235(3, 10), 237(3), 238, 239, 239, 240, 270(16), 271(19), 272(22), 275(22), 276, 283(10), 285(19), 288(29), 290(29), 291, 292, 390
Jenkinson, D. A., 391
Johann, H. H., 12, 16, 165, 201
Johansen, O. O., 122, 322(154), 348
Johansson, G., 208(7,8), 217(7, 8), 221(7,8), 222
Johansson, S. A. E., 124(2), 128(2), 161, 256(5), 276, 404(11), 406(19a), 410, 411

AUTHOR INDEX

Johansson, T. B., 12, 16, 124, 128(2), 133, 134(9), 138(12), 159(9), 161, 256(5), 276, 404(11), 406(19a), 410, 411
Johnson, J. L., 270(14), 271(14), 276
Johnson, W. C., 221(36), 223
Jolly, R. K., 142, 162
Jones, R. W., 290(71), 293
Joshi, A., 221(36), 223
Joy, D. C., 166(20), 201
Joyce, J. M., 62(2), 68, 127(6), 133(6), 136(6), 161
Jundt, F. C., 125(4), 161

K

Karlsson, S. E., 208(7), 217(7), 221(7), 222
Kasemsanta, S. P., 390
Kaufman, L., 35(9), 56
Kaufmann, H. C., 152(16), 162
Kaufmann, R., 173(27), 202
Kaye, M. J., 320(123), 325(123), 331(123), 347
Kehl, G. L., 283(17), 291
Keil, K., 183(36), 198(84), 202, 204, 325(197), 339(242), 340(242,243), 341(269), 342(197), 349, 351, 352
Kemp, A., 324(181), 349
Kemp, J. W., 321(136), 347
Kemp, K., 125(5), 161
Kemper, M. A., 283, 290(9), 291
Kennett, T. J., 55(26), 57
Keponen, M., 290(83), 294
Kerr Del Grande, N., 26(4), 56
Kessler, H., 290(43), 292
Kessler, J. E., 404(15), 411
Khan, J. M., 290(46), 293
Kharchenko, A. M., 323(175), 348
Kienberger, C. A., 290(41), 292
Kiessling, R., 197, 204
Kikkert, J. N., 388
Kilday, B. E., 283, 291
Kim, Y. S., 339(233), 351
Kindbert, B., 208(7), 217(7), 221(7), 222
Kinnunen, J., 290(83), 294

Kissin, S. A., 341(262), 352
Knapp, K. T., 405(17), 411
Knoke, D. R., 376(30), 380, 387
Kodoma, H., 313(94), 314(94), 321(94), 327(94), 346
Kometani, T. Y., 404(15), 411
Kraeft, U., 314(103,104), 333(104), 346
Kreula, S., 389
Kriege, O. H., 290(93), 294
Krieger, A. S., 485(4), 487(4), 495
Kubo, H., 125(4), 161
Kulin, S. A., 213(25), 223
Kunzendorf, H., 331(215), 350
Kurtz, A. D., 213(25), 223
Kyser, D. F., 193(64), 198(64), 204

L

Laby, T. H., 394, 410
Lachance, G. R., 274(21), 276, 299, 323(158,159,160), 326, 348, 373, 385
Ladle, G. H., 342(274,275), 352
Lamothe, R., 342(277), 353
Landis, D. A., 56, 160(17), 162, 232(7), 240
Lange, N., 197, 204
Langheinrich, A. P., 331(214), 350, 365(5), 374(5,12,13,14), 375(13), 376(14), 377(12,36), 379(36), 385, 387, 391
Larribau, E. P., 300(14), 329(14), 343
Larson, R. R., 339(241), 341(263), 351, 352
Läuchli, A., 197, 204
Laughon, R. B., 342(275), 352
Law, S. L., 253(2), 275, 377(34), 384(34), 387
Lawrence, D., 122
Leake, B. E., 324(181), 349
Lee, J. D., 219(32), 223
Leitner, J. W., 326(198), 327(198), 349
Leltouillier, R., 313(92), 315(92), 345

Lentz, A. J., 285(22) 291
Leoni, L., 320(124,125), 323(124, 125), 324(125), 325(124,125, 188), 328(125), 331(124,125), 347, 349
Lewis, C. L., 290(58,59), 293, 375(24), 386
Lewis, G. J., 321(128), 347
Lewis, R., 60(1), 68
Leyden, D. E., 253(3), 275, 311(66,67,68,69), 331(69), 345, 408(22), 411
Lichtinger, R. W., 35(7), 56
Lidstrom, L. J., 391
Liebhafsky, H. A., 84(6), 87(7), 99(7,9), 102(7), 106(7), 107(7), 108(7), 110, 310(59), 320(118), 323(118), 344, 346
Lifshin, E., 165(15), 192(50), 201, 203
Lignon, D. T., 341(263,270,271), 352
Lin, I. J., 305(27), 343
Lincoln, A. J., 290(85), 294
Lindemann, F. A., 12(10), 16
Lindgren, I., 208(7), 217(7), 221(7), 222
Lindgren, J. L., 401(7), 404(7), 410
Lindsay, J. R., 341(263), 352
Lindsey, K., 209(21), 223
Linn, T. A., Jr., 374(12), 375(20), 377(12,20,33,36), 379(36), 382, 385, 386, 387
Loch, G., 301(17), 343
Lochte-Holtgreven, W., 489(12), 496
Loftus-Hills, G. D., 339(235), 351
Longobucco, R. J., 308(44), 309(44), 310(44), 313(95), 344, 346
Loo, B. W., 54(23), 57
Loomis, T. C., 404(15), 411
Lorimer, G. W., 192(53), 203
Loseva, L. E., 315, 346
Louis, R., 270(11), 271(11), 276, 407(20), 408(20c), 411
Lovering, J. F., 316(227), 350
Lowman, P., 342(277), 353

Lubozynski, M. F., 408(24), 409(24), 411
Lucas-Tooth, H. J., 122, 323(163), 327, 348, 373, 385
Lucas-Tooth, H. S., 106, 110
Luke, C. L., 253(4), 275, 290(97,100), 294, 295, 311(70), 345, 383, 387, 394(3), 404(15), 410, 411
Lund, P. K., 408(23), 411
Lundgren, F. O., 391
Lunel, T., 324(181), 349
Luschow, H. M., 122
Lynch, D. R., 218(31), 220(31), 223
Lytle, F. W., 216(34), 223

M

Maassen, G., 383, 388
Machaček, V., 321(131), 331(131), 347
MacHenzie, W. S., 341(265), 352
Madden, M. L., 374(17), 379, 385
Madlem, K. W., 299(13), 306(31), 307(31), 308(13), 312(13), 329(13), 343
Maeda, F., 290(99), 295
Magnusson, T., 492(13), 496
Malanify, J. J., 406(19b), 407(19), 411, 494(16), 496
Malinify, J. J., 138(12), 162
Mallet, J. H., 26(4), 56
Mallett, J. H., 279(3), 291
Malone, D. F., 49(22), 52(22), 57, 405(18), 411
Malyukov, B. A., 310(57), 344
Mamuso, T., 194(67), 204
Mander, J. E., 326(198), 327(198), 349
Manne, R., 208(8), 217(8), 221(8), 222
Manners, V. J., 283, 290(88), 291, 294
Manuszewski, R. C., 198(83), 204
Mapper, D., 341(267), 352
Marchant, G. R., 374(15), 378(37), 385, 387

Marr, H. E., III, 376(27), 378, 386
Marsh, R. H., 388
Maschin, B., 290(57), 293
Mathies, J. C., 408(23), 411
Mathiesen, J. M., 311(72), 345
Mathieson, J. M., 390
Matocha, C. K., 299(105), 308(40), 310(40), 314(105), 315(105), 344, 346
Matthews, F. W. H., 270(13), 271(13), 276
Mattson, R. A., 329, 350
Maurice, F., 189(47), 203
McGarry, P. E., 376(31), 387
McGeorge, J. C., 41(13), 56
McIntyre, D. B., 299(13), 308(13), 312(13), 329(13), 343
McKaveney, J. P., 290(61,98), 293, 294
McKay, D. S., 342(274,275), 352
McKee, K. H., 290(60), 293
McKinney, C. N., 327(201), 333(201), 349
McMaster, W. H., 26(4), 56, 67(9), 68, 279, 291
McNelles, L. A., 133(8), 142(8), 161
McNely, D. J., 375(25), 377(25), 386
Mero, J. L., 375(25), 377(25), 386
Meyer, C., 341(268), 352
Michaelis, R. E., 195(71,72,73, 74,75,76), 204, 283(12), 288(28.31), 291, 292
Mihura, J., 300(14), 329(14), 343
Millard, R. C., 312(85), 329(85), 345
Mitchell, B. J., 323(166,167, 176), 327(166,167), 348
Mitchell, R., 487(7), 496
Mitchell, R. J., 388
Mizuike, A., 384(42), 388
Mokler, P. H., 495(18), 496
Montgomery, C., 455, 460, 481
Moore, G. A., 195(71), 204
Moore, W. J., 307(39), 308(39), 343

Mori, T., 316(228), 350
Morteani, G., 328(206), 339(206), 349
Moschen, B., 290(74), 294
Moseley, H. G. J., 1, 2(1), 15, 394(1), 410
Muir, M. D., 171(23), 202
Muller, L. D., 304(25), 343
Muller, R. O., 51(18), 57, 290, 292
Mulligan, B. W., 290(79,101), 294, 295
Mulvey, T., 164(7), 201
Murata, K., 193(64), 198(64), 204
Murata, M., 311(65), 344
Myers, A. T., 319(112), 331(217), 346, 350
Myers, R. H., 290(84), 294, 307(35), 323(164,165), 343, 348
Myklebust, R. L., 62(4), 68, 186(39), 189(39,46), 192(39, 46), 195(72), 202, 203, 204, 328(229,230), 350

N

Nadalin, R. J., 290(93), 294
Nadiv, S., 305(27), 343
Nagaoka, M., 316(228), 350
Nagel, D. J., 208(6), 216(6), 221, 489(11), 490(11), 492(11), 496
Naggar, M. H., 339(232), 351
Nakajima, K., 193(62), 203
Nathanson, B., 404(15), 411
Nelson, I. V., 326(195), 349
Nelson, J. W., 133(9), 134(9), 159(9), 161
Nelson, R. O., 52(20), 57
Nettles, P. H., 138(12), 161
Newbury, D. E., 193(63), 194(65), 203, 204
Newbury, M. L., 308(47), 320(47), 324(47), 326(47), 331(47), 344
Nguyen, A. D., 326(199), 333(199), 349

Nichiporuk, W., 341(257), 351
Nichols, I. A., 316, 329(109), 346
Nicholson, J. B., 209(20), 222
Nicholson, W. A. P., 197(82), 204
Nicolas, J., 328(203), 349
Niles, W. W., 319(112), 346
Nix, D., 41(13), 56
Nixon, W. C., 165(19), 201
Noddack, W., 2(3), 15
Noguchi, M., 311(65), 344
Nordberg, R., 208(7), 217(7), 221(7), 222
Nordling, C., 208(7,8), 217(7, 8), 221(7,8), 222
Norrish, K., 299, 308(12,49), 312(12,49,83), 313(12), 314(12,83), 315, 320(49), 323(49), 325(12,49,83,190), 327(12), 329(49,83), 342, 344, 345, 349
Numata, M., 333(224), 350

O

Oatley, C. W., 165, 201
Oda, Y., 193(62), 203
Ogilvie, R. E., 174(28), 184, 202
Ohata, M., 290(68), 293
O'Neill, M. J., 314(97), 346
O'Nions, R. K., 322(192), 325(192), 333(192), 349
Orr, B. H., 133(8), 142(8), 161
Ott, W. L., 290(59), 293
Ottoline, A. C., 270(14), 271(14), 276
Owen, L. B., 322(152), 347
Oyama, K., 339(237), 351

P

Paganelli, M., 290(63), 293
Pankhurst, R. J., 322(192), 325(192), 333(192), 349
Papouchado, K., 415(1), 481

Parker, A., 312(84), 315(84), 321(133), 331(133), 345, 347
Patrick, T. J., 487(6), 495
Pease, R. F. W., 165(19), 201
Peck, L. C., 303(23), 305(23), 343
Peckerar, M. C., 67(8), 68, 407(20), 408(20b), 411
Pehl, R. H., 35(6), 56, 310(61), 311(61), 331(61), 344, 404(13), 410
Pellett, J. R., 376(31), 387
Perlman, M. L., 217(35), 223
Peterman, Z. E., 333(218), 350
Peters, P. D., 193(56), 203
Peterson, L. E., 487(9), 488(9), 496
Petrasso, R., 485(4), 487(4), 495
Pfeiffer, H. G., 84(6), 87(7), 99(7,9), 102(7), 106(7), 107(7), 108(7), 110, 320(118), 346
Pherson, F. L., 374(13), 375(13), 385
Philibert, J., 188(43), 189, 202, 203
Phillips, G. C., 395(5), 410
Phillips, R., 341(264), 352
Pickard, E. D., 122
Pickles, W. L., 37(10), 56
Piskunova, L. F., 323(175), 348
Plant, A. G., 324(181), 349
Plesch, R., 333(225), 350
Pluchery, M., 290(77), 294
Poole, A. B., 122, 327(200), 328(200), 349
Porsche, F. W., 270(10), 271(10), 276
Porter, D. E., 404(14), 411
Post, H. R., 282(8), 283(8), 291
Powell, J. L., 333(220), 350
Pradzynski, A. H., 384(44), 388, 390, 401(7), 404(7,12), 410
Prestwich, W. V., 55(26), 57
Price, B. J., 122
Price, N. B., 310(58), 344
Prinz, M., 341(269), 352
Purdham, J. T., 395(5), 410
Purser, K. H., 125(4), 161
Pyne, C., 323(163), 327, 348,

[Pyne, C.]
  373, 385
Pyne, E. C., 106, 110

Q

Quintin, M., 323(161), 328(203), 348, 349

R

Rammernsee, H. Z., 290(43), 292
Randle, K., 322(155), 348
Rasberry, S. D., 67(13,14), 68, 195(72), 204, 226, 239, 288(31), 290(79,81,86,101), 292, 294, 295, 299, 323(7), 327, 342, 373, 385
Rasmussen, N. C., 54(25), 57
Rautavalta, P., 290(83), 294
Rawling, B. S., 375(23), 386
Ray, L., 327, 349
Reed, D. J., 388, 389
Reed, S. J. B., 176(31), 178(31), 188, 189, 202, 203, 349(252), 341(259), 351, 352
Regler, F., 489(10), 496
Reid, A. M., 341(268,272), 352
Reidy, W. P., 485(5), 487(5), 495
Rendic, D., 395(5), 410
Rengan, K., 388
Reymont, T. M., 311(71), 345
Reynolds, R. C., 321(137), 324, 347
Rhines, F. N., 280, 281(5), 291
Rhodes, J. R., 32(5), 51(19), 56, 57, 307(33), 311(74,75), 343, 345, 374(11), 375(11), 376(11), 377(11), 384(43,44), 385, 388, 390, 391, 401(7), 402(8), 404(7,12), 410
Riccardi, R., 323(179), 348
Richter, P., 313(86), 314(86), 329(86), 345
Riding, J. R., 378(37), 387
Ridley, W. I., 341(272), 352

Rittmayer, G., 212(24), 213(24), 223
Riviere, J. C., 221(37), 223
Robertson, A., 55(26), 57
Roering, C., 299, 342
Rose, A. W., 339(234), 351
Rose, C. V., 311(76), 345
Rose, H. J., Jr., 299(11), 308(11), 309(11), 313(11), 314(11), 315(107), 320(119, 121), 324(121), 325(197), 329(11), 333(121), 339(11, 121,239,240), 340(256), 341(263,270,271), 342(197), 342, 346, 347, 349, 351, 352, 377(32), 387
Rosenberg, A. S., 327(201), 333(201), 349
Ross, D. C., 339(238), 351
Rotter, R., 390
Rouseau, R., 323(162), 348
Rowe, E. M., 217(35), 223
Rudd, P., 122
Rudolph, J. S., 290(93), 294
Ruff, A. W., Jr., 195(76), 197(78), 204
Russ, J. C., 35(7), 56, 62(3), 68, 193(54,58), 203, 333(226), 350
Russell, B. G., 312(82), 314(82), 321(82), 323(82), 329(82), 345
Russell, D. S., 333(221), 350
Ryabukhin, V. A., 311(64), 344

S

Sahores, J. J., 300(14), 329(14), 343
Saito, A., 290(52,55,67,72), 293
Saitta, M., 320(124,125), 323(124,125), 324(125), d25(124,125,188), 328(125), 331(124,125), 347, 349
Salmon, M. L., 325, 349
Samson, C., 282(7), 291, 307(32), 343
Samuels, L. E., 282, 291
Sandborg, A. O., 35(7), 56

Saubermann, A., 173(26), 202
Scarborough, J. B., 84(5), 95(5), 110
Schaefer, K., 122
Schamadabek, R., 342(277), 353
Schenk, E. A., 125(4), 161
Schiel, E., 209(19), 214(19), 215(19), 222
Schmitt-Ott, W. D., 41(13), 56
Schneider, H., 290(39), 292
Schnopper, H. W., 484(2), 485(2), 487(2), 495
Schonwals, D., 290(39), 292
Schreiber, T. P., 270(14), 271(14), 276
Schumann, H., 290(39), 292
Schwarts, S., 290(47), 293
Schwartz, H. P., 341(262), 352
Scott, F. H., 283(11), 290(88), 291, 294
Scott, R. W., 285(22), 291
Scott, S. D., 341(262), 352
Scribner, B. F., 290(79,101), 294, 295
Seaman, G. G., 131, 161
Seebold, R. E., 193(60), 203
Sellers, B., 375(26), 386
Semeniuk, P., 333(221), 350
Shabason, L., 135, 138(10), 162
Shafroth, S. M., 138(12), 161
Shane, K. C., 131, 161
Sheather, P. H., 487(6), 495
Sherman, J., 67(7), 68, 323(156, 157), 348
Shrimizu, I., 290(68), 293
Sieberg, R. D., 390
Siegbahn, K., 208(7,8), 217(7, 8), 221(7,8), 222
Siegbahn, M., 208(3), 209(18), 213(18), 215(18,28), 221, 222, 223, 492(13), 496
Silk, J. K., 485(4), 487(4), 495
Sine, N. M., 290(58,59), 293
Sisefsky, J., 194(66), 204
Sjösström, M., 197, 204
Sjurseth, J. R., 404(11), 406(11b), 407(11b), 410
Smales, A. A., 341(267), 352
Smallbone, A. H., 391
Smith, A. L., 333(218), 350

Smith, D. L., 290(81), 294
Smith, K. C. A., 165, 201
Smith, T. K., 312(81), 314(81), 345
Snyder, R. L., 164(2), 175(2), 201
Soderquist, C. E., 35(7), 56
Solomon, M., 339(235), 351
Solt, M. W., 331(217), 350
Spano, E. F., 102(10), 110, 290(42), 292, 310(56), 311(56), 344, 394(4), 410
Sparks, C. J., 286, 292
Speer, R. J., 209(21), 223
Spencer, J. P., 166(20), 201
Spitz, H. D., 308(41), 344
Stacey, G. S., 389
Stein, D. F., 221(36), 223
Stein, H. J., 495(18), 496
Steinnes, E., 322(154), 348
Stephenson, D. A., 313(90), 314(90), 315(90), 321(90), 323(168), 326(168,193), 345, 348, 349
Stern, W. B., 313(88), 314(88), 329(88), 345
Stewart, J. E., 232(6), 239
Stone, B. C., 313(94), 314(94), 321(94), 327(94), 329(94), 346
Stoner, G. A., 290(40), 292
Stoops, R. F., 290(60), 293
Strasheim, A., 312(78), 314(78), 326(78), 329(78), 345
Strasz, O. P., 395(5), 410
Strausz, K. I., 395(5), 410
Street, K., Jr., 299, 309(4), 314(4), 315(4), 327(4), 329, 342, 407(21), 411
Strelow, F. W. E., 322(153), 348
Stroganova, N. S., 311(64), 344
Stump, I. G., 409(25), 411
Stump, N. W., 365(3,4), 375(3), 376(3), 384, 389
Stumpel, E. F., 341(265), 352
Suchan, H. L., 208(10), 217(10), 221(10), 222
Suhr, N. H., 322(151), 331(151), 347
Sullivan, D. C., 290(44), 292
Sundkvist, G. J., 391

Sutarno, R., 322(144), 347
Sweatman, T. R., 376(29), 387
Sweeney, W. E., 193(60), 203
Swinne, R., 206, 221
Switzer, P., 306(28,29), 322(28, 29), 343

T

Tacke, I., 2(3), 15
Tan, F. C., 322(151), 331(151), 347
Taylor, B. L., 290(95), 294
Taylor, D. M., 302(22), 308(22), 312(22), 314(22), 315(22), 327(22), 329(22), 343
Taylor, G. J., 341(261), 352
Taylor, R. M., 325(190), 349
Tefa, O., 290(90), 294
Tertian, R., 285(20), 291, 314(98,99), 323(99,169,170, 171,172,178), 326(98,99,169, 170,171,172,194), 329(98), 346, 348, 349
Tester, M. A., 209(13), 222
Thatcher, J. W., 407(20), 408(20a), 411
Thomas, T. L., 311(69), 331(69), 345, 408(22), 411
Thomas, W. W., 341(260), 352
Thomsen, J. S., 213(27), 223
Thomson, J. A., 133(8), 142(8), 161
Thorne, G. N., 272(22), 275(22), 276
Thornley, R. F. M., 165(18), 201
Timm, W. H. A., 375(24), 386
Tixier, R., 188(43), 202
Toft, R. W., 270, 271(20), 276
Togel, K., 310(55), 344
Tolmie, R. W., 374(16), 385
Toong, K. S., 376(29), 387
Toulmin, P., III, 325(197), 342(197), 349
Townsend, J. E., 346
Traill, R. J., 274(21), 276, 299, 323(158,159,160), 326, 348, 373, 385
Trombka, J. I., 342(276,277), 352, 353

Tsukamoto, A., 290(68), 293
Tuddenham, W. M., 365(5), 374(5,13), 375(13), 385
Turmel, S., 313(92), 315(92), 345

U

Uchikaw, H., 333(224), 350
Ukrainskii, Yu. M., 310(57), 344
Umbarger, C. J., 138(12), 161, 406(19b), 407(19), 411
Unus, I., 41(13), 56
Urbanoski, L., 374(16), 385
Urdy, C. E., 326(195), 349

V

Vaiana, G. S., 485(4), 487(4), 495
Valkovic, V., 395(5), 410
Van Greiken, R. E., 133(9), 134(9), 159(9), 161
Van Lehn, A. L., 404(12), 410
Vanninen, P., 389
van Nordstrand, R. A., 270(9), 271(9), 276
Vassilaros, G. L., 290(61,98), 293, 294
Vaughn, R. W., 374(14), 376(14), 385
Veigele, W. J., 138(13), 162
Verghese, K., 402(8), 410
Verheijke, M. L., 64(6), 68
Vernon, M. J., 325(191), 333(191, 219), 341(219), 349, 350
Vieth, D. L., 182(34), 195(73, 75), 202, 204
Vincent, H. A., 308(51), 320(51), 326(51), 328(51), 331(51), 344
Vincent, P., 495(17), 496
Violet, C. E., 492(14), 496
Volborth, A. V., 303(79), 304(26), 307(26), 308(26,46, 50,51), 309(46), 310(46), 316(26,50), 320(50,51), 326(50,51), 328(26,46,51,205),

[Volborth, A. V.]
 329(210,213), 331(51),
 343, 344, 345, 349, 350
Volkov, N. G., 54(24), 57
Volrath, J. D., 321(138),
 331(138), 347
von Ardenne, M., 164, 201
von Hevesy, G., 2(2), 15, 278,
 291

W

Wagman, J., 405(17), 411
Wagner, F., 290(45,62), 293
Wahlberg, J. S., 331(217), 350
Waldron, H. F., 376(30),
 380(30), 387
Walsh, C. J., 35(7), 56
Walter, R. L., 62(2), 68, 127,
 133, 136(6), 161
Ware, N. G., 340(252), 351
Warner, J. L., 341(272), 352
Waterbury, G. R., 290(66,92),
 293, 294
Waterstrat, R. M., 198(83), 204
Watling, J., 122
Watson, R. E., 217(35), 223
Watson, R. L., 404(11), 406(11b),
 407(11b), 410
Watt, J. S., 365(2), 374(18),
 375(3,21,22), 376(3,18,21),
 384, 386, 389, 390
Webb, M. S. W., 341(267), 352
Webber, G. R., 308(47), 320(47),
 321(138), 324(47), 326(47),
 331(47,138), 344, 347
Webster, R. K., 341(267), 352
Wedberg, G. H., 138(14), 162
Weihrauch, J. H., 194(68), 204
Weisskopf, M. C., 487(7), 496
Welday, E. E., 299, 308(13),
 312(13), 329(13), 343
Wells, O. C., 169(21), 202
Wenk, G. J., 375(22), 386, 388,
 389
Werme, L. O., 208(8), 217(8),
 221(8), 222
West, N. G., 375(19), 381, 386
Westwood, W., 331(216), 350

White, E. W., 183(35), 202
White, G., 122
White, H. B., 142, 162
Whitehead, H. R., 290(76), 294
Whitman, A., 313(87,89,93),
 314(87,89,93), 315(93),
 329(87), 345
Wiech, G., 216(30), 223
Wilband, J. J., 324(212), 350
Wilkinson, L. R., 375(22), 386,
 388, 389
Willis, J. P., 341(258), 351
Willis, R. D., 62(2), 68,
 127(6), 133(6), 136(6), 138,
 162
Wilson, J. D., 341(267), 352
Wilson, J. R., 324(181), 349
Winchester, J. W., 133(9),
 134(9), 159(9), 161
Winslow, E. H., 87(7), 99(7),
 102(7), 106(7), 107(7),
 108(7), 110, 320(118), 346
Witmer, A. W., 64(6), 68
Wittry, D. B., 164(4), 175(4),
 201, 209(20), 222
Woldseth, R., 178(32), 202,
 226(1), 239
Wolfe, I. A., 312(139), 321(139),
 347
Wolff, R. S., 487(7), 496
Womeldoph, D., 290(84), 294
Womeldorph, D., 307(35), 343
Wones, D. R., 340(247), 351
Wong, T. C., 376(29), 387
Wonsidler, D. R., 192(51), 203
Wood, R. E., 320, 321(120), 347
Wronka, G., 290(69), 293

Y

Yakowitz, H., 170(22), 182(34),
 187(41), 189(45,46), 192(46),
 193(59,63), 194(65), 195(71,
 73,74,75,76), 196(77), 202,
 203, 204, 288(31), 292,
 328(230), 350
Yanak, N. M., 383, 387
Yao, T. C., 270(10), 271(10), 276
Yelland, J. V., 122

Yezer, A., 290(86), 294
Yin, L., 342(277), 353
Yolken, H. T., 387, 292

Z

Zanin, S. J., 290(87), 294
Zeitlin, H., 312(139), 321(139), 347

Zemany, P. D., 84(6), 87(7), 99(7,9), 102(7), 106(7), 107(7), 108(7), 109(13), 110, 310(59), 320(118), 344, 346 253(1), 275, 310(59), 320(118), 344, 346
Zemskoff, A., 189(47), 203
Ziebold, T. O., 174(28), 184, 202
Zimmerman, J. B., 326(196), 349
Zulliger, H. R., 232(6), 239
Zussman, J., 341(265), 352
Zworykin, V. K., 164, 175(2), 201

SUBJECT INDEX

A

Absorption coefficient, 188
Absorption correction, 325
Absorption edge, 5
Absorption effect, 188
Absorption spectra, 216, 217
Absorption spectroscopy, 217
Accuracy, 72, 249
Aluminum alloys, 290
Alloys, 277ff, 436, 437
Assaying, 356
Astronomy, 483ff
Atomic-number effect, 187
Attenuation, see absorption
Auger effect, 23, 34
Auger electron microscope, 170
Auger electron spectroscopy, 218, 221

B

Background, 54, 61ff, 100, 126, 174, 323, 371, 397, 398
Backscatter factor, 188
Backscatter loss, 187
Beam current, 129
Beam penetration, 130
Beam scanning, 153
Binding energy, 206
Biological samples, 140, 144, 146, 151, 158, 198, 408
Blood, 132, 409
Bore hole, 377
Boron, 114, 119
Botany, 133, 155
Brass, 118, 426, 427
Bremsstrahlung, 5, 24, 33
Briquetting, 308
Bronze, 419, 425-432, 437-439, 440

C

Calibration, 46ff, 66, 103, 118, 136, 183, 266, 287, 318, 370, 401
Carbon in steel, 119
Carbonate-phosphate rocks, 333
Castings, 455
Cathodoluminescence, 171
Cement, 257
Ceramics, 478-480
Characteristic energy, 21
Characteristic fluorescence, 189
Characteristic lines, 2
Coating, 172
Coincidence, counting correction, 74, 76
Collimator, 3, 23
Commercial instruments, 253
Compton scatter, 27
Concentration, 396
Concentration, normal, 361
Concentration range, 302
Contamination, 316, 317
Continuum, 5
  fluorescence, 190
Copper alloys, 290, 424
Coprecipitation, 253
Core level, see electron energy level
Correction methods, 272, 323ff
Counting precautions, 71
Counting rate, 44
Counting strategy, 247
Counting technique, 70
Counting time, 247
Critical excitation voltages, 190
Cross section for X-ray generation, 27

Crystals, 210
Crystals, curved, 11, 175
Crystals, long spacing, 165, 215
Crystal size, 10
Crystal spacing, 3, 9, 10
Crystal spectrometer, 6, 10
Crystal type, 11
Curie unit, 31
Cyclotron, 125

D

Data interpretation, 59ff, 69ff, 144, 149, 150, 187, 270, 271, 370
Dead time, 42, 60ff, 75, 174
de Broglie optics, 7
Detection limit, see limit of detection
Detectors, 12, 13, 36
Detector efficiency, 40
Detector resolution, 41
Diffraction efficiency, 6, 8
Diffraction order, 3
Diffuser, 128
Dilution, 251
Double-crystal spectrometer, 213

E

Elastic scatter, 27, 28
Electron channeling, 171
Electron energy level, 206, 207
Electron excitation, 111ff
Electron probe analysis, 163ff
Electron spectroscopy, 217
Electron temperature, 490
Energy dispersion, 17ff, 176
Energy resolution, 41
Enhancement, see matrix effects
Errors, 71
ESCA, 206
Escape peaks, 39
Excitation efficiency, 108, 113

F

Figure of merit, 247

Filters, 398
Flow cell, 365
Flow proportional counter, 14
Fluorescent yield, see x-ray yield
Fluxes, 312
Focal circle (Rowland circle), 11, 176
Forensic analysis, 261
Fundamental parameters, 68, 185, 325
Fusion techniques, 253, 311ff

G

Gaussian distribution, 84
Geometry for energy dispersion, 32
Germanium detector, 37
Glass, 460-477
Gold alloys, 290
Gratings, ruled, 215
Grinding samples, 304, 368
Guard-ring detector, 36, 39

H

Heterogeneity, 306, 369, 423
Homogeneity, 51, 304, 305

I

Inelastic scatter, 27, 28
Instrument parameters, 262
Instruments, commercial, 254
Interelement effect, see matrix effect
Interference, line, 22, 62, 118, 178, 263, 264, 323, 399
Internal standard, 415
Ion exchange resins, 253, 399, 407
Ion excitation, 25, 123ff
Ionization cross section, 187
Iron alloys, 290, 408
Iteration, 190

J

Johann optics, 6
Johansson optics, 11

L

Laboratory design, 227
Lead, 119
Limit of detection, 35, 96ff, 130, 160, 404ff
Line shape, 214
Liquid samples, 132, 133, 260
Lithium drifted detector, see silicon detector
Lunar samples, 340

M

Magnesium alloys, 290
Maintenance of equipment, 367
Mass absorption coefficients, 25
Matrix effects, 174, 185, 245, 250, 269, 323
Meteorites, 198, 360
Methods of analysis, comparison of, 243
Microanalysis, 163ff, 393ff
Mineral, 297ff
Mineral exploration, 373
Mining, 377
Monte Carlo calculations, 194
Moseley's law, 1
Multiple-crystal spectrometer, 254

N

Nickel alloys, 290, 408
Nonconducting samples, 172
Nuclear structure, 494

O

Oil, 260
Oil, mineral, 408
On-stream analysis, 358, 361
Ores, 355ff
Oxygen in cryolite, 119

P

Particle size effect, 51, 118, 402
Particulate samples, 172, 194, 259, 306ff, 394, 399
Pewter, 454-460
Phases, alloy, 281
Photoelectric absorption, 27
Photoelectron, 207
Photon excitation, 25
Pile-up, pulse, 44
PIXEA, 152
Plasma diagnostics, 488
Poisson distribution, 79
Polarized x-rays, 35
Pollution analysis, 259, 393ff, 404
Powder samples, 306, 368, 369
Precision, 72
Preconcentration, 251
Proportional counter, 13
Proton excitation, 123ff

Q

Quantitative analysis, 181, 370, 397

R

Radiation hazards, 235ff
Radioisotope sources, 33, 420
Rayleigh scattering, 27
Refining, mineral, 382
Reflection coefficient, integral, 9
Refractive index, 215
Regression equations, 66
Relative x-ray intensity, 66, 185
Resolution, 9, 20, 205, 213, 264
Rocking curve, crystal, 9
Rowland circle, see focal circle

S

Safety, 234ff
Safety, legal requirements, 236
Sample geometry, 107
Sample heating, 11, 142
Sample preparation, 131ff, 250ff, 278ff, 303ff, 368ff, 395
Sample thickness, 64, 192, 193
Sample type, 64, 65, 131, 132, 153, 256, 285
Scanning electron microscope, 166, 169
Scattered background, 397
Scintillation detector, 13
Secondary electrons, 168
Secondary fluorescer, 32
Semiconductor detector, see Silicon detector
Sensitivity, 26, 137
Signal processing, 42
Silicate rocks, 328
Silicon detector, 36
Silver alloys, 420, 444-454
Slurry, 362, 365
Smelting, 380
Sodium alloys, 290
Soil samples, 154, 156, 157
Solid-state detector, see Silicon detector
Spectra, generation, 5
Spectrometer, crystal, 3
Spectrometer error, 99
Spectrometer geometry, 3
Standard deviation, 83
Standards, 195, 265, 287, 318, 370, 371
Standards, internal, 321
Statistics, counting, 69ff
Steel, 258, 286, 408
Stopping power for electrons, 187
Sulfur compounds, 209, 211
Surface preparation, 114, 115

T

Tailings, 379
Takeoff angle, 171
Telescope, x-ray, 488
Titanium alloys, 290
Total reflection, 215, 487
Trace analysis, 31, 329, 393ff, 415
Tungsten alloys, 290

V

Van de Graaff generator, 125
Valence effects, 205ff, 245
Variance, 87, 93

W

Water analysis, 148
Wavelength, 2
Wavelength dispersion, 1ff, 174

X

X-ray tubes, 4, 31, 230
X-ray yield, 174

Y

Yield, fluorescence, 34
Yield, X-ray, 174

Z

ZAF method, 187
Zirconium alloys, 290